Lead Compounds from Medicinal Plants for the Treatment of Neurodegenerative Diseases

A volume in the *Pharmaceutical Leads from Medicinal Plants* series

"Cogito, ergo sum."

René Descartes
French mathematician, natural scientist, and philosopher (1596–1650)

Lead Compounds from Medicinal Plants for the Treatment of Neurodegenerative Diseases

Christophe Wiart, PharmD, PhD, ACS

Ethnopharmacologist

ELSEVIER

AMSTERDAM • BOSTON • HEIDELBERG • LONDON • NEW YORK • OXFORD
PARIS • SAN DIEGO • SAN FRANCISCO • SINGAPORE • SYDNEY • TOKYO

Academic Press is an imprint of Elsevier

Academic Press is an imprint of Elsevier
32 Jamestown Road, London NW1 7BY, UK
225 Wyman Street, Waltham, MA 02451, USA
525 B Street, Suite 1800, San Diego, CA 92101-4495, USA

First edition 2014

British Library Cataloguing-in-Publication Data
A catalogue record for this book is available from the British Library

Library of Congress Cataloging-in-Publication Data
A catalog record for this book is available from the Library of Congress

ISBN: 978-0-12-398373-2

For information on all Academic Press publications
visit our website at elsevierdirect.com

Typeset by MPS Limited, Chennai, India www.adi-mps.com

Contents

Foreword

By Atta-ur-Rahman

The tremendous advances in natural product chemistry in the last few decades have been triggered by spectacular developments in NMR, mass spectroscopy, and various hyphenated techniques that allow rapid separation and identification of the individual compounds in complex mixtures. The development of high throughput screening methods has greatly facilitated the discovery of new bioactive compounds.

Alzheimer's disease (AD) is the most common of neurodegenerative diseases, affecting almost 30 million people globally, and it has been estimated that with growing old-age populations, 1 in 85 people on our planet will be affected by 2050. The medicines developed include acetylcholinesterase inhibitors and an NMDA antagonist. However, they are only of marginal benefit. Parkinson's disease (PD) is another neurodegenerative disorder resulting from the death of dopamine-generating cells in the brain. There is no known cure for PD, although some relief may be provided by levodopa, dopamine agonists, and MAO-B inhibitors. Amyotrophic lateral sclerosis (ALS) is one of five motor neuron diseases that results in muscle weakness and atrophy. In a small percentage (about 5%), the causes have been attributed to genetic defects, but in the majority of cases the causes are not known. Again there is no known cure.

There is an urgent need of finding new compounds that can attack the underlying mechanisms involved in PD, AD, and ALS, not only to block the progression of the disease with age but also offer a cure. Natural products offer a vast reservoir of compounds that present a huge structural diversity. This is accompanied by a corresponding span of biological activities of various types. They can serve as a treasure chest when searching for such novel lead compounds.

This book is concerned with the medicinal chemistry of those natural products that have been found to have potential for the treatment of these neurodegenerative disorders. I would like to compliment Dr. Christophe Wiart for writing an excellent book that comprehensively covers various classes of natural products that can be potentially employed directly or that can offer interesting pharmacophores for the treatment of these diseases.

The book should be of great interest to a large community of medicinal chemists working in this challenging area.

Atta-ur-Rahman, FRS
International Center for Chemical & Biological Sciences
University of Karachi
Karachi, Pakistan

 Professor Atta-ur-Rahman obtained his PhD in organic chemistry from Cambridge University (1968). He has 910 publications in several fields of organic chemistry including 701 research publications, 27 international patents, 117 books and 65 chapters in books published largely by major U.S. and European presses. He is the Editor-in-Chief of 12 European Chemistry journals and the Editor of *Studies in Natural Product Chemistry*—37 volumes of which have been published by Elsevier (The Netherlands) under his editorship during the last two decades. Seventy-six students have completed their PhD degrees under his supervision.

Professor Rahman is the first scientist from the Muslim world to have won the prestigious UNESCO Science Prize (1999) in the 35-year-old history of the Prize. He was elected as Fellow of the Royal Society (London) in July 2006. He has been awarded honorary doctorate degrees by many universities including the degree of Doctor of Science (ScD) by Cambridge University (UK) (1987), Honorary degree of Doctor of Education by Coventry University (UK) (2007), Honorary DSc degree by Bradford University (UK) (2010), Honorary PhD by the Asian Institute of Technology (2010) and Honorary Doctorate by the University of Technology, Mara, (2011) (bestowed by the King of Malaysia). He was elected Honorary Life Fellow of King's College, Cambridge University, UK in 2007. Professor Atta-ur-Rahman was awarded the TWAS Prize for Institution Building in Durban, South Africa in October 2009 in recognition of his contributions for bringing about revolutionary changes in the higher education sector in Pakistan. The Austrian government also honored him with its highest civil award (Grosse Goldene Ehrenzeischen am Bande) (2007) in recognition of his eminent contributions. Successive Governments of Pakistan have conferred on him four civil awards, Tamgha-i-Imtiaz (1983, President General Ziaul Haq), Sitara-i-Imtiaz (1991, Prime Minister Mohtarma Benazeer Bhutto), Hilal-i-Imtiaz (1998, Prime Minister Nawaz Sharif), and the highest national civil award Nishan-i-Imtiaz (2002, President General Musharraf).

He is President of the Network of Academies of Sciences of Islamic Countries (NASIC) and the Vice-President (Central & South Asia) of the Academy of Sciences for the Developing World (TWAS) Council, Foreign Fellow of Korean Academy of Sciences, and Foreign Fellow of the Chinese Chemical Society. Professor Atta-ur-Rahman was the President of the Pakistan Academy of Sciences (2003–2006). He was again elected as the President of the Academy from 1st January 2011 and continues in that capacity.

Professor Atta-ur-Rahman was the Federal Minister for Science and Technology (14th March, 2000—20th November, 2002), Federal Minister of Education (2002), and Chairman of the Higher Education Commission with the status of a Federal Minister from 2002 to 2008.

Professor Atta-ur-Rahman was the Coordinator General of COMSTECH, an OIC Ministerial Committee comprising the 57 Ministers of Science & Technology from 57 OIC member countries, from 1996 to 2012. He is Distinguished National Professor as well as Professor Emeritus at Karachi University. He is also the Patron-in-Chief of the International Center of Chemical and Biological Sciences (which comprises a number of institutes, including the Husein Ebrahim Jamal Research Institute of Chemistry and the Dr. Panjwani Center of Molecular Medicine and Drug Development) at Karachi University.

Foreword

By Derek J. McPhee

As Editor-in-Chief of the MDPI journals *Molecules* and *Pharmaceuticals*, I came to know Dr. Wiart as a result of a letter he had sent to the *Molecules* Editorial Office alerting us of a misidentified plant species in one of our published papers. During the subsequent email exchanges, I became aware of his profound knowledge of all matters related to plant pharmacognosy, so it is with great pleasure that I learn he has now chosen to follow up his numerous other well-received books in this area with one dedicated entirely to the topic of Lead Compounds from Medicinal Plants for the Treatment of Neurodegenerative Diseases.

This is indeed a timely subject, for neurodegenerative diseases constitute an area of pressing interest given the expectation that the number of people afflicted worldwide by these diseases will rapidly expand with the increasing aging population, while it remains a therapeutic area where there is currently a dearth of approved drugs. As several of these approved drugs are plant-derived natural products or close analogs, the expectation that plants will provide additional leads for such drugs seems entirely reasonable. Confirming my view on the timeliness of the topic, only a few days before I received the publisher's kind invitation to pen this foreword, I had seen a May 2013 conference announcement for a New York Academy of Sciences sponsored meeting on the topic of Translating Natural Products into Drugs for Alzheimer's and Neurodegenerative Diseases, and I was already aware of several journal reviews and some chapters in the *Springer Handbook of Natural Compounds* that have appeared in the past few years and cover different aspects of this field.

To this body of literature we can now add Dr. Wiart's most recent tome, where every major chemical class is covered in one of its three chapters (Chapter 1: Alkaloids, further divided according to skeleton into Amide, Piperine, and Pyridine Alkaloids; Indole Alkaloids; Isoquinoline Alkaloids and Derivatives; and Terpenoid Alkaloids; Chapter 2: Terpenes, with subchapters dedicated to Monoterpenes, Sesquiterpenes, Diterpenes, and Triterpenes; and Chapter 3: Phenolics, with headings covering Benzopyrones, Quinones, and Lignans). Within each chapter all the medical plant species containing these chemical entities are listed, with extensive critically evaluated coverage of chemical structures, detailed structure–activity relationship information, biological activity targets, and mechanisms of action. All this is complemented by an exhaustive listing of the primary literature sources and valuable cross-referenced indexes by Natural Product, Pharmacological Terms, and Plants.

I have no doubt that this book's readers, which will include both the experienced scientist and the novice in the field seeking background to guide a search for novel entities with biological activity in this therapeutic area, will soon come to consider this the definitive "go-to" book for comprehensive information in this area for many years to come, and Dr. Wiart is to be congratulated for another success in his lengthy and distinguished publishing history.

Derek J. McPhee
Senior Director of Technology Strategy
Amyris Inc.
Emeryville, California, USA

 Derek J. McPhee is currently the Senior Director of Technology Strategy at Amyris Inc., a publicly traded biotechnology company based in Emeryville (California, USA) focused on the production of renewable alternatives to petroleum-based fuels and specialty chemicals. A native of Scotland, he has a LicC degree in Applied Chemistry from the Universidad de Málaga (Spain) and a PhD in Organic Chemistry from the University of Calgary (Canada). After pursuing a NSERC Postdoctoral Research Associateship at the Division of Chemistry of the Canadian National Research Council in Ottawa, the remainder of his career has been in industry.

Following a period of 14 years with Uniroyal Chemical (now part of Chemtura Corporation) working on the discovery, process development, and manufacture of agricultural, rubber, and specialty chemicals, he has worked in the generic pharmaceutical industry as a Senior Scientist at Brantford Chemicals (now Apotex Pharmachem), and as Director of Chemistry and Vice President of a U.S./Canada-based custom synthesis company. Before joining Amyris in March 2005, he was a self-employed consultant to several chemical and pharmaceutical startups.

In addition to his current position at Amyris, he has been that company's Director and Senior Director of Chemistry, leading a team that, with funding from the Bill and Melinda Gates Foundation, developed a novel low-cost route to the antimalarial drug precursor artemisinin using a raw material produced by fermentation of genetically engineered yeast. The launch by Sanofi of an ACT drug made using this technology was officially announced a few days prior to World Malaria Day in March 2013.

During the period 2000–2005 he was the Managing Editor of MDPI's online chemistry journal *Molecules*, and since 2005 he has served as its Editor-in-Chief. He has also been the Editor-in-Chief of the MDPI journal *Pharmaceuticals* since its launch in 2004. He is the sole author/coauthor of 23 patents, 19 papers in peer-reviewed scientific journals, and two book chapters.

Foreword

By Cornelis J. Van der Schyf

Neurodegenerative diseases are a group of disorders with complex pathoetiological pathways leading to neuronal cell death. These disorders constitute an emerging epidemic as the aging cohort of the world population expands and their burden on society grows inexorably, with enormous economic and human costs. Of note—and of great concern—is the fact that neurodegenerative diseases lack effective treatment options for patients. Although some *de novo* "designed" agents show significant promise in preclinical studies as neuroprotective and disease-modifying agents, the need to discover unique organic molecules to serve as design leads for drug discovery programs is a growing concern. In this regard, natural products have served exquisitely as design templates for several complex drug design studies. Plant-derived secondary metabolites have long served as an important resource for the development of small-molecule therapeutics due primarily to their combination of unique chemical features and potent bioactivities. Accumulating evidence suggests that phytochemicals themselves may potentially mitigate neurodegeneration, and improve memory and cognitive and neuromotor function. Ironically, nutraceutical products (most of which are derived directly from plants) may offer a viable short-term option for many patients suffering from neurodegenerative disorders since these products are subject to fewer regulations than traditional pharmaceuticals and therefore could be made available to patients much more expeditiously than newly developed prescription drugs.

It is in these contexts that the value of Christophe Wiart's second volume in the series *Pharmaceutical Leads from Medicinal Plants* can be truly appreciated. This volume, titled *Lead Compounds from Medicinal Plants for the Treatment of Neurodegenerative Diseases*, collates data from the peer-reviewed literature that present pharmacological evidence, structure-activity relationships, cellular targets, and mechanisms of action in a very compelling way.

Chapters follow a primary arrangement using chemical structure types rather than plant species or specific pharmacological mechanism of action as indexing mechanisms. This makes perfect sense in view of the target audience, which would draw extensively from the drug development and drug discovery community. For example, listed under the title of Chapter 1, "Alkaloids," detailed descriptions follow that describe the amide, piperine, pyridine, indole, isoquinoline, and terpenoid subclass alkaloids. The same pattern is used for Chapters 2, "Terpenes," and 3, "Phenolics." Sublisted below these descriptors, the individual plant species that actually produce the chemical classes and subclasses are described individually. The elegance of this approach can be exemplified by the stilbene scaffold that has become popular in particular due to the neuroprotective effects of the non-flavonoid natural product resveratrol, and compounds derived from the xanthine scaffold that afford neuroprotection in Parkinson's disease through mechanisms that include dual adenosine A_{2A} receptor antagonism and MAO-B inhibition. Both the stilbene and xanthene scaffolds are present in a number of related and unrelated plant species, and it is extremely useful to have these species listed after the primary desired chemical scaffold has been identified in the index.

Natural products derived from medicinal plants in particular are widely anticipated to play a significant and increasing role in the development of new therapeutic leads for neurodegenerative disease. It is my contention that this volume by Dr. Wiart will play a core role in this evolving era, and that every discovery in this arena will prove to have been inspired, at least in part, by the monumental work of this author.

Cornelis J. (Neels) Van der Schyf
Dean of the Graduate School & Professor of Biomedical and Pharmaceutical Sciences
Idaho State University
Pocatello, Idaho, USA

Dr. Van der Schyf is Dean of the Graduate School and Professor of Biomedical and Pharmaceutical Sciences at Idaho State University (ISU). Before joining ISU, Van der Schyf was Associate Dean for Research and Graduate studies in the College of Pharmacy and Professor of Neurobiology in the College of Medicine at Northeast Ohio Medical University. He earned his BPharm, MSc, DSc (PhD), and DTE degrees from Potchefstroom University (now North-West University) in South Africa and completed a postdoctoral fellowship in the Department of Medicinal Chemistry at the University of Connecticut, during which time he did research at the Francis Bitter National Magnet Lab at MIT in Cambridge, Massachusetts.

Cornelis J. Van der Schyf is considered as one of the leading contributors to the concept of designed multiple ligands in the treatment of neurodegenerative diseases, and has published more than 110 peer-reviewed research and review articles, more than 200 abstracts and presentations—many of these as invited keynote speaker, seven book chapters, several reports to industry, and journal editorials. He holds 14 patents. He is the Editor-in-Chief of the *Journal of Biophysical Chemistry* and is or has been a member of the editorial advisory boards for *BMC Pharmacology and Toxicology, Expert Opinion on Drug Discovery, Molecules, International Journal of Brain and Cognitive Sciences, Medicinal Chemistry Research, The Open Medicinal Chemistry Journal, Open Medicinal Chemistry Letters, Neurotoxicity Research*, and *Pharmaceutics & Novel Drug Delivery Systems: Current Research*, and serves on the International Advisory Board of the *South African Journal of Chemistry*. He was an invited guest editor for the January 2009 issue of the journal *Neurotherapeutics*. As visiting professor in Australia (University of Queensland, Brisbane), Belgium (FUNDP, Namur), USA (Virginia Tech), and currently as Emeritus Extraordinary Professor at North-West University (South Africa), he remains active internationally.

Besides serving or having served *ad hoc* and as a chartered member on several NIH Study Sections and many other national and international granting agencies, he is a member of the Phi Beta Delta Honor Society, Sigma Xi, and The Scientific Research Society, and has received several honors, including "Most Cited Paper" awards, the APSSA Upjohn Achievement Award and South Africa's highest honor in drug discovery research, the FARMOVS Prize for Pharmacology and Drug Development.

Preface

During the Tertiary period, some little apes gained optical, olfactory, and locomotor abilities which allowed feeding not only on seeds, fruits, and leaves from trees but also herbs, mosses, and mushrooms, the phytonutrients of which may have in fact possibly contributed to the subsequent birth of humanity. This possibility raises interesting questions regarding how plant natural products may have induced or facilitated the constitution of the bewildering web of neuronal connections required to form the human brain. This question is yet unanswered, but acetylcholine, dopamine, and serotonin, which account for neurotransmission, occur in plants which indeed appeared in the Tertiary period.

Today, the accelerating speed of aging comes with an increased number of patients diagnosed with neurodegenerative diseases for which there is no robust treatment despite intensive research. In effect, the progression of Alzheimer's disease and Parkinson's disease can be slowed down but not stopped or reversed. Besides this, other critical conditions such as amyotrophic lateral sclerosis and spinal cord injuries remain completely untreatable. In this light, we present in this volume evidence that natural products are not only able to protect neurons and to boost their activities, but also to induce neuritogenesis, raising the captivating possibility that the flowering plants that helped to lead apes to evolve into *Homo sapiens* may in the near future allow not only complete victory over neurodegenerative diseases and neuronal injuries but also the boosting of human intelligence.

Christophe Wiart, PharmD, PhD, ACS

About the Author

Dr. Christophe Wiart was born August 12, 1967 in Saint Malo, France. He obtained a Doctorate of Pharmacy from the University of Rennes in 1996 and was a pupil of the pharmacognosist Professor Loic Girre and the botanist Lucile Allorge from the Botanical section of the Museum of Natural History in Paris. Dr. Wiart has been studying the medicinal plants of India, Southeast Asia, China, Korea, Japan, Australia, and the Pacific Islands for the last 20 years. He has collected, identified, classified, and made botanical plates of about 2000 medicinal plants. Dr. Wiart is regarded as the most prominent living authority in the field of Asian ethnopharmacology, chemotaxonomy, and ethnobotany. His research team currently works on the identification and pharmacological evaluation of Asian medicinal plants at the University of Nottingham. He has authored numerous bestselling books devoted to the medicinal plants of Asia and their pharmacological and cosmetological potentials.

Alkaloids

■ INTRODUCTION

Neurons convey information by synthesizing and secreting neurotransmitters which bind to cytoplasmic membrane proteins or receptors. The ability of a neurotransmitter to bind to a receptor depends biochemically on strict structural requirements which define several neuronal routes including the nicotinic, dopaminergic, adrenergic, and serotoninergic pathways, whereas acetylcholine, dopamine, noradrenaline, and serotonin bind to and activate cholinergic, dopaminergic, adrenergic, and serotoninergic receptors, respectively. The cholinergic system, for instance, encompasses the synthesis and secretion of acetylcholine, the depletion of which, as a result of cholinergic neuron progressive degeneration, results in cognitive decline, decrease in brain weight, and dementia as described by Alois Alzheimer in 1907 and known since as Alzheimer's disease (AD). Since the pathophysiology of AD involves a steady decline in cholinergic neurotransmission in the cortex, alkaloids from plants able to bind to cholinergic receptors, namely nicotinic and muscarinic receptors, thus mimicking acetylcholine, can conceptually be viewed as agents or starting points for the synthesis of leads to fight senile dementia. Nicotinic receptors are the target of alkaloids, such as, notably, nicotine from *Nicotiana tabacum* L. (1.1.3), lobeline from *Lobelia inflata* L. (1.1.3), lupanine from *Sophora flavescens* Aiton (1.1.3), ibogaine from *Tabernanthe iboga* Baill. (1.2.6), pteleprenine from *Ptelea trifoliata* L. (1.4.1), and methyllycaconine from *Aconitum* L. (1.4.2), which in fact share structural similarities with acetylcholine, including the presence of a nitrogen atom. Other cholinergic types of receptors are the muscarinic receptors that are targeted, for instance, by scopolamine from *Atropa belladonna* L., arecoline from *Areca catechu* L. (1.1.3), 2-β-hydroxy-6β-acetoxy-nortropane from *Erycibe obtusifolia* Benth. (1.1.3), himbacine from *Galbulimima baccata* F.M. Bailey (1.1.3), cryptolepine from *Cryptolepis sanguinolenta* (Lindl.) Schltr. (1.2.4), and rhynchophylline from *Uncaria sinensis* (Oliv.) Havil. (1.2.6). The dopaminergic system involves the synthesis and secretion of dopamine, the

C. Wiart: Lead Compounds from Medicinal Plants for the Treatment of Neurodegenerative Diseases.
DOI: http://dx.doi.org/10.1016/B978-0-12-398373-2.00001-7

depletion of which, as a result of dopaminergic neuron progressive degeneration, results in locomotor and cognitive decline, or *paralysis agitens*, as described by James Parkinson in 1817 and called Parkinson's disease (PD) by Jean-Martin Charcot in 1888. Since the pathophysiology of PD involves a collapse in dopaminergic neurotransmission in the *substantia nigra*, alkaloids from plants able to bind to dopaminergic receptors, thus mimicking dopamine, can conceptually be viewed as agents or starting points for the synthesis of leads to fight this neurodegenerative disease. Of note, dopamine itself is a catecholamine which originates from the amino acid tyrosine, which in plants is the precursor of alkaloids that bind to dopaminergic receptors. Such alkaloids are, for instance, the isoquinolines norreticuline and reticuline from *Papaver somniferum* L. (family Papaveraceae Juss.) (1.2.6), stepholidine from *Stephania intermedia* H.S. Lo (1.3.1), boldine from *Peumus boldus* Molina (1.3.3), glaucine from *Glaucium flavum* Crantz (1.3.3), and nantenine *Nandina domestica* Thunb. (1.3.3). The adrenergic system in the brain includes the synthesis and secretion of noradrenaline from the *locus coeruleus*, which modulates cognition in the prefrontal cortex, and a decline and dysfunction of this system occur during AD and PD. Noradrenaline derives from dopamine, thus allowing alkaloids such as raubasine and corynanthine from *Catharanthus roseus* (L.) G. Don (1.2.6), tabersonine from *Melodinus fusiformis* Champ. ex Benth. (1.2.6), norreticuline and reticuline from *Papaver somniferum* L. (family Papaveraceae Juss.) (1.2.6), xylopine from *Annona rugulosa* (Schltdl.) H. Rainer (1.3.1), and dicentrine *Lindera macrophylla* Boerl. (family Lauraceae Juss.) (1.3.3). The serotoninergic system encompasses the synthesis and secretion of an indolic neurotransmitter known as serotonin, the levels of which decrease in the brain of patients with AD and PD, thus accounting for mood disorders. Serotoninergic receptors are the target of alkaloids and particularly indole alkaloids from plants, such as geissoschizine methyl ether from *Uncaria sinensis* (Oliv.) Havil. (1.1.3), 12-methoxy-1-methyl-aspidospermidine from *Geissospermum vellosii* Allemão (1.1.3), psychollatine from *Psychotria umbellata* Thonn. (1.1.3), or other types of alkaloids such as asimilobine from *Nelumbo nucifera* Gaertn. (1.3.3) and skimmianine from *Adiscanthus fusciflorus* Ducke (1.4.1). By docking into and stimulating receptors, alkaloids from plants not only mimic the agonistic properties of neurotransmitters, thus allowing the maintenance of neurotransmission within a neuropathological context, but also initiate cascades on biochemical pathways which favor the survival or even the growth of neurons as explained in this chapter. Furthermore, alkaloids from plants have the ability to abrogate the enzymatic activity of enzymes involved in the catabolism of acetylcholine or dopamine in the synthesis of chimeric peptides, neuroinflammation, or to interfere with protein kinases, thus allowing neuroprotection. Taking into account their unique ability to mimic neurotransmitters, to hinder or stimulate the activity of enzymes to the benefit of neuronal viability, one can view plant alkaloids as robust candidates for the treatment of AD, PD, and amyotrophic lateral sclerosis (ALS). In this light, this chapter proposes the plant alkaloids *N*-trans-feruloyltyramine, piperlonguminine, matrine, scorodocarpines, 9-methyl-harmanes, dehydroevodiamine, cryptolepine, carbazoles, tabersonine and plumeran, berberine, sanguinarine, nandenine, tetrandrine, hydrastine, sinomenine, 1-*O*-acetylambelline, aristolacatam BII, skimmianine, dendrobane sesquiterpenes, songorine, and gagaminine are identified as leads and sources of synthetic derivative alkaloids for the treatment of these neurodegenerative diseases.

Topic **1.1**

Amide, Piperine, and Pyridine Alkaloids

1.1.1 *Polyalthia suberosa* (Roxb.) Thwaites

History The plant was first described by George Henry Kendrick Thwaites in *Enumeratio Plantarum Zeylaniae* published in 1864.

Synonyms *Guatteria suberosa* (Roxb.) Dunal, *Uvaria suberosa* Roxb.

Family Annonaceae Juss., 1789

Common Name An luo (Chinese)

Habitat and Description This tree grows to a height of 5 m in the forests of India, Sri Lanka, Laos, Burma, Thailand, China, Vietnam, Malaysia, and the Philippines.

The stems are dark reddish-brown and lenticelled. The leaves are simple. The petiole is 0.5 cm long. The leaf blade is obovate, 5–10 cm × 2–5 cm, glossy, acute at the base, round or acute at the apex, and with inconspicuous secondary nerves. The flowers are solitary and cauliflorous. The pedicel is slender and up to 3 cm long. The calyx comprises 3 sepals, which are broadly triangular and 0.3 cm long. The corolla comprises 6 petals, which are yellowish green, lanceolate, leathery, and 1 cm long. The fruit consists of numerous ripe carpels, which are globose, 0.5 cm across, glossy, and fleshy (Figure 1.1).

Medicinal Uses In the Philippines, the plant is used to abort.

Phytopharmacology The plant produces the lanostanes triterpenes suberosol,[1] the azaanthracene alkaloid kalasinamide,[2] *N*-trans-feruloyltyramine, and *N*-trans coumaroyltyramine.[3,4]

Proposed Research Pharmacological study *N*-trans-feruloyltyramine and synthetic derivatives for the treatment of neurodegenerative diseases.

Rationale Small molecules inhibiting the enzymatic activity of acetylcholinesterase (AChE) (CS 1.1) have been developed to delay the progression of Alzheimer's disease (AD). One such molecule is the amide rivastigmine (CS 1.1), which sustains the synaptic levels of acetylcholine

■ **FIGURE 1.1** *Polyalthia suberosa* (Roxb.) Thwaites.

(CS 1.2) and therefore cholinergic neurotransmission between neurons that inexorably subside to apoptosis due to increasing aggregation of β-amyloid peptide and neurofibrillary tangles.[5] Of note, the amide alkaloid N-trans-feruloyltyramine (CS 1.3) isolated from *Polyalthia suberosa* (Roxb.) Thwaites (family Annonaceae Juss.) at a dose of $250\,\mu$M protected cortical neurons against reactive oxygen species (ROS), subsequent activation of pro-apoptotic Bcl-2-associated X protein (Bax), mitochondrial insults, and caspase 3 activation provoked by β-amyloid ($A\beta_{1-42}$) peptide.[4]

The precise mechanism underlying the neuroprotective effects of N-trans-feruloyltyramine is yet undeciphered, but the inhibition of ROS might be a cardinal event because this amide alkaloid hindered the production of nitric oxide (NO) by inducible nitric oxide synthetase (iNOS) in macrophages[6] challenged with lipopolysaccharide (LPS).[7] In fact, one could frame the hypothesis that N-trans-feruloyltyramine abrogates β-amyloid peptide-induced generation of ROS such as NO, thus dampening mitochondrial insults either directly[8] or via Fas stimulation, hence of pro-apoptotic Bax translocation into the mitochondria, release of cytochrome c, and activation of caspase 9 and 3.[9] Note that pro-apoptotic Bax is over-expressed in neurons of Alzheimer patients,[10] and agents able to negate pro-apoptotic Bax are of therapeutic interest.

The pathophysiology of AD, frontotemporal dementia, and Parkinson's disease (PD) is characterized by microtubule associated protein tau, which upon phosphorylation polymerizes into tangles.[11] Inhibitors of microtubule-associated protein tau aggregation via dephosphorylation or direct dockage[12] are of therapeutic interest. One such inhibitor is the synthetic amide derivative BSc3094 (CS 1.4), which reacted with and inhibited the polymerization of microtubule-associated protein tau with an

■ **CS 1.1** Rivastigmine.

■ **CS 1.2** Acetylcholine.

■ **CS 1.3** *N*-trans-Feruloyltyramine.

■ **CS 1.4** BSc3094.

■ **CS 1.5** Compound 23.

IC_{50} value equal to $1.6\,\mu M$.[13] PD involves the oxidation of DJ-1 protein, which physiologically protects dopaminergic neurons[14] against 6-hydroxydopamine (6-HODA)-induced apoptosis[15] and stroke-induced neurodegeneration,[16] and Compound 23 (CS 1.5) reacted with DJ-1 protein and protected SH-SY5Y cells against hydrogen peroxide or 6-HODA at a dose of $1\,\mu M$.[17] Furthermore, Compound 23 prevented dopaminergic neuron 6-HODA-induced apoptosis in the *substantia nigra* of rodents.[17]

Other amide alkaloids of remarkable neuroprotective interest are the *N*-acyl ethanolamines (NAE) and their synthetic derivatives. In neurophysiological conditions, the central nervous system produces NAEs such as arachidonylethanolamide (anandamide, CS 1.6), which is involved in the control of manifold brain function such as pain, appetite, locomotion, and neuronal fate by direct interaction with cannabinoid (CB1) receptor and the transient receptor potential vanilloid subtype 1 (TRPV1). In effect, the binding of anandamide to neuronal cannabinoid receptors subtype 1 (CB1) inhibits adenylate cyclase (AC) followed by a collapse of cyclic adenosine monophosphate (cAMP), protein kinase A (PKA) inhibition, rectifying K^+ channels stimulation,[18] inhibition of voltage-sensitive Ca^{2+} channels (VSCC), and activation of kinases, which include c-Jun N-terminal kinase (JNK) and mitogen-activated protein kinase (MAPK) p38.[19] Furthermore, anandamide binds to the TRPV1, the activation of which results in neuroapoptosis resulting from excessive free cytoplasmic Ca^{2+}, activation of PKA, activation of the arachidonate cascade, mitochondrial insults, release of cytochrome c, and activation of caspase 3.[20]

Structure activity relationship evidence points to the fact that the activation of the CB1 receptor can be achieved only by an amide alkaloid comprising

■ **CS 1.6** Anandamide (NAE 20:4).

■ **CS 1.7** Palmytoylethanolamine (NAE 16:0).

■ **CS 1.8** Linoleoylethanolamine (NAE 18:2).

an aliphatic chain of 20–22 carbons with a minimum of three non-conjugated *cis* double bonds,[21] and therefore, other *N*-acyl ethanolamines, including NAE 16:0 and NAE 18:2, do not activate the CB1 receptor but are yet strikingly neuroprotective. Strangely, palmytoylethanolamine (CS 1.7) (NAE 16:0) at a dose of $100 \mu M$ protected mouse hippocampal cells (HT-22) cells against oxidative insults via activation of protein kinase B (Akt).[22] Likewise, neurons were protected against glutamate insults by $120 \mu M$ of linoleoylethanolamine (CS 1.8) (NAE 18:2) via a mechanism probably involving fatty acid amide hydrolase (FAAH) inhibition[23] by possible TRPV1 antagonism. Note that the phosphorylation of extracellular signal-regulated kinase (ERK1/2) is commonly observable in neurons exposed to glutamate,[24] the release of which is inhibited by the inhibition of voltage-sensitive Ca^{2+} channels (VSCC) by anandamide.[25,26]

1.1.2 *Piper kadsura* (Choisy) Ohwi

History The plant was first described by Jisaburo Ohwi in *Acta Phytotaxonomica et Geobotanica* published in 1934.

Synonyms *Ipomoea kadsura* Choisy, *Piper arboricola* C. DC., *Piper futokadsura* Siebold, *Piper subglaucescens* C. DC.

Family Piperaceae Giseke, 1792

Common Name Feng teng (Chinese)

Habitat and Description This plant is a climber that grows wild in the forests of China, Taiwan, Korea, and Japan. The stem is woody, articulated, and sparsely hairy. The leaves are simple. The petiole is 1–1.5 cm long, hairy, and channeled. The leaf blade is broadly lanceolate, 5–12.5 cm × 3.5–10 cm, asymmetrical, and subcordate at the base, acute at the apex, and wavy at the margin. It presents 2 pairs of secondary nerves, emerging from the base. The inflorescence is a spike opposite to the leaf. The male spike is slender and 10 cm long on a hairy stem. The fruit is a drupe, which is yellow and 0.5 cm across (Figure 1.2).

Medicinal Uses In Taiwan, Korea, and Japan, it is used as carminative and to draw phlegm from the lungs. In China, the plant is used to treat inflammation and rheumatic diseases.

Phytopharmacology The plant contains aristolactam alkaloids, including piperlactam S[27] and aristololactam AIIIa,[28] cohorts of lignans including kadsurenin C and H,[29] piperkadsin A and B, futoquinol, kadsurenone,[28] piperkadsin C, wallichinine, denudatin A, kadsurenin L, isofutoquinol A, futokadsurin C,[30] and N-pcoumaroyl, tyramine.[28]

■ **FIGURE 1.2** *Piper kadsura* (Choisy) Ohwi.

Proposed Research Pharmacological study of piperlonguminine (CS 1.9) and synthetic derivatives for the treatment of neurodegenerative diseases.

Rationale The amyloid precursor protein (APP) is pivotal in the pathophysiology of Alzheimer's disease since its abnormal cleavage by β-secretase and γ-secretase generates β-amyloid peptide, which aggregates into neurotoxic amyloid plaques in the brain tissues. Therefore, agents capable of mitigating APP have the potential to withhold the progression of AD, and such agents may come from medicinal plants. In fact, such agents have been identified by Xia et al.[31] as the amide alkaloids piperlonguminine and dihydropiperlonguminine (CS 1.10) from *Piper kadsura* (Choisy) Ohwi, which inhibited the level of APP in human SK-N-SH neuroblastoma cells at a dose of 13.1 μg/mL[31] and raised the exiting possibility of isolating amide alkaloids from members of the vast genus *Piper* L. for the stabilization of AD.

The neuroprotective potential of amide alkaloids from the genus *Piper* L. is further demonstrated with piperine (CS 1.11) from *Piper nigrum* L. (family Piperaceae Giseke), which protected rodents against ethylcholine aziridinium (AF64A) ion-induced dementia and neuronal loss.[32] The biochemical mechanism underlying the neuroprotective effects of piperine

■ **CS 1.9** Piperlonguminine.

■ **CS 1.10** Dihydropiperlonguminine.

■ **CS 1.11** Piperine.

is still elusive, but one could reasonably speculate that the inhibition of monoamine oxidase (MAO) is involved.[33] Indeed, piperine inhibited the enzymatic activity of monoamine oxidases A (MAO-A) and B (MAO-B) with an IC_{50} value equal to $0.4\,\mu M$ and $0.2\,\mu M$, respectively,[34] and the structure activity relationship demonstrates that the ketone moiety, the two double bonds, and the piperine heterocycle are necessary for activity against MAO-A.[34] MAO-A is a critical enzyme in the pathophysiology of Parkinson's disease because it interacts with the endogenous dopamine-derived neurotoxin N-methyl-(R)-salsolinol, which causes cell death in dopaminergic neurons.[35] Indeed, the binding of N-methyl-(R)-salsolinol to MAO-A favors the opening of mitochondrial permeability transition pore (MPT), hence mitochondrial insult, release of cytochrome c, activation of caspase 3, and apoptosis.[36] Furthermore, the piperine derivative (E,E)-1-[5-(3,4-dihydroxyphenyl)-1-oxo-2,4-pentadienyl]piperidine (HU0622, CS 1.12) at a dose of $7\,\mu M$ activated ERK1/2 in rat pheochromocytoma (PC12) cells and induced neurite outgrowth[37] via probable activation of cAMP response element binding protein (CREB).[38] Note that the amide alkaloid bastadin 9 (CS 1.13), isolated from the sponge *Ianthella basta* Pallas (family Lanthellidae), inhibited the enzymatic activity of β-secretase against APP with an IC_{50} value equal to $0.3\,\mu M$.[39] Such compounds also

■ **CS 1.12** HU0622.

■ **CS 1.13** Bastadin 9.

have the captivating potential to block the neurotoxic accumulation of Ca^{2+}, for example NP04634 (CS 1.14), the derivative of 11,19-dideoxyfis-tularin from the sponge *Aplysina cavernicola* Vacelet (family Aplysinidae), which, at a concentration $10\,\mu M$, protected bovine chromaffin cells exposed to $0.3\,\mu M$ of L-type Ca^{2+} channel activator.[40]

In the past few decades, observations were made that a series of 1-benzyl-4-ethylpiperidine derivatives elicited potent anti-AChE activities, such as 1-benzyl-4-[2-(*N*-benzoylamino)ethyl]piperidine, and 3g, 3h, and 3p (CS 1.15–1.18).[41] These exciting activities heralded a rapid increase in struc-ture activity relationship and synthesis of analogs and resulted in the dis-covery of donepezil (CS 1.19), which inhibited AChE with an IC_{50} value equal to 5.7 nM and was developed as a drug for the treatment of AD.[42]

■ **CS 1.14** NP04634.

■ **CS 1.15** 1-benzyl-4-[2-(*N*-benzoylamino)ethyl]piperidine.

■ **CS 1.16** 3g.

■ **CS 1.17** 3 h.

■ **CS 1.18** 3p.

■ **CS 1.19** Donezepil.

1.1.3 *Sophora flavescens* **Aiton**

History The plant was first described by William Aiton in *Hortus Kewensis* published in 1789.

Synonyms *Sophora angustifolia* Siebold & Zucc., *Sophora angustifolia* var. *stenophylla* Makino & Nemoto, *Sophora flavescens* fo. *angustifolia* (Siebold & Zucc.) Yakovlev, *Sophora flavescens* var. *stenophylla* Hayata, *Sophora macrosperma* DC., *Sophora tetragonocarpa* Hayata

Family Fabaceae Lindl., 1836

Common Names Kurara (Japanese), ku shen (Chinese)

Habitat and Description This herb grows to 2 m in the wastelands of India, China, Russia, Korea, and Japan. The stem is terete and subglabrous; it develops from a fleshy, anthropomorphic root. The leaves are imparipinnate and stipulate. The stipules are stipules lanceolate and 1 cm long, acuminate. The rachis is 20–25 cm long and supports 6–12 pairs of leaflets plus a terminal one, lanceolate; 3–4 cm × 0.5–2 cm; membranaceous; and

■ **FIGURE 1.3** Sophora flavescens Aiton.

subglabrous beneath. The inflorescence is a terminal, 25 cm long raceme of numerous yellowish-white flowers. The calyx is campanulate, subglabrous, and presents 5 lobes which are inconspicuous. The corolla is 1.5 cm long and comprises 5 petals, which include a spatulate standard. The androecium comprises 10 stamens. The fruit is a pod that is 5–10 cm long and irregularly constricted around 1–5 seeds (Figure 1.3).

Phytopharmacology The plant produces the pyridine alkaloids kuraramine[43] and isokuraramine[44]; a broad array of quinolizidine alkaloids including matrine, oxymatrine, sophocarpine, sophoramine, sophoranol, 5α,9α-dihydroxymatrine, 9α-hydroxysophoramine, matrin-*N*-oxide, sophocarpine-*N*-oxide, sophoranol-*N*-oxide, lupanine, anagyrine, baptifoline, *N*-methylcytisine, rhombifoline, mamanine[44]; and cohorts of prenylated flavonoids, which include kosamol A,[45] sophoraflavanone G,[46] sophoflavescenol, kurarinol, kushenols K and H,[47] sophoraflavanones K and L, and 8-lavandulylkaempferol.[48]

■ **CS 1.20** Nicotine.

■ **CS 1.21** Nornicotine.

■ **CS 1.22** Anabaseine.

■ **CS 1.23** *N*-methylanabasine.

■ **CS 1.24** Anabasine.

Medicinal Uses In China, the plant is used to expel worms from the intestines, to treat fatigue and jaundice, and to break fever.

Proposed Research Pharmacological study of matrine and synthetic derivatives for the treatment of neurodegenerative diseases.

Rationale Neuronal nicotinic receptors consist of 5 proteins $\alpha2$–$\alpha10$ and $\beta2$–$\beta4$, which form a transmembrane ion channel, which is responsible for synaptic cholinergic transmission and modulation of the secretion of other types of neurotransmitters.[49] Of compelling interest is the fact that the $\alpha7$ and $\alpha4\beta2$ subtypes of neuronal nicotinic receptors are critically involved in cognition and are of considerable importance in the pathophysiology and possible treatment of dementia. Indeed, Wang et al.[50] provided evidence that β-amyloid peptide binds to $\alpha7$ nicotinic receptors expressed by neurons with an IC_{50} value equal to 0.01 pM, resulting in neurotoxic complexes impairing the cholinergic neurotransmission and stimulating the formation of plaques.[51,52] Furthermore, the stimulation of $\alpha7$ and $\alpha4\beta2$ nicotinic mitigate dementia[53,54] and such agonists often exposes a piperidine or a pyridine framework.

One notable example of pyridine alkaloid agonist at the $\alpha4\beta2$ nicotinic receptor agonist is nicotine (CS 1.20) from *Nicotiana tabacum* L. (family Solanaceae Juss.), which binds to this receptor with an IC_{50} value equal to 3.5 μM.[55] Nicotine improved the cognition of rodents at a dose of 0.2 mg/kg in the one-way active avoidance test, Lashley III maze test, and 17-arm radial maze test[56]; and nornicotine (CS 1.21) inhibited β-amyloid peptide aggregation,[57] suggesting that nicotine and derivatives could be of value to treat AD. It is noteworthy to mention here that nicotine fights Parkinson neurodegeneration by protecting dopaminergic neurons against the pro-apoptotic ROS generated by MAO-B during dopamine catabolism and by stimulating $\alpha7$ nicotinic receptors and nicotinic receptors containing $\alpha4$ or $\alpha6$ sub-units.[58-62] Consonant with the beneficial effect of nicotine and derivatives against PD was the observation that nicotine, nornicotine, anabaseine (CS 1.22), *N*-methylanabasine (CS 1.23), and anabasine (CS 1.24) from *Nicotiana tabacum* L. (family Solanaceae Juss.) induced the secretion of dopamine in striatal preparations.[63] Along these lines, the nicotinic congener and dreadful poison epibatidine (CS 1.25) from the frog *Epipedobates tricolor* is a robust $\alpha4\beta2$ nicotinic partial agonist, which at a dose of 0.6 μg/kg incurred dopamine secretion in caudate putamen rodents.[64] Additional support to the contention that pyridine alkaloid nicotinic receptor agonists are beneficial against dementia is provided by the finding that anabaseine produced by marine worms including

Paranemertes peregrina[65] is an agonist of $\alpha 7$ nicotinic receptors and partial agonist of $\alpha 4\beta 2$ nicotinic, which at a dose equal to $10\,\mu M$ prolonged the survival of PC12 cells deprived of both serum and nerve growth factor (NGF).[66] In addition, anabaseine derivative GTS-21 (CS 1.26) given at a dose of 1 mg/kg enhanced the cognitive function of aged rodents in the one-way active avoidance test and 17-arm radial maze test.[56] In an effort to magnify $\alpha 4\beta 2$ nicotinic receptors agonist selectivity, several synthetic derivatives of nicotine have been produced, such as A-84543 (CS 1.27) and the piperidine derivative 2b (CS 1.28), which showed potent agonist activity against $\alpha 4\beta 2$ nicotinic receptors with IC_{50} values equal to $0.7\,\mu M$[67] and to $0.01\,\mu M$,[68] respectively. Some of these derivatives underwent preclinical studies, such as altinicline (CS 1.29) and TC-1734 (CS 1.30) for the treatment of cognitive dysfunction.[69–72] The piperidine alkaloid lobeline (CS 1.31) from *Lobelia inflata* L. (family Campanulaceae Juss.) is a partial agonist and human $\alpha 4\beta 2$ nicotinic expressed in the human epithelial cell line (SH-EP1) and an $\alpha 7$ nicotinic receptor antagonist that improved the cognitive function of rodents at a dose of $19\,\mu mol/kg$.[73] However, lobeline could potentially be seen as a dopaminergic neurotoxin, because at a dose of $3\,\mu M$, it boosted the release of cytoplasmic dopamine in striatal preparation with inhibition of dopamine uptake by synaptosomes and synaptic vesicles with IC_{50} values equal to $80\,\mu M$ and $0.8\,\mu M$, respectively[74] via the inhibition of vesicular monoamine transporter 2 (VMAT2).[75] In neurophysiological conditions VMAT2 accumulates dopamine into synaptic vesicles, thus preventing dopamine-induced cellular insults and allowing exocytosis and dopaminergic transmission; hence, VMAT2 inducers are of interest for the treatment of PD.[76] Therefore, it is not difficult to imagine that a series of synthetic derivatives of lobeline could be engineered and assessed for their ability to increase the activity of VMAT2.

Members of the family Fabaceae Lindl., including members of the genus *Cytisus* Desf. and *Lupinus* L., engineer a bewildering array of quinolizidine alkaloids that bind to nicotinic receptors. One such compound is lupanine (CS 1.32), which binds to and antagonizes nicotinic receptors

■ **CS 1.25** Epibatidine.

■ **CS 1.26** GTS-21.

■ **CS 1.27** A-84543.

■ **CS 1.28** 2b.

■ **CS 1.29** Altinicline.

■ **CS 1.30** TC-1734.

■ **CS 1.31** Lobeline.

■ **CS 1.32** Lupanine.

■ **CS 1.33** Cytisine.

■ **CS 1.34** Arecoline.

in the brain with an IC_{50} value equal to $5\,\mu M$.[77] It therefore seems plausible that the systematic study of fabaceous quinolizidine alkaloids and derivatives may result in the development of nicotinic receptor agonists that could be used as neuroprotective agents. This is shown with the quinolizidine alkaloid cytisine (CS 1.33), which is a partial agonist of $\alpha 4\beta 2$ nicotinic receptor with an IC_{50} value equal to $2\,\mu M$, a full agonist for the $\alpha 7$ nicotinic receptor,[55] and an agonist at muscarinic receptors.

In the brain, acetylcholine binds to the muscarinic receptor subtype 1 (M_1), coupled with the guanosine triphosphate binding protein Gq/11, hence causing phospholipase C (PLC) activation and membrane phospholipid phosphatidylinositol 4,5-bisphosphate (PIP2) conversion into inositol-3-phosphate (IP3) and diacylglycerol (DAG).[78] DAG activates the protein kinase C (PKC).[78] In parallel, inositol-1,4,5-trisphosphate (IP3) binds to and compels the aperture of inositol-1,4,5-trisphosphate (IP3)-receptor Ca^{2+} channels expressed by the endoplasmic reticulum (ER), which results in increased levels of cytoplasmic Ca^{2+}, which activates calmodulin and results in the closure of K^+ channels and abrogation of K^+ efflux, hence depolarization[78] and enhancement of cognition. An example of M_1 agonist is the piperidine alkaloid arecoline (CS 1.34) from *Areca catechu* L. (family Arecaceae Bercht. & J. Presl), or betel nut, which incurred, at low doses, beneficial effects in Alzheimer patients.[79] Of particular interest is the fact that selective agonists of post-synaptic cortical and hippocampal M_1 receptors are of value for the treatment of AD by virtue of their ability to both enhance cognition[80] and to compromise β-amyloid peptide aggregation.[81]

Note that baogongteng A (2-β-hydroxy-6β-acetoxy-nortropane, CS 1.35) from *Erycibe obtusifolia* Benth. (family Convolvulaceae Juss.) displayed muscarinic effect and was modified into the synthetic derivative 6-β-acetoxynortropane (CS 1.36), which exhibited robust muscarinic agonist properties.[82] The arecoline derivative xanomeline induced the release of neurotrophic *b*-amyloid precursor protein (*b*-APP) by CHOM$_1$ cells, a secreted form of APP (sAPPα), by Chinese hamster ovary (CHO) cells via M$_1$ activation with an IC$_{50}$ equal to 10 nM and enhanced the cognitive function of rodents at a dose of 3 mg/kg.[81,83] The remedial effect of xanomeline (CS 1.37) against dementia was further shown in a 6-month double-blind placebo-controlled clinical study conducted by Bodick et al.[84] In this context, it is important to mention that muscarinic agonists, such as oxotremorine, increased the levels of dopamine into the tegmental area.[84] In fact, xanomeline at a dose of 10 mg/kg increased extracellular levels of dopamine in the prefrontal cortex and *nucleus accumbens* by 400% and 170%, respectively.[85] Along the same lines, the synthetic pyrazine derivative WAY-132983 (CS 1.38) nullified scopolamine-impaired performance in rodents at a dose of 3 mg/kg.[69] Further efforts to develop highly selective M$_1$ receptor agonists resulted in the synthesis of 1-((4-cyano-4-(pyridin-2-yl)

■ **CS 1.35** 2-β-hydroxy-6β-acetoxy-nortropane.

■ **CS 1.36** 6-β-acetoxynortropane.

■ **CS 1.37** Xanomeline.

■ **CS 1.38** WAY-132983.

■ **CS 1.39** PQCA.

■ **CS 1.40** Oxotremorine.

■ **CS 1.41** Matrine.

■ **CS 1.42** Oxymatrine.

piperidin-1-yl)methyl)-4-oxo-4H-quinolizine-3-carboxylic acid, or PQCA (CS 1.39), which at a dose of 10 mg/kg improved the recognition memory of rodents poisoned with scopolamine.[70] Evidence suggests some correlation between M_1 and nuclear factor kappa-light-chain-enhancer of activated B cells (NF-κB). For instance, the synthetic pyrrolidine alkaloid oxotremorine (CS 1.40) is an M_1 agonist, which at a dose of 200 μM induced the generation of ROS in PC12 M1 cells, hence activation of ERK1/2 and an increase of AP-1 and NF-κB transcriptional activities.[86] Likewise, the stimulation of muscarinic receptors with 10 μM of acetylcholine compelled the activation of PKC, ERK1/2, nuclear factor of kappa light polypeptide gene enhancer in B-cells inhibitor α (IκBα) phosphorylation, and NF-κB activation in human bronchial epithelial cell line (16HBE) cells.[87] It is therefore conceivable that M_1 agonists may fight neurodegeneration. This assertion is supported by the demonstration that the quinolizidine alkaloid matrine (CS 1.41) from *Sophora flavescens* Aiton (family Fabaceae Lindl.), at a dose of 200 μmol/L protected astrocytes and cortical neurons against oxygen glucose deprivation by increasing the levels of IκBα hence inhibition of NF-κB-induced transcription of proapoptotic protein (p53) and c-Myc.[88] Reciprocally, M_1 antagonists or weak agonists may very well inhibit NF-κB because the inhibition of NF-κB was observed in brain tissues of rodents with permanent middle cerebral artery occlusion treated with 120 mg/kg of quinolizidine alkaloid oxymatrine (CS 1.42).[89] Furthermore, oxymatrine protected rodents against intracerebral hemorrhage with NF-κB inhibition at 120 mg/kg.[90] The cardinal role of NF-κB in neurodegeneration is further exemplified with the pyridine derivative pioglitazone (CS 1.43), which attenuated the effects of experimental ischemia in rodents by activating the peroxisome proliferator-activated receptor-γ (PPAR-γ).[91] Therefore, agonists of PPAR-γ are of interest for the treatment of neurodegenerative diseases. Note that the activation of nuclear factor kappa-light-chain-enhancer of activated B cells (NF-κB) in neurons during cerebral ischemia compels the transcription of pro-apoptotic protein c-Myc.[92] Likewise, the pyridine alkaloid anatabine from a member of the family Solanaceae inhibited β-amyloid peptide production by Chinese

■ **CS 1.43** Pioglitazone.

■ **CS 1.44** Pyrroloquinoline quinine.

hamster ovary (CHO) cells with an IC_{50} equal to $640 \mu g/mL$ together with inhibition β-cleavage of APP, inhibition of NF-κB, hence β-secretase inhibition[93] confirming that compounds which inhibit NF-κB have the compelling attribute to mitigate amyloidogenesis. The possibility of pyridine alkaloids preventing neurodegeneration by abating intraneuronal levels of ROS is exemplified with pyrroloquinoline quinone (CS 1.44), which is a powerful bacterial antioxidant that formed a complex with α-synuclein, thus nullifying the formation amyloid fibrils at a dose of $140 \mu M$[94] and protected human dopaminergic human SH-SY5Y neuroblastoma cells against the oxidative insults incurred by 6-HODA or H_2O_2.[95] Likewise, in neurons of rodents, fetal ventral mesencephalon pyrroloquinoline quinone nullified the oxidative insults incurred by H_2O_2.[95] Pyrroloquinoline quinone at a dose of $25 \mu M$ inhibited the generation of ROS by 25% in human SH-SY5Y neuroblastoma cells exposed to amyloid $\beta_{(25-35)}$ peptide and prevented apoptosis by reducing pro-apoptotic Bax, mitochondrial insult and caspase 3 activation.[96] Furthermore, pyrroloquinoline quinone at a dose of $100 \mu M$ protected hippocampal neurons against glutamate insults by counteracting glutamate-induced pro-apoptotic Bax upregulation, cytoplasmic Ca^{2+} increase, caspase 3 activation, decrease of superoxide dismutase (SOD) activity, and increase in ROS via a mechanism implying the activation of Akt.[97] It is therefore tempting to speculate that pyrroloquinoline quinone may be beneficial for the treatment of amyotrophic lateral sclerosis (ALS). In this context, it is of interest to note that $10 \mu M$ of the synthetic pyrazine PHID (CS 1.45) engineered by Kim et al.[98] protected human SH-SY5Y neuroblastoma cells against the pro-apoptotic

■ **CS 1.45** PHID.

■ **CS 1.46** (+)-*N*-deoxymilitarinone A.

■ **CS 1.47** Militarinone A.

effects of 1-methyl-4-phenylpyridinium ion (MPP$^+$) by attenuating caspase 3 activation and ROS accumulation. Likewise, the activation of Akt accounted for the development of neurites in PC12 cells exposed to 100 μM of (+)-*N*-deoxymilitarinone A (CS 1.46) isolated from the fungi *Paecilomyces farinosus*.[99] Furthermore, PC12 cells exposed to 40 μM of militarinone A (CS 1.47) increased cytoplasmic levels in phosphorylated Akt and increased phosphorylated ERK1/2.[100]

Note that pyridine alkaloids such as 12′-hydroxy-7′-multijuguinol (CS 1.48), 12′-hydroxy-8′-multijuguinol (CS 1.49), methyl multijuguinate (CS 1.50), 7′-multijuguinol (CS 1.51), and 8′-multijuguinol (CS 1.52) from *Senna multijuga* (Rich.) H.S. Irwin & Barneby (family Fabaceae Lindl.) inhibited the enzymatic activity of AChE by 51%, 28%, 40%, 52%, and 19% respectively at a dose of 350 μM.[101] In fact, the synthetic pyridine

■ **CS 1.48** 12′-hydroxy-7′-multijuguinol.

■ **CS 1.49** 12′-hydroxy-8′-multijuguinol.

■ **CS 1.50** Methyl multijuguinate.

■ **CS 1.51** 7′-multijuguinol.

■ **CS 1.52** 8′-multijuguinol.

derivative tacrine (CS 1.53) inhibited the enzymatic activity of AChE with an IC_{50} value equal to 125 nM and has been marketed in the United States for the treatment of AD[102] but was discontinued because of serious hepatotoxic effect.[103,104] Huperzine A (CS 1.54) from *Huperzia serrata* (Thunb.) Trevis. (family Lycopodiaceae P. Beauv. ex Mirb.) inhibited the enzymatic activity of acetylcholinesterase (AChE) with an IC_{50} value equal to 0.08 μM[105] and was used in China to treat AD.[106] The seeds of *Coffea*

■ **CS 1.53** Tacrine.

■ **CS 1.54** Huperzine A.

■ **CS 1.55** Caffeine.

■ **CS 1.56** DMPX.

arabica L. (family Rubiaceae Juss.) contain the pyridine alkaloid trigonelline, which at a dose of $30 \mu g$ induced the formation of dendrites by human SK-N-SH neuroblastoma cells cultured *in vitro*.[107]

A compelling body of evidence has led to the generally accepted view that caffeine (CS 1.55) intake decreases the probability of developing PD.[108–110] Jones et al.[111] first reported that the brain of rodents exposed to kainic acid was protected with injection of adenosine analogue at a dose of $25 \mu g/kg$ by a mechanism involving blockade of the metabotropic adenosine$_{2A}$ receptors. Furthermore, caffeine at a dose of $10 mg/kg$ sustained the nigrostriatal activity of rodents poisoned with 1-methyl-4-phenyl-1,2,3,6-tetrahydropyridine (MPTP) by 40%, by blocking adenosine$_{2A}$ receptors.[112] The anti-apoptotic property of caffeine was further demonstrated, whereas at a dose of $100 \mu M$ this alkaloid preserved the viability of human SH-SY5Y neuroblastoma cells exposed to 1-methyl-4-phenylpyridinium ion (MPP$^+$) by 80% via activation of phosphoinositide 3-kinase (PI3K), and thus Akt and inhibition of caspase 3,[113] suggesting mitochondrial protection. Akt is in fact inhibited by cAMP,[114] so one could speculate that the cytoplasmic depletion of cAMP and the resulting inhibition of PKA may account for neuroprotection. Indeed, caffeine blocks adenoside$_{2A}$ receptors expressed by striatopallidal neurons and reduces γ-amino butyric acid (GABA) transmission with the *globus pallidus*.[115] PD involves apoptosis of neurons located in the *substantia nigra* which then are unable to secrete enough dopamine to stimulate dopaminergic subtype 1 (D$_1$) and subtype 2 (D$_2$) receptors in the striatum.[115] The *globus pallidus pars externa* is inhibited by striatopallidal GABA, hence release of glutamate by the subthalamic nucleus, which activates *globus pallidus pars interna/substantia nigra pars reticulata* that inhibit the pedunculopontine nucleus and excitants of motoneurons.[115] Thus, derivatives of caffeine were prepared such as DMPX (CS 1.56), istradefylline (CS 1.57), and MSX3 (CS 1.58), culminating with preladenant, which is currently being evaluated in clinical trials for the treatment of PD.[115] Theophylline and theobromine (CS 1.59–1.60) block adenosine receptor A$_{2A}$R with Ki values equal to 1710nM.[116]

A number of other miscellaneous alkaloids of neurological value are worth being mentioned here. One such alkaloid is leonurine (CS 1.61), from *Leonurus heterophyllus* Sweet (family Lamiaceae Martinov), which at a dose of $60 mg/kg$ reduced the size of necrosis by 10% in rodent stroke models with reduction of ROS, protection of mitochondrial integrity, and therefore reduction of cytochrome c.[117] The securinega-type alkaloid (+)-securinine (CS 1.62), from *Phyllanthus amarus* Schumach. & Thonn. (family Phyllanthaceae Martinov), is a γ-amino butyric acid subtype $_A$(GABA$_A$) antagonist, which at a dose of $40 mg/kg$ protected rodents against the cognitive disturbances imposed by intracerebral ventricle

■ **CS 1.57** Istradefylline.

■ **CS 1.58** MSX3.

■ **CS 1.59** Theophylline.

■ **CS 1.60** Theobromine.

■ **CS 1.61** Leonurine.

injection of β-amyloid (Aβ_{25-35}) in rodents, with a decrease of AChE activity and neuroinflammation.[118] Juliflorine (CS 1.63), from *Prosopis juliflora* (Sw.) DC. (family Fabaceae Lindl.), inhibited the enzymatic activity of AChE with an IC$_{50}$ value of 0.4 μM and relaxed the K$^+$-induced contraction of rabbit jejunum.[119] Himbacine (CS 1.64) is a piperidine alkaloid

■ **CS 1.62** Securinine.

■ CS 1.63 Juliflorine.

■ CS 1.64 Himbacine.

isolated from *Galbulimima baccata* F.M. Bailey (family Himantandraceae Diels), which is a potent inhibitor of M2 muscarinic receptor antagonist with a Ki value equal to 4.5.[120] Note the stimulation of cerebral muscarinic receptors subtype 2 (M_2) results in the activation of guanosine triphosphate binding protein $G\alpha i$, which inhibits adenylate cyclase (AC), resulting in cAMP collapse, opening of K^+ channels, hyperpolarization, and paralysis of neurons; therefore, M_2 antagonists might be useful to enhance neurotransmission.[121]

REFERENCES

[1] Li HY, Sun NJ, Kashiwada Y, Sun L, Snider JV, Cosentino L, et al. Anti-AIDS agents, 9. Suberosol, a new C 31 lanostane-type triterpene and anti-HIV principle from *Polyalthia suberosa*. J Nat Prod 1993;56(7):1130–3.

[2] Tuchinda P, Pohmakotr M, Munyoo B, Reutrakul V, Santisuk T. An azaanthracene alkaloid from *Polyalthia suberosa*. Phytochemistry 2000;53(8):1079–82.

[3] Tuchinda P, Pohmakotr M, Reutrakul V, Thanyachareon W, Sophasan S, Yoosook C, et al. 2-Substituted furans from Polyalthia suberosa. Planta Med 2001;67(6):572–5.

[4] Thangnipon W, Suwanna N, Kitiyanant N, Soi-ampornkul R, Tuchinda P, Munyoo B, et al. Protective role of N-trans-feruloyltyramine against β-amyloid peptide-induced neurotoxicity in rat cultured cortical neurons. Neurosci Lett 2012;513(2):229–32.

[5] Polinsky RJ. Clinical pharmacology of rivastigmine: a new-generation acetylcholinesterase inhibitor for the treatment of Alzheimer's disease. Clin Ther 1998;20(4):634–47.

[6] Yokozawa T, Wang TS, Chen CP, Hattori M. *Tinospora tuberculata* suppresses nitric oxide synthesis in mouse macrophages. Biol Pharm Bull 1999;22:1306–9.

[7] Moncada S, Palmer RM, Higgs EA. Nitric oxide: physiology, pathophysiology and pharmacology. Pharmacol Rev 1991;43:109–42.

[8] Cecchi C, Fiorillo C, Baglioni S, Pensalfini A, Bagnoli S, Nacmias B, et al. Increased susceptibility to amyloid toxicity in familial Alzheimer's fibroblasts. Neurobiol Aging 2007;28:863–76.

[9] Sharpe C, Arnoult D, Youle RJ. Control of mitochondrial permeability by Bcl-2 family members. Biochim Biophys Acta 2004;1644:107–13.

[10] Wei MC, Zong WX, Cheng EH, Lindten T, Panoutsakopoulou V, Ross AJ, et al. Proapoptotic Bax and Bak: a requisite gateway to mitochondrial dysfunction and death. Science 2001;292:727–30.

[11] Arima K. Ultrastructural characteristics of tau filaments in tauopathies: immuno-electron microscopic demonstration of tau filaments in tauopathies. Neuropathol 2006;26:475–83.

[12] Iqbal K, Grundke-Iqbal I. Pharmacological approaches of neurofibrillary degeneration. Curr Alzheimer Res 2005;2:335–41.

[13] Pickhardt M, Larbig G, Khlistunova I, Coksezen A, Meyer B, Mandelkow EM, et al. Phenylthiazolyl-hydrazide and its derivatives are potent inhibitors of tau aggregation and toxicity *in vitro* and in cells. Biochem 2007;46:10016–10023.

[14] Bandopadhyay R, Kingsbury AE, Cookson MR, Reid AR, Evans IM, Hope AD, et al. The expression of DJ-1 (PARK7) in normal human CNS and idiopathic Parkinson's disease. Brain 2004;127:420–30.

[15] Miyazaki S, Yanagida T, Nunome K, Ishikawa S, Inden M, Kitamura Y, et al. DJ-1–binding compounds prevent oxidative stress-induced cell death and movement defect in Parkinson's disease model rats. J Neurochem 2008;105:2418–34.

[16] Yanagisawa D, Kitamura Y, Inden M, Takata K, Taniguchi T, Morikawa S, et al. DJ-1 protects against neurodegeneration caused by focal cerebral ischemia and reperfusion in rats. J Cereb Blood Flow Metab 2007;28:563–78.

[17] Kitamura Y, Watanabe S, Taguchi M, Takagi K, Kawata T, Takahashi-Niki K, et al. Neuroprotective effect of a new DJ-1–binding compound against neurodegeneration in Parkinson's disease and stroke model rats. Mol Neurodegener 2011;6(1):48.

[18] Howlett AC, Qualy JM, Khachatrian LL. Involvement of Gi in the inhibition of adenylate cyclase by cannabimimetic drugs. Mol Pharmacol 1986;29:307–13.

[19] Guzman M, Sanchez C, Galve-Roperh I. Control of the cell survival/death decision by cannabinoids. J Mol Med 2001;78:613–25.

[20] Maccarrone M, Lorenzon T, Bari M, Melino G, Finazzi-Agro A. Anandamide induces apoptosis in human cells via vanilloid receptors. Evidence for a protective role of cannabinoid receptors. J Biol Chem 2000;275:31938–31945.

[21] Reggio PH, Traore H. Conformational requirements for endocannabinoid interaction with the cannabinoid receptors, the anandamide transporter and fatty acid amidohydrolase. Chem Phys Lipids 2000;108:15–35.

[22] Duncan RS, Chapman KD, Koulen P. The neuroprotective properties of palmitoylethanolamine against oxidative stress in a neuronal cell line. Mol Neurodegener 2009;4:50.

[23] Duncan RS, Xin H, Goad DL, Chapman KD, Koulen P. Protection of neurons in the retinal ganglion cell layer against excitotoxicity by the *N*-acylethanolamine, *N*-linoleoylethanolamine. Clin Ophthalmol 2011;5(1):543–8.

[24] Luo Y, DeFranco DB. Opposing roles for ERK1/2 in neuronal oxidative toxicity: distinct mechanisms of ERK1/2 action at early versus late phases of oxidative stress. J Biol Chem 2006;281:16436–42.

[25] Mackie K, Hille B. Cannabinoids inhibit N-type calcium channels in neuroblastoma-glioma cells. Proc Natl Acad Sci U S A 1992;89:3825–9.

[26] Shen M, Piser TM, Seybold VS, Thayer SA. Cannabinoid receptor agonists inhibit glutamatergic synaptic transmission in rat hippocampal cultures. Res J Neurosci 1996;16:4322–34.

[27] Tsai JY, Chou CJ, Chen CF, Chiou WF. Antioxidant activity of *piperlactam* S: prevention of copper-induced LDL peroxidation and amelioration of free radical-induced oxidative stress of endothelial cells. Planta Med 2003;69(1):3–8.

[28] Lin LC, Shen CC, Shen YC, Tsai TH. Anti-inflammatory neolignans from *Piper kadsura*. J Nat Prod 2006;69(5):842–4.

[29] Jiang RW, Mak TCW, Fung KP. Molecular structures of two bicyclo-(3.2.1)-octanoid neolignans from *Piper kadsura*. J Mol Struct 2003;654(1–3):177–82.

[30] Kim KH, Choi JW, Ha SK, Kim SY, Lee KR. Neolignans from *Piper kadsura* and their anti-neuroinflammatory activity. Bioorg Med Chem Lett 2010;20(1):409–12.

[31] Xia W, Zeng JP, Chen LB, Jiang AL, Xiang L, Xu J, et al. Inhibition of beta-amyloid precursor protein gene in SK-N-SH cells by piperlonguminine/dihydropiperlonguminine components separated from Chinese herbal medicine Futokadsura stem. Chin J Physiol 2007;50(4):157–63.

[32] Chonpathompikunlert P, Wattanathorn J, Muchimapura S. Piperine, the main alkaloid of Thai black pepper, protects against neurodegeneration and cognitive impairment in animal model of cognitive deficit like condition of Alzheimer's disease. Food Chem Toxicol 2010;48:798–802.

[33] Lee SA, Hong SS, Han XH, Hwang JS, Oh GJ, Lee KS, et al. Piperine from the fruits of *Piper longum* with inhibitory effect on monoamine oxidase and antidepressant-like activity. Chem Pharm Bull 2005;53(7):832–5.

[34] Mu LH, Wang B, Ren HY, Liu P, Guo DH, Wang FM, et al. Synthesis and inhibitory effect of piperine derivates on monoamine oxidase. Bioorg Med Chem Lett 2012;22(9):3343–8.

[35] Naoi M, Maruyama W, Akao Y, Yi H. Mitochondria determine the survival and death in apoptosis induced by an endogenous neurotoxin, *N*-methyl(R)salsolinol, and neuroprotection by propargylamines. J Neural Transm 2002;109:607–21.

[36] Naoi M, Maruyama W, Akao Y, Yi H, Yamaoka Y. Involvement of type A monoamine oxidase in neurodegeneration: regulation of mitochondrial signaling leading to cell death or neuroprotection. J Neural Transm 2006;71:67–77.

[37] Uwabe KI, Iwakawa T, Matsumoto M, Talkahashi K, Nagata K. HU0622: a small molecule promoting GAP-43 activation and neurotrophic effects. Neuropharmacol 2006;51(4):727–36.

[38] Lonze BE, Riccio A, Cohen S, Ginty DD. Apoptosis, axonal growth defects, and degeneration of peripheral neurons in mice lacking CREB. Neuron 2002;34:371–85.

[39] Williams P, Sorribas A, Liang Z. New methods to explore marine resources for Alzheimer's therapeutics. Curr Alzheimer Res 2010;7(3):210–13.

[40] Valero T, del Barrio L, Egea J, Cañas N, Martínez A, García AG, et al. NP04634 prevents cell damage caused by calcium overload and mitochondrial disruption in bovine chromaffin cells. Eur J Pharmacol 2009;607(1–3):47–53.

[41] Sugimoto H, Tsuchiya Y, Sugumi H, Higurashi K, Karibe N, Iimura Y, et al. Novel piperidine derivatives. Synthesis and anti-acetylcholinesterase activity

of 1-benzyl-4-[2-(*N*-benzoylamino)ethyl]piperidine derivatives. J Med Chem 1990;33(7):1880–7.

[42] Sugimoto H. Donepezil hydrochloride: a treatment drug for Alzheimer's disease. Chem Rec 2001;1(1):63–73.

[43] Murakoshi I, Kidoguchi E, Haginiwa J, Ohmiya S, Higashiyama K, Otomasu H. (+)-Kuraramine, a possible metabolite of (−)-*N*-methylcytosine in flowers of *Sophora flavescens*. Phytochemistry 1981;20(6):1407–9.

[44] Murakoshi I, Kidoguchi E, Haginiwa J, Ohmiya S, Higashiyama K, Otomasu H. Isokuraramine and (−)-7,11-dehydromatrine, lupin alkaloids from flowers of *Sophora flavescens*. Phytochemistry 1982;21(9):2379–84.

[45] Ryu SY, Kim SK, No Z, Ahn JW. A novel flavonoid from *Sophora flavescens*. Planta Med 1996;62(4):361–2.

[46] Kim YK, Min BS, Bae KH. A cytotoxic constituent from *Sophora flavescens*. Arch Pharmacol Res 1997;20(4):342–5.

[47] Woo ER, Jong HK, Hyoung JK, Park H. A new prenylated flavonol from the roots of *Sophora flavescens*. J Nat Prod 1998;61(12):1552–4.

[48] Shen CC, Lin TW, Huang YL, Wan ST, Shien BJ, Chen CC. Phenolic constituents of the roots of *Sophora flavescens*. J Nat Prod 2006;69(8):1237–40.

[49] Dajas-Bailador F, Wonnacott S. Nicotinic acetylcholine receptors and the regulation of neuronal signalling. Trends Pharmacol Sci 2004;25(6):317–24.

[50] Wang HY, Lee DHS, Davis CB, Shank RP. Amyloid peptide Abeta(1–42) binds selectively and with picomolar affinity to alpha7 nicotinic acetylcholine receptors. J Neurochem 2000;75(3):1155–61.

[51] Auld DS, Kar S, Quirion K. Beta-amyloid peptides as direct cholinergic neuromodulators: a missing link? Trends Neurosci 1998;21:43–9.

[52] Gouras GK, Tsai J, Nasland J, Vicent B, Edgar M, Checler F, et al. Intraneuronal Ab42 accumulation in human brain. Am J Pathol 2000;156:15.

[53] Newhouse PA, Potter A, Lenox RH. Effects of nicotinic agents on human cognition: possible therapeutic applications in Alzheimer's and Parkinson's diseases. Med Chem Res 1993;2:628–42.

[54] Sahakian B, Jones G, Levy R, Gray J, Warburton D. The effects of nicotine on attention, information processing, and short-term memory in patients with dementia of the Alzheimer type. Br J Psychiatry 1989;154:797–800.

[55] Chavez-Noriega LE, Gillespie A, Stauderman KA, Crona JH, Claeps BO, Elliott KJ, et al. Characterization of the recombinant human neuronal nicotinic acetylcholine receptors alpha3beta2 and alpha4beta2 stably expressed in HEK293 cells. Neuropharmacol 2000;39(13):2543–60.

[56] Arendash GW, Sengstock GJ, Sanberg PR, Kem WR. Improved learning and memory in aged rats with chronic administration of the nicotinic receptor agonist GTS-21. Brain Res 1995;674:252–9.

[57] Dickerson TJ, Janda KD. Glycation of the amyloid β-protein by a nicotine metabolite: a fortuitous chemical dynamic between smoking and Alzheimer's disease. Proc Natl Acad Sci U S A 2003;100:8182–7.

[58] Fowler JS, Volkow ND, Wang GJ, Pappas N, Logan J, MacGregor R, et al. Inhibition of monoamine oxidase B in the brains of smokers. Nature 1996;379:733–6.

[59] Obata T, Aomine M, Inada T, Kinemuchi H. Nicotine suppresses 1-methyl-4-phenylpyridinium ion-induced hydroxyl radical generation in rat striatum. Neurosci Lett 2002;330:122–4.

[60] Ryan RE, Ross SA, Drago J, Loiacono RE. Dose-related neuroprotective effects of chronic nicotine in 6-hydroxydopamine treated rats, and loss of neuroprotection in a4 nicotinic receptor subunit knockout mice. Br J Pharmacol 2001;132:1650–6.

[61] Toulorge D, Guerreiro S, Hild A, Maskos U, Hirsch EC, Michel PP. Neuroprotection of midbrain dopamine neurons by nicotine is gated by cytoplasmic Ca^{2+}. FASEB J 2011;25:2563–73.

[62] Gronier B, Perry KW, Rasmussen K. Activation of the mesocorticolimbic dopaminergic system by stimulation of muscarinic cholinergic receptors in the ventral tegmental area. Psychopharmacology 2000;147(4):347–55.

[63] Dwoskin LP, Teng L, Buxton ST, Ravard A, Deo N, Crooks PA. Minor alkaloids of tobacco release [3H]dopamine from superfused rat striatal slices. Eur J Pharmacol 1995;276(1–2):195–9.

[64] Janhunen SL, Ahtee L. Comparison of the effects of nicotine and epibatidine on the striatal extracellular dopamine. Eur J Pharmacol 2004;494:167–77.

[65] Kern WR. A study of the occurrence of anabaseine in *Paranemertes* and other nemertines. Toxicon 1971;9:23–32.

[66] Martin EJ, Panickar KS, King MA, Deyrup M, Hunter BE, Wang G, et al. Cytoprotective actions of 2,4-dimethoxybenzylidene anabaseine in differentiated PC12 cells and septal cholinergic neurons. Drug Dev Res 1994;31(2):135–41.

[67] Lin NH, Gunn DE, Li Y, He Y, Bai H, Ryther KB, et al. Synthesis and structure-activity relationships of pyridine-modified analogs of 3-[2-((S)-pyrrolidinyl)methoxy]pyridine, A-84543, a potent nicotinic acetylcholine receptor agonist. Bioorg Med Chem Lett 1998;8(3):249–54.

[68] Albrecht BK, Berry V, Boezio AA, Cao L, Clarkin K, Guo W, et al. Discovery and optimization of substituted piperidines as potent, selective, CNS-penetrant alpha4beta2 nicotinic acetylcholine receptor potentiators. Bioorg Med Chem Lett 2008;18(19):5209–12.

[69] Bartolomeo AC, Morris H, Buccafusco JJ, Kille N, Rosenzweig-Lipson S, Husbands MG, et al. The preclinical pharmacological profile of WAY-132983, a potent M1 preferring agonist. J Pharmacol Exp Ther 2000;292(2):584–96.

[70] Uslaner JM, Eddins D, Puri V, Cannon CE, Sutcliffe J, Chew CS, et al. The muscarinic M1 receptor positive allosteric modulator PQCA improves cognitive measures in rat, cynomolgus macaque, and rhesus macaque. Psychopharmacology 2013;225(1):21–30.

[71] Cosford ND, Bleicher L, Vernier JM, Chavez-Noriega LE, Rao TS, Siegel RS, et al. Recombinant human receptors and functional assays in the discovery of altinicline (SIB-1508Y), a novel acetylcholine-gated ion channel (nAChR agonist). Pharm Acta Helv 2003;74:125–30.

[72] Gatto GJ, Bohme GA, Caldwell WS, Letchworth SR, Traina VM, Obinu MC, et al. TC-1734: an orally active neuronal nicotinic acetylcholine receptor modulator with antidepressant, neuroprotective and long-lasting cognitive effects. CNS Drug Rev 2004;10(2):147–66.

[73] Decker MW, Majchrzak MJ, Arneric SP. Effects of lobeline, a nicotinic receptor agonist, on learning and memory. Pharmacol Biochem Behav 1993;45(3):571–6.

[74] Teng L, Crooks PA, Sonsalla PK, Dwoskin LP. Lobeline and nicotine evoke [3H]overflow from rat striatal slices preloaded with [3H]dopamine: differential

inhibition of synaptosomal and vesicular [3H]dopamine uptake. J Pharmacol Exp Ther 1997;280(3):1432–44.

[75] Teng L, Crooks PA, Dwoskin LP. Lobeline displaces [3H]dihydrotetrabenazine binding and releases [3H]dopamine from rat striatal synaptic vesicles: comparison with d-amphetamine. J Neurochem 1998;71(1):258–65.

[76] Zheng G, Dwoskin LP, Crooks PA. Vesicular monoamine transporter 2: role as a novel target for drug development. AAPS J 2006;8(4):682–92.

[77] Schmeller T, Sauerwein M, Sporer F, Wink M, Muller WE. Binding of quinolizidine alkaloids to nicotinic and muscarinic acetylcholine receptors. J Nat Prod 1994;57(9):1316–19.

[78] Berstein G, Blank JL, Smrcka AV, Higashijima T, Sternweis PC, Exton JH, et al. Reconstitution of agonist stimulated phosphatidylinositol 4,5-bisphosphate hydrolysis using purified m1 muscarinic receptor, Gq/11, and phospholipase C-beta. J Biol Chem 1992;267:8081–8.

[79] Soncrant TT, Raffaele KC, Asthana S, Berardi A, Morris PP, Haxby JV. Memory improvement without toxicity during chronic, low dose intravenous arecoline in Alzheimer's disease. Psychopharmacol (Berl) 1993;112(4):421–7.

[80] Volpicelli LA, Levey AI. Muscarinic acetylcholine receptor subtypes in cerebral cortex and hippocampus. Prog Brain Res 2004;145:59–66.

[81] Eckols K, Bymaster FP, Mitch CH, Shannon HE, Ward JS, DeLapp NW. The muscarinic M1 agonist xanomeline increases soluble amyloid precursor protein release from Chinese hamster ovary-M1 cells. Life Sci 1995;57:1183–90.

[82] Pei XF, Gupta TH, Badio B, Padgett WL, Daly JW. 6beta-Acetoxynortropane: a potent muscarinic agonist with apparent selectivity toward M2-receptors. J Med Chem 1998;41(12):2047–55.

[83] Bymaster FP, Whitesitt CA, Shannon HE, Delapp N, Ward JS, Calligaro DO, et al. Xanomeline: G selective muscarinic agonist for the treatment of Alzheimer's disease. Drug Dev Res 1997;40(2):158–70.

[84] Bodick NC, Offen WW, Levey AI, Cutler NR, Gauthier SG, Satlin A, et al. Effects of xanomeline, a selective muscarinic receptor agonist, on cognitive function and behavioral symptoms in Alzheimer disease. Arch Neurol 1997;54(4):465–73.

[85] Perry KW, Nisenbaum LK, George CA, Shannon HE, Felder CC, Bymaster FP. The muscarinic agonist xanomeline increases monoamine release and immediate early gene expression in the rat prefrontal cortex. Biol Psychiatry 2001;49(8):716–25.

[86] Mangelus M, Kroyter A, Galron R, Sokolovsky M. Reactive oxygen species regulate signaling pathways induced by M1 muscarinic receptors in PC12M1 cells. J Neurochem 2001;76(6):1701–11.

[87] Profita M, Bonanno A, Siena L, Ferraro M, Montalbano AM, Pompeo F, et al. Acetylcholine mediates the release of IL-8 in human bronchial epithelial cells by a NFκB/ERK1/2-dependent mechanism. Eur J Pharmacol 2008;582(1–3):145–53.

[88] Xu M, Yang L, Hong LZ, Zhao XY, Zhang HL. Direct protection of neurons and astrocytes by matrine via inhibition of the NF-κB signaling pathway contributes to neuroprotection against focal cerebral ischemia. Brain Res 2012;1454:48–64.

[89] Liu Y, Zhang X, Yang C, Fan H. Oxymatrine protects rat brains against permanent focal ischemia and downregulates NF-kappaB expression. Brain Res 2009;1268:174–80.

[90] Huang M, Hu YY, Dong XQ, Xu QP, Yu WH, Zhang ZY. The protective role of oxymatrine on neuronal cell apoptosis in the hemorrhagic rat brain. J Ethnopharmacol 2012;143(1):228–35.

[91] Zhang HL, Xu M, Wei C, Qin AP, Liu CF, Hong LZ, et al. Neuroprotective effects of pioglitazone in a rat model of permanent focal cerebral ischemia are associated with peroxisome proliferator-activated receptor gamma-mediated suppression of nuclear factor-κB signaling pathway. Neurosci 2011;176:381–95.

[92] Zhang W, Potrovita I, Tarabin V, Herrmann O, Beer V, Weih F, et al. Neuronal activation of NF-kappaB contributes to cell death in cerebral ischemia. J Cereb Blood Flow Metab 2005;25(1):30–40.

[93] Paris D, Beaulieu-Abdelahad D, Bachmeier C, Reed J, Ait-Ghezala G, Bishop A, et al. Anatabine lowers Alzheimer's Aβ production *in vitro* and *in vivo*. Eur J Pharmacol 2011;670(2–3):384–91.

[94] Kobayashi M, Kim J, Kobayashi N, Han S, Nakamura C, Ikebukuro K, et al. Pyrroloquinoline quinone (PQQ) prevents fibril formation of alpha-synuclein. Biochem Biophys Res Commun 2006;349(3):1139–44.

[95] Nunome K, Miyazaki S, Nakano M, Iguchi-Ariga S, Ariga H. Pyrroloquinoline quinone prevents oxidative stress-induced neuronal death probably through changes in oxidative status of DJ-1. Biol Pharm Bull 2008;31(7):1321–6.

[96] Zhang JJ, Zhang RF, Meng XK. Protective effect of pyrroloquinoline quinone against Abeta-induced neurotoxicity in human neuroblastoma SH-SY5Y cells. Neurosci Lett 2009;464(3):165–9.

[97] Zhang Q, Shen M, Ding M, Shen D, Ding F. The neuroprotective action of pyrroloquinoline quinone against glutamate-induced apoptosis in hippocampal neurons is mediated through the activation of PI3K/Akt pathway. Toxicol Appl Pharmacol 2011;252(1):62–72.

[98] Kim IS, Koppula S, Kim BW, Song MD, Jung JY, Lee G, et al. A novel synthetic compound PHID (8-Phenyl-6a, 7, 8, 9, 9a, 10-hexahydro-6H-isoindolo [5, 6-g] quinoxaline-7, 9-dione) protects SH-SY5Y cells against MPP(+)-induced cytotoxicity through inhibition of reactive oxygen species generation and JNK signaling. Eur J Pharmacol 2011;650(1):48–57.

[99] Cheng Y, Schneider B, Riese U, Schubert B, Li Z, Hamburger M. (+)-N-Deoxymilitarinone A, a neuritogenic pyridone alkaloid from the insect pathogenic fungus *Paecilomyces farinosus*. J Nat Prod 2006;69(3):436–8.

[100] Riese U, Ziegler E, Hamburger M. Militarinone A induces differentiation in PC12 cells via MAP and Akt kinase signal transduction pathways. FEBS Lett 2004;577(3):455–9.

[101] Francisco W, Pivatto M, Danuello A, Regasini LO, Baccini LR, Young MCM, et al. Pyridine alkaloids from *Senna multijuga* as acetylcholinesterase inhibitors. J Nat Prod 2012;75(3):408–13.

[102] Snape MF, Misra A, Murray TK, De Souza RJ, Williams JL, Cross AJ, et al. A comparative study in rats of the *in vitro* and *in vivo* pharmacology of the acetylcholinesterase inhibitors tacrine, donepezil and NXX-066. Neuropharmacology 1999;38(1):181–93.

[103] Watkins PB, Zimmerman HJ, Knapp MJ, Gracon SI, Lewis KW. Hepatotoxic effects of tacrine administration in patients with Alzheimer's disease. JAMA 1994;271:992–8.

[104] Blackard WG, Sood GK, Crowe DR, Fallon MB. Tacrine. A cause of fatal hepatotoxicity? J Clin Gastroenterol 1998;26(1):57–9.

[105] Tang XC, Han YF. Pharmacological profile of huperzine A, a novel acetylcholinesterase inhibitor from Chinese herb. CNS Drug Rev 1999;5(3):281–300.

[106] Little JT, Walsh S, Aisen PS. An update on huperzine A as a treatment for Alzheimer's disease. Expert Opin Investig Drugs 2008;17(2):209–15.

[107] Tohda C, Nakamura N, Komatsu K, Hattori M. Trigonelline-induced neurite outgrowth in human neuroblastoma SK-N-SH cells. Biol Pharm Bull 1999;22(7):679–82.

[108] Costa J, Lunet N, Santos C, Santos J, Vaz-Carneiro A. Caffeine exposure and the risk of Parkinson's disease: a systematic review and meta-analysis of observational studies. J Alzheimer's Dis 2010;20(Suppl. 1):S221–38.

[109] Ascherio A, Zhang SM, Hernan MA, Kawachi I, Colditz GA, Speizer FE, et al. A prospective study of caffeine consumption and risk of Parkinson's disease in men and women. Ann Neurol 2001;50(1):56–63.

[110] Ross GW, Abbott RD, Petrovitch H, Morens DM, Grandinetti A, Tung KH, et al. Association of coffee and caffeine intake with the risk of Parkinson disease. JAMA 2000;283:2674–9.

[111] Jones PA, Smith RA, Stone TW. Protection against hippocampal kainite excitotoxicity by intracerebral administration of an adenosine A2A receptor antagonist. Brain Res 1998;800:328–35.

[112] Chen JF, Xu K, Petzer JP, Staal R, Xu YH, Beilstein M, et al. Neuroprotection by caffeine and A(2A) adenosine receptor inactivation in a model of Parkinson's disease. J Neurosci 2001;21:RC143.

[113] Nakaso K, Ito S, Nakashima K. Caffeine activates the PI3K/Akt pathway and prevents apoptotic cell death in a Parkinson's disease model of SH-SY5Y cells. Neurosci Lett 2008;432:146–50.

[114] Bernstein D, Fajardo G, Zhao M. The role of β-adrenergic receptors in heart failure: differential regulation of cardiotoxicity and cardioprotection. Prog Pediatr Cardiol 2011;31(1):35–8.

[115] Morelli M, Di Paolo T, Wardas J, Calon F, Xiao D, Schwarzschild MA. Role of adenosine A2A receptors in parkinsonian motor impairment and l-DOPA–induced motor complications. Prog Neurobiol 2007;83(5):293–309.

[116] Armentero MT, Pinna A, Ferré S, Lanciego JL, Müller CE, Franco R. Past, present and future of A(2A) adenosine receptor antagonists in the therapy of Parkinson's disease. Pharmacol Ther 2011;132(3):280–99.

[117] Qi J, Hong ZY, Xin H, Zhu YZ. Neuroprotective effects of leonurine on ischemia/reperfusion-induced mitochondrial dysfunctions in rat cerebral cortex. Biol Pharm Bull 2010;33(12):1958–64.

[118] Lin X, Jun-Tian Z. Neuroprotection by D-securinine against neurotoxicity induced by beta-amyloid (25–35). Neurol Res 2004;26(7):792–6.

[119] Choudhary MI, Nawaz SA, Zaheer-Ul-Haqa Azim MK, Ghayur MN, Lodhi MA, et al. Juliflorine: a potent natural peripheral anionic-site-binding inhibitor of acetylcholinesterase with calcium-channel blocking potential, a leading

candidate for Alzheimer's disease therapy. Biochem BiophysRes Commun 2005;332(4):1171–9.

[120] Doller D, Chackalamannil S, Czarniecki M, McQuade R, Ruperto V. Design, synthesis, and structure-activity relationship studies of himbacine derived muscarinic receptor antagonists. Bioorg Med Chem Lett 1999;9(6):901–6.

[121] Egan TM, North RA. Acetylcholine hyperpolarizes central neurones by acting on an M2 muscarinic receptor. Nature 1986;319(6052):405–7.

Topic **1.2**

Indole Alkaloids

1.2.1 *Scorodocarpus borneensis* (Baill.) Becc.

History The plant was first described by Odoardo Beccari in *Nuovo Giornale Botanico Italiano* published in 1877.

Synonym *Ximenia borneensis* Baill.

Family Olacaceae R. Br., 1818

Common Name Kulim (Malay, Indonesian)

Habitat and Description This massive and magnificent timber grows to 60m tall in the rainforest of Thailand, Malaysia, Indonesia, and the Philippines. The tree is easily spotted because it produces an intense smell of garlic. The trunk is straight and covered with a conspicuously fissured grayish bark. The inner bark is sappy and purplish red. The leaves are simple, spirally arranged, and without stipules. The petiole is 1–2cm long. The blade is broadly elliptic, leathery, dark green glossy above, round at the base, acuminate at the apex, 7–15cm × 3–8cm, and it presents 4–5 pairs of discrete secondary nerves. The inflorescence is an axillary raceme of 20cm in length. The calyx is minute, cupular, and produces 5 inconspicuous lobes. The corolla consists of 5 oblong white petals, which are 0.8–1.5cm × 0.2–0.5cm, hairy at the throat, and reflexed. The androecium includes 8–10 stamens developing from the petal. The fruit is a greenish drupe that is globose and 5cm across. The seed is concealed in a crustaceous shell and exudes a yellowish oil which has an intense smell of garlic (Figure 1.4).

■ **FIGURE 1.4** *Scorodocarpus borneensis* (Baill.) Becc.

Medicinal Uses In Malaysia, the seeds are used to treat skin infection and as a substitute for garlic.

Phytopharmacology The few phytochemical studies devoted to this plant allowed the isolation and characterization of flavonoid glycosides such as lucenin-2; vicenin-2; isoschaftoside; some megastigmanes which include grasshopper ketone and blumenol B[122]; series of alkyl-sulfurs[123]; the sesquiterpene scodopin[124]; and the tryptamine derivatives scorodo-carpines A–C.[124]

The medicinal property of the plant and its smell are due to the alkyl-sulfurs.

Proposed Research Pharmacological study of scorodocarpines A–C and derivatives for the treatment of neurodegenerative diseases.

Rationale The powerful pharmacotoxicological impact of plant indole alkaloids on the brain is based on the mere fact that both the

■ **CS 1.65** Serotonin.

■ **CS 1.66** Fluoxetine.

■ **CS 1.67** 2,5,6-tribromogramine.

neurotransmitter serotonin (5-hydroxytryptamine) and the neuroendocrine hormone melatonin (*N*-acetyl-5-methoxytryptamine) are indoles themselves. Serotonin mediates numerous cerebral functions including mood, and a deficit of serotonin (CS 1.65) in the brain induces anxiety disorders and depression. A way to control depression is to increase the synaptic concentrations of serotonin by inhibiting presynaptic reuptake.

Increased synaptic concentrations of serotonin in the brain are beneficial, as α-synucleopathic MBP-hasyn mice treated with 18 mg/kg of fluoxetine had improved motor performance. The fluoxetine restored neuronal density in the frontal *cortex* and *hippocampus;* induced neurogenesis; reduced the amounts of α-synuclein in the *corpus callosum;* and boosted the levels of neurotrophic factors (NTFs), glial-derived neurotrophic factors (GDNFs), and brain-derived neurotrophic factors (BDNFs) from astrocytes.[125] Furthermore, the stimulation of serotoninergic receptors including serotoninergic receptor subtype 2A (5-HT2A), 2B (5-HT2B), and 2C (5-HT2C) resulted in neuronal G protein Ras activation and therefore extracellular signal-regulated kinase (ERK1/2) phosphorylation and cAMP response element binding protein (CREB) activation, hence neuron survival and neuritogenesis.[126] Along the same lines, fluoxetine (CS 1.66) given daily at a dose of 20 mg/kg to R6/1 transgenic Huntington's disease mice improved spatial memory and enhanced the survival of hippocampal neurons.[127] Indeed, treatment with 10 mg/kg of fluoxetine significantly attenuated the loss of dopaminergic neurons in the substantia nigra and increased the striatal dopamine levels of mice poisoned with 1-methyl-4-phenyl-1,2,3,6-tetrahydropyridine (MPTP), via mitigation of microglia production of reactive oxygen species (ROS) by NADPH oxidase, reduction of interleukin-1β (IL-1β), tumor necrosis factor-α (TNF-α), and inducible nitric oxide synthetase (iNOS).[128] In addition, serotonin stimulates 5-HT2B astrocytic receptors, hence ERK1/2 phosphorylation[129] and secretion of BDNF.[130] It is therefore tempting to speculate that indole alkaloids by virtue of their serotoninergic properties represent a vast source of leads for the treatment of neurodegenerative diseases. Note that the brominated indole alkaloid 2,5,6-tribromogramine (CS 1.67) from the marine organism *Zoobotryon pellucidum* at a dose of 30 μM induced the translocation of protein kinase C (PKC), hence the activation of nuclear factor kappa-light-chain-enhancer of activated B cells (NF-κB), leading to the production of nerve growth factor (NGF) by astrocytoma cells.[131] Melatonin (CS 1.68) is produced in the brain by the pineal gland and is involved in the circadian rhythm and the secretion of hormones. At a dose of 10–5 M, melatonin protected C6 cells against β-amyloid (Aβ_{1-42}) peptide-induced nitric oxide (NO) production and apoptosis and against Aβ_{1-40}-induced Ca^{2+}

■ **CS 1.68** Melatonin.

■ **CS 1.69** *N*-hexadecanyl tryptamine.

■ **CS 1.70** *N*-octadeca-9Z,12Z-dienyl tryptamine.

accumulation.[132] In fact, Schmidt et al.[133] developed series of tryptamine derivatives including *N*-hexadecanyl tryptamine (CS 1.69), *N*-octadeca-9Z,12Z-dienyl tryptamine (CS 1.70), and *N*-15-hydroxypentadecanyl tryptamine (CS 1.71), which at a dose of 10 nM induced neuritogenesis in mesencephalic dopaminergic neurons by 100%, 55%, and 65%, respectively, with parallel and profound activity against ROS[133] on probable account of the tryptamine moiety.[134] Furthermore, the antioxidant indole fatty alcohols engineered by Coowar et al.[135]—5-methoxyindole-3-hexadecanol (CS 1.72) and 5-methoxyindole-3-octadecanol—increased by 70% the differentiation of neural stem cell derived neurospheres into neurons of 10 nM via possible decrease in Notch4 level expression. In this context, it is important to mention that the stimulation of Notch receptors by Delta-like protein compels γ-secretase complex-induced cleavage and cytoplasmic release of the intracellular portion of Notch, which then enters the nucleus, where it binds to CBF1 protein, where upon it imposes the transcription of basic helix-loop-helix (bHLH) proteins and the inhibition of neurogenesis by Ascl1 and Neurog1/2.[136] Therefore, Notch inhibitors

■ **CS 1.71** *N*-15-hydroxypentadecanyl tryptamine.

■ **CS 1.72** 5-methoxyindole-3-hexadecanol.

■ **CS 1.73** Scorodocarpine A.

including indole alkaloids with a long alkyl chain in position 3 are of considerable interest for the development of neurotrophic agents; they include scorodocarpines A–C (CS 1.73–1.75) from *Scorodocarpus borneensis* (Baill.) Becc. (family Olacaceae R. Br.).[124] Methylation of tryptamine at C3 and subsequent nucleophilic attack to the iminium ion create physostigmine (CS 1.76) (eserine) produced by *Physostigma venenosum* Balf. f. (family Fabaceae Lindl.), which mitigates Alzheimerian dementia by inhibiting the enzymatic activity of acetylcholinesterase (AChE).[137,138] From the same plant, the oxidized derivative of physostigmine, gesenerine (CS 1.77), inhibited the enzymatic activity of AChE with an IC_{50} value equal to $1\,\mu M$ and has been synthetically modified into ganstigmine (CS 1.78), which protected cortical neurons against β-amyloid 25–35 peptide insults at a dose of $3\,\mu M$ by a mechanism strikingly independent of

■ **CS 1.74** Scorodocarpine B.

■ **CS 1.75** Scorodocarpine C.

■ **CS 1.76** Physostigmine.

■ **CS 1.77** Gesenerine.

cholinergic activity.[139] Another compelling example of a neuroprotective indole alkaloid is indirubin (CS 1.79) from *Indigofera tinctoria* L. (family Fabaceae Lindl.) as well as the indirubin-3′-oxime which inhibited the enzymatic activity of cyclin dependent kinase 1(CDK1)-cyclin B,

■ **CS 1.78** Ganstigmine.

■ **CS 1.79** Indirubin.

■ **CS 1.80** Indirubin-3′-monoxime.

■ **CS 1.81** 30A.

CDK-2-cyclin A, CDK-2-cyclin E, CDK-4-cyclin D1, and CDK5-p35 with IC50 values equal to $10\,\mu M$, $2.2\,\mu M$, $7.5\,\mu M$, $12\,\mu M$, $5.5\,\mu M$ and $0.1\,\mu M$, $0.4\,\mu M$, $0.2\,\mu M$, $3.3\,\mu M$, $0.1\,\mu M$, respectively, by docking into cyclin dependent kinase ATP catalytic pockets.[140] Furthermore, indirubin-3′-monoxime (CS 1.80) inhibited the enzymatic activity of glycogen synthase kinase 3-β (GSK-3β) and cyclin dependent kinase (CDK5) with IC50 values equal to $0.02\,\mu M$ and $0.1\,\mu M$, which account for the phosphorylation of the microtubule-associated protein tau.[141] Additionally, $10\,\mu M$ of indirubin-3′-monoxime protected cerebellar granule neurons from K$^+$-induced apoptosis by inhibiting the phosphorylation of c-Jun N-terminal kinase (JNK),[142,143] as it activates caspase 8 and therefore apoptosis. The synthetic indole 30A (CS 1.81) developed by Stoit et al.[144] is a partial agonist activity against $\alpha 4\beta 2$ nicotinic receptor with a K_i value equal to $125\,nM$, making it an attracting candidate for the treatment of dementia[145] and showing that indole alkaloids may fight neurodegeneration by targeting not only the serotoninergic receptor but also others such as the nicotinic receptors and probably opioid, adrenergic, and muscarinic receptors.

1.2.2 *Peganum nigellastrum* Bunge

History The plant was first described by Alexander Andrejewitsch Bunge in *Mémoires Presentes a l'Académie Impériale des Sciences de St.-Pétersbourg par divers savants et lus dans ses Assemblées* published in 1835.

Synonym *Ximenia borneensis* Baill.

Family Nitrariaceae Lindl., 1830

Common Name Luo tuo hao (Chinese)

Habitat and Description This little perennial herb grows in open areas in China, Mongolia, and Russia. The stems are angled. The leaves are simple, sessile, spiral, and without stipules. The blade is deeply incised, to 0.5–2 cm long, and presents several filiform lobes. The flowers are showy, axillary, and terminal. The calyx includes 5 sepals, which are deeply incised and 1.5 cm long. The corolla comprises 5 petals, which are obovate, yellow, 1.5 cm long, and nerved. The androecium consists of 15 stamens. The filaments are inflated at the base, and the anthers are cylindrical. The gynoecium is made of 3 carpels united to form a 3-lovular and globose ovary. The fruit is a capsule, which is 3-locular, 0.5–1 cm across, and contains numerous tiny linear seeds (Figure 1.5).

Medicinal Uses   In China and Mongolia, the plant is used to heal putrefied wounds.

Phytopharmacology   The few phytochemical studies devoted to this plant allowed the characterization of a series of b-carboline alkaloids such as harmine; the canthin-6-one alkaloids luotonins C and D; and a series of quinoline alkaloids including 3-phenylquinoline,

■ **FIGURE 1.5** *Peganum nigellastrum* Bunge.

■ **CS 1.82** *N*-methyl β-carbolinium cation.

■ **CS 1.83** 1-methyl-4-phenylpyridinium ion (MPP⁺).

■ **CS 1.84** 1-methyl-4-phenyl-1,2,3,6-tetrahydropyridine (MPTP).

■ **CS 1.85** 9-methyl-β-carboline.

■ **CS 1.86** Dopamine.

3-(4-hydroxyphenyl)quinoline, 3-(1H-indol-3-yl)quinoline; as well as the phenolics dihydrosinapyl ferulate and dihydroconiferyl ferulate.[146]

Proposed Research Pharmacological study of 9-methyl-harmane alkaloid derivatives for the treatment of Parkinson's disease (PD).

Rationale The cyclization of tryptamine produces the neurotoxin β-carboline indole alkaloids, which are found in grilled meat, cigarettes, and numerous medicinal plants and account for the development of Parkinson's disease. Indeed, β-carboline indole alkaloids are, in the brain, metabolized by β-carboline *N*-methyltransferases and heme peroxidases into *N*-methyl β-carbolinium cations (CS 1.82) which, by virtue of their structural similitude with 1-methyl-4-phenylpyridinium ion (MPP⁺) (CS 1.83), induce apoptosis in dopaminergic neurons.[147] MPP⁺ is obtained from the conversion of 1-methyl-4-phenyl-1,2,3,6-tetrahydropyridine (MPTP) (CS 1.84) by monoamine oxidase B (MAO-B), which imposes dopamergic apoptosis via mitochondrial insults.[147] Thus, one could reasonably draw an inference that the β-carboline framework may very well be used as a starting point for the synthesis of β-carboline *N*-methyltransferases, heme peroxidases, and/or MAO-B inhibitors as a therapeutic strategy to slow down the progression of PD.

Consonant with the usefulness of the β-carboline framework to fight Parkinson's disease is the compelling fact that rodents treated with 2 μmol/100 g of 9-methyl-β-carboline (CS 1.85) had increased levels of hippocampal dopamine (CS 1.86) and dendrite growth.[148] Furthermore, 9-methyl-β-carboline lowered the production of α-synuclein and protected dopaminergic neurons against 1-methyl-4-phenylpyridinium ion (MPP⁺) and rotenone and mitigated the microglial activation and secretion of cytokines induced by lipopolysaccharide (LPS).[149] Likewise, 9-methyl-β-carboline at a dose of 0.1 mg/kg/day protected rodents against MPP⁺-induced Parkinsonism by promoting the survival of dopaminergic neurons in the *substantia nigra* by inducing the transcription of GDNF.[149] Along the same lines, 9-methyl-β-carboline imposed an increase in dopaminergic neurons by 48% at a dose of 90 μM via cellular incorporation through dopamine transporter (DAT), tyrosine hydroxylase, protein kinase A (PKA), and PKC activation.[150]

Notwithstanding the neurotoxicological limitations of natural β-carboline indole alkaloids, attention has been focused on their biological properties. Notably, harmine (CS 1.87), harmaline (CS 1.88), harmol (CS 1.89), and harman (CS 1.90), from *Peganum nigellastrum* Bunge (family Nitrariaceae Lindl.) were found to inhibit the activity of AChE at doses equal to 0.01 μg, 0.01 μg, 0.01 μg, and 0.1 μg, respectively.[146]

These results support the theory that 9-methyl-harmane alkaloid derivatives might be of value to fight PD by virtue of their unique synergistic neurodopaminergic properties and β-carboline N-methyltransferases, heme peroxidases, and/or MAO-B and acetylcholinesterase (AChE) inhibition.

■ CS 1.87 Harmine.

1.2.3 *Tetradium daniellii* (Benn.) T.G. Hartley

History The plant was first described by Thomas Hartley Gordon in *The Gardens' Bulletin: Singapore* published in 1981.

Synonyms *Ampacus danielli* (Benn.) Kuntze, *Euodia daniellii* (Benn.) Hemsl., *Euodia delavayi* Dode, *Euodia henryi* Dode, *Euodia hupehensis* Dode, *Euodia labordei* Dode, *Euodia sutchuenensis* Dode, *Euodia velutina* Rehder & E.H. Wilson, *Euodia vestita* W.W. Sm., *Zanthoxylum bretschneideri* Maxim., *Zanthoxylum daniellii* Benn.

Family Rutaceae Juss., 1789

Common Name Chou tan wu yu (Chinese)

■ CS 1.88 Harmaline.

Habitat and Description This ornamental tree grows wild, to 15 m tall, in the forests of China and Korea. The leaves are imparipinnate, exstipulate, and opposite. The rachis supports 2–4 pairs of folioles, which are somewhat glossy, asymmetrical, broadly lanceolate, 5–15 cm × 2–10 cm, with 6–15 pairs of secondary nerves and wavy margins. The inflorescence is a cyme, which is 20 cm across, showy, somewhat fleshy, and terminal. The calyx consists of 5 sepals, which are minute. The corolla comprises 5 petals, which are white, 0.5 cm long, and broadly lanceolate. Male flowers expose an androecium of 5 stamens that protrude out of the corolla and present conspicuous orange anthers. The female flowers present a gynoecium of 5 carpels partially fused into a 5-locular ovary. The fruit is a dehiscent capsule concealing 2 black and glossy seeds per carpel (Figure 1.6).

■ CS 1.89 Harmol.

■ CS 1.90 Harman.

Medicinal Uses In China and Korea, the plant is used to assuage headache, to heal gastric ulcer, and to sooth dermatitis.

Phytopharmacology The plant has not raised enough interest in the natural product community. The few phytochemical studies devoted to this plant allowed the characterization of a series of flavonoids, including flavaprin, evodioside B, vitexin, and hesperidin; the coumarins bergapten, xanthotoxin, and isopimpinellin[151]; and probably evodiamine and congeners.

■ **FIGURE 1.6** *Tetradium daniellii* (Benn.) T.G. Hartley.

■ **CS 1.91** Dehydroevodiamine.

Proposed Research Pharmacological study of dehydroevodiamine (CS 1.91) and derivatives for the treatment of Alzheimer's disease (AD).

Rationale The *β*-carboline indole alkaloid dehydroevodiamine from *Tetradium ruticarpum* (A. Juss.) T.G. Hartley (Rutaceae Juss.) inhibited the enzymatic activity of AChE with an IC_{50} value equal to $6.3\,\mu M$ and protected rodents against scopolamine-induced amnesia at a dose of $6.25\,mg/kg$.[152] This neuroprotective effect was further confirmed by Wang et al.[153]

whereby amnesia induced in rodents by administration of 3 nmol of β-amyloid (Aβ$_{25-35}$) was corrected by a treatment with 0.7 mg/kg of dehydroevodiamine. The precise neuroprotective activity of dehydroevodiamine and congeners is yet to be deciphered, but there is increasing evidence to support the notion that NO is somewhat involved.

Indeed, dehydroevodiamine at a dose of 5 μM mitigated the release of glutamate by cerebellar glial cells challenged with *N*-Methyl-D-aspartate (NMDA),[154] and at 200 μmol/L inhibited the hyperphosphorylation of tau protein in brain slices exposed to 0.1 μmol/L of the protein phosphatase inhibitor calyculin,[155] which are downstream effects of NO production in astrocytes. In normal physiological conditions, activated astrocytes accumulate massive amounts of cytoplasmic NO, resulting in secretion of glutamate[156] and hyperphosphorylation of tau protein.[157]

In fact, the indole alkaloids dehydroevodiamine (CS 1.91), evodiamine (CS 1.92), and rutaecarpine (CS 1.93) from *Tetradium ruticarpum* (A. Juss.) T.G. Hartley (family Rutaceae Juss.) inhibited the enzymatic activity of NADPH oxidase at doses of 50 μM and at a dose of 20 μM abated the production of NO and the expression of iNOS by microglial cells exposed to LPS, whereas dehydroevodiamine that inhibited the production of ROS by neutrophils challenged the PKC activator phorbol 12-myristate 13-acetate (PMA) with an IC$_{50}$ value equal to 22.4 μM.[158] In an attempt to magnify the neuroprotective effect of dehydroevodiamine, several derivatives were engineered, such as compounds 17 (CS 1.94) and 5c (CS

■ **CS 1.92** Evodiamine.

■ **CS 1.93** Rutaecarpine.

■ **CS 1.94** Compound 17.

■ **CS 1.95** Compound 5c.

■ **CS 1.96** 10-hydroxy-infractopicrin.

■ **CS 1.97** Infractopicrin.

1.95), which inhibited the enzymatic activity of AChE with IC_{50} values equal to $3.4\,\mu M$ and $10\,nM$, respectively.[159,160] Note that indole alkaloids 10-hydroxy-infractopicrin (CS 1.96) and infractopicrin (CS 1.97) isolated from the mushroom *Cortinarius infractus* Berk. (family Cortinariaceae R. Heim ex Pouzar) inhibited the enzymatic activity of AChE with IC_{50} values equal to $12.7\,\mu M$ and $9.7\,\mu M$, respectively.[161]

1.2.4 *Cryptolepis sinensis* (Lour.) Merr.

History The plant was first described by Elmer Merrill Drew in *Philippine Journal of Science* published in 1920.

Synonyms *Aganosma edithae* Hance, *Cryptolepis edithae* (Hance) Benth. & Hook. f. ex Maxim., *Cryptolepis elegans* Wall. ex G. Don, *Emericia sinensis* (Lour.) Roem. & Schult., *Pergularia sinensis* Lour., *Periploca chinensis* (Lour.) Spreng., *Periploca sinensis* (Lour.) Steud., *Streptocaulon chinense* (Lour.) G. Don, *Vallaris sinensis* (Lour.) G. Don

Family Apocynaceae Juss., 1789

Common Name Bai ye teng (Chinese)

Habitat and Description This climber grows in the rainforest of India, Southeast Asia, China, and Taiwan. The main stem is woody and exudes a milky latex. The leaves are simple, opposite, and exstipulate. The petiole is 0.5–1 cm long. The blade is lanceolate, 1.5–8 cm × 1–2.5 cm, cordate, apiculate with 5–9 pairs of secondary nerves, and glaucous below. The inflorescence is a terminal cyme, which is 10 cm across. The calyx consists of 5 minute glandiferous sepals. The corolla is salver shaped and yellow. The corolla tube is 0.5 cm long and develops 5 contorted lobes that are linear and 1.5 cm long. The androecium consists of 4 stamens, the anthers of which are attached to the stigma, which forms a hood. The gynoecium comprises 2 carpels fused into an ovary. The fruit is a pair of follicles, which are cylindrical, 12 cm long, and stuffed with numerous comose seeds. The seeds are minute, and the coma is 2 cm long.

■ **CS 1.98** Cryptolepine.

Medicinal Uses In China, the plants produce a topical antidote for snake bites.

Phytopharmacology The plant probably shelters gamma carboline indole alkaloids such as cryptolepine (CS 1.98).

Proposed Research Pharmacological study of cryptolepine and derivatives for the treatment of AD.

Rationale Shifting the nitrogen atom from position 2 to 3 on the carboline framework produces gamma carbolines. One such compound is cryptolepine, from *Cryptolepis sanguinolenta* (Lindl.) Schltr. (family Apocynaceae Juss.), which at a dose of 30 μM blocked muscarinic receptors subtype 1 (M_1).[152] In addition, cryptolepine inhibited the aggregation of platelets induced by adenosine diphosphate (ADP) with an IC_{50} value equal to 2.7 × 10^{-5} M by stimulating adenylate cyclase (AC), hence the formation of cyclic adenosine monophosphate (cAMP).[153] Likewise, cryptolepine at a dose of 10 μM reduced by 50% the production of NO by RAW 264.7 challenged with LPS by inhibiting NF-κB and therefore inhibiting the synthesis of iNOS.[154] Note that 11-hydroxycryptolepine (CS 1.99) inhibited the enzymatic activity of xanthine oxidase and neutralized superoxide radicals with IC_{50} values equal to 32.3 μM and 28 μM, respectively.[155] This evidence raises the exciting hypothesis that gamma carboline alkaloids may produce new chemical entities from which could be developed neuroprotective agents acting via muscarinic M_1 receptors, platelet aggregation, and antioxidative capacities. Another inference to be drawn is that gamma carboline may be used to develop agents to induce neurite growth by virtue of their ability to increase cAMP and therefore PKA and CREB.

■ **CS 1.99** 11-hydroxycryptolepine.

An example of synthetic neuroprotective gamma carbolines is stobadine (CS 1.100), which at dose of 2 mg/kg protected rodents with brain ischemia against the ROS and subsequent lipid peroxidation.[162] Stobadine scavenged hydroxyl, peroxyl, and alkoxyl and superoxy radicals due to its ability to form a nitrogen-centered stable radical on indolic nitrogen,[163] hence its ability to protect endoplasmic reticulum (ER) against $FeSO_4$/EDTA/hydrogen peroxide insults with an IC_{50} value equal to 2.5×10^{-5} mol/L,[164] and mitochondrial insults by Fe^{2+}/ascorbate at a dose of 10^{-3} mol/L.[165] In addition, stobadine mitigated the deleterious effects of ROS on Na^+,K^+/ATPase, and Na^+-dependent Ca^{2+} uptake activity to 73.3% and 100%, respectively, in rodent synaptosomes exposed to Fe^{3+}/EDTA.[165] Dimebon (CS 1.101) neutralized amyloid ($A\beta_{25-35}$) peptide insults against neurons at a dose of 25 μM via inhibition of AChE, Ca^{2+} influx, and NMDA receptor antagonistic effects with IC_{50} values equal to 42 μM, 57 μM, and 42 mg/kg.[166] Along the same lines, dimebon at a dose of 1 mg/kg/day protected rodents against ethylcholine aziridinium ion-induced dementia.[166] Furthermore, dimebon given at a dose of 60 mg per day to Alzheimer's patients improved cognition.[167] In fact, dimedon underwent a phase III clinical trial,[168] which showed no convincing therapeutic effects,[169] exemplifying further the concept that the current standards in pre-clinical evaluation of lead compounds for the treatment of neurodegenerative disease need robust improvements. An exhaustive neuropharmacological study of members of the genus *Cryptolepis* R. Br. is warranted.

1.2.5 *Clausena lansium* (Lour.) Skeels

History The plant was first described by Homer Collar Skeels in the *U.S. Department of Agriculture Bureau of Plant Industry Bulletin* published in 1909.

■ **CS 1.100** Stobadine.

■ **CS 1.101** Dimebon.

Synonyms *Aulacia punctata* (Sonn.) Raeusch., *Clausena punctata* (Sonn.) Rehder & E.H. Wilson, *Clausena wampi* (Blanco) Oliv., *Cookia punctata* Sonn., *Cookia wampi* Blanco, *Quinaria lansium* Lour., *Sonneratia punctata* (Sonn.) J.F. Gmel.

Family Rutaceae Juss., 1789

Common Name Huang pi (Chinese)

Habitat and Description This tree native to China grows to 10 m high. The leaves are imparipinnate and alternate. The rachis bears 5–11 leaflets, which are asymmetrical, lanceolata, 5 cm–15 cm × 3 cm–6 cm, cuneate at the base, crenulated at the margin, and apiculate. The inflorescence is a terminal panicle. The calyx comprises 5 ovate lobes, which are minute. The corolla comprises 5 petals, which are white, oblong, and 0.5 cm long. The androecium includes 10 stamens, which are of unequal length. The receptacle presents a disc. The ovary presents 5 lodges. The fruit is a berry that is 2 cm across, has 1–5 seeds, and is edible (Figure 1.7).

■ **FIGURE 1.7** *Clausena lansium* (Lour.) Skeels.

Medicinal Uses In China, the plant is used to treat inflammation and bronchitis.

Phytopharmacology The plant produces the pyrrolidine clausenamide[170]; the triterpene lansiol[171]; and coumarins including wampetin[172] and lansiumarin A, B, and C.[173] It accumulates cohorts of carbazole indole alkaloids including 3-formyl-6-methoxycarbazole, methyl 6-methoxycarbazole-3-carboxylate, 3-formyl-1,6-dimethoxycarbazole, 3-formyl carbazole, methyl carbazole-3-carboxylate, murrayanine, glycozoline, and indizoline[174]; mafaicheenamine A–C[174]; mafaicheenamines D and E[175]; and claulansines A–J.[176]

Proposed Research Pharmacological study of rutaceous carbazoles and derivatives for the treatment of neurodegenerative diseases.

Rationale There is a bulk of captivating evidence to support the notion that carbazole indole alkaloids are first-line candidates for the treatment of neurodegenerative disease. In effect, carbazole alkaloids fight neurodegeneration via several yet undelineated mechanisms, but it is agreed that the carbazole framework *per se* is antioxidant and stops lipid (CS 1.102) peroxidation via the NH moiety.[177] Notably, the synthetic carbazole alkaloid carvedilol (CS 1.103) protected cerebellar granule cells against glutamate and oxidative insults with IC_{50} values equal to of $1.1\,\mu M$ and $5\,\mu M$, respectively, due to its antioxidant virtues.[178] Along the same lines, β-carvedilol mitigated NMDA/glycine-stimulated boost of cytoplasmic Ca^{2+} by blocking calcium channel and NMDA receptors with a Kd value equal to $29.4\,\mu M$.[179] Additionally, carvedilol abated the Na^+ induced veratridine-mediated neuronal death by blocking Na^+ receptors and therefore the release of $^3[H]$aspartate with an IC_{50} of $1.7\,\mu M$, suggesting that it inhibits the Na^+-dependent glutamate transporter.[180] Note that the aforementioned steroidal alkaloid veratridine from *Veratrum album* L. (family Melanthiaceae Batsch ex Borkh.) is a dreadful poison which binds to the neurotoxin receptor site 2 of voltage-dependent Na^+ channels (VDSC), resulting in the engulfment of Na^+ into neurons and therefore Ca^{2+} accumulation and apoptosis.[181] Furthermore, carvedilol given at a dose of 5 mg/kg improved the cognition of rodents poisoned with aluminium chloride by boosting the enzymatic activity of neuronal superoxide dismutase (SOD), catalase (CAT), and glutathione S-transferase (GST),[182] which are stimulated by NO.[183]

Is carvedilol a neglected candidate for the treatment of amyotrophic lateral sclerosis (ALS)? The neuroprotective virtue of carbazole alkaloids was

■ **CS 1.102** Lipid.

■ **CS 1.103** Carvedilol.

further shown by Kotoda et al.[185] because carbazomadurins A and B (CS 1.104–1.105) isolated from the pathogenic fungi *Actinomadura madurae* Lechevalier and Lechevalier (family Thermomonosporaceae Rainey et al.) nullified the detrimental effects of glutamate against N18-RE-105 cells with EC_{50} values 97 nM and 84 nM, respectively, probably due to the antioxidant property of the hydroxylated carbazole framework.

Additional support for the contention that the carbazole framework protects neurons because it is antioxidant was provided by the finding that synthetic carbazole P7C3 (CS 1.106) protected cultured U2OS cells from the calcium ionophore A23187-induced mitochondrial insults at 10 nM, enhanced neuron formation in the *hippocampus* of rodents at a dose of 40 mg/kg, improved the cognition of terminally aged rodents with

■ **CS 1.104** Carbazomadurin A.

■ **CS 1.105** Carbazomadurin B.

■ **CS 1.106** P7C3.

hippocampal neurogenesis, and prevented caspase 3-mediated apoptosis at a dose of 10 mg/kg.[186] Oddly enough, the PKC inhibitor and neurotoxin staurosporine (CS 1.107) from the bacterium *Streptomyces staurosporeus*, which is used to induce experimental neuroapoptosis at a dose of 10 ng, prevented the death of CAl subfield neurons of rodents subjected to ischemia by 87% and 42%, respectively, by inhibiting the enzymatic activity of PKC and PKA with IC_{50} values equal to 1.8 nM and 23 nM, respectively.[187]

Additionally, staurosporine at a dose of 100 pM rescued hippocampal neurons against Aβ-amyloid (Aβ_{1-40}) peptide by stabilizing intracellular concentrations of free Ca^{2+}.[188] However, staurosporine at a dose of 100 nM abated the survival-promoting effects of epidermal growth factor (EGF) on hippocampal and cerebellar neurons *in vitro* via PKC inhibition,[189] hence the development of synthetic analogs like CEP-1347 (CS 1.108), which at a dose of 500 nM prolonged the survival of motoneurons by inactivation of JNK and underwent clinical trials for the treatment of PD, the result of which were inconclusive.[190,191] In flowering plants, carbazole indole alkaloids are mainly found in members of the family Rutaceae Juss.

■ **CS 1.107** Staurosporine.

■ **CS 1.108** CEP-1347.

including the genus *Clausena* Burm. f., *Glycosmis* Corrêa, and *Murraya* J. König ex L. It is therefore tempting to contemplate members of the family Rutaceae Juss. as an untapped source of neuroprotective agents. Indeed, incorporation of leaflets of *Murraya koenigii* (L.) Spreng. to 8% in the diet of rodents mitigated scopolamine-induced dementia with a reduction of AChE activity by 20%[192] due to carbazole alkaloids,[193] which reduced brain NO and lipid peroxidation levels, and prompted the activity of protective antioxidants such as glutathione peroxidase (GPx), glutathione reductase (GRD), SOD, and CAT.[194] The actual carbazole alkaloids involved here are yet to be identified, but euchrestine B (CS 1.109), bis-murrayafoline E (CS 1.110), and (+)-mahanine (CS 1.111) from *Murraya*

■ **CS 1.109** Euchrestine B.

■ **CS 1.110** Bismurrayafoline E.

■ **CS 1.111** (+)-mahanine.

koenigii (L.) Spreng. (Rutaceae Juss.) are antioxidant *in vitro*, with IC_{50} values equal to $21.7\,\mu M$, $6.8\,\mu M$, and $21.9\,\mu M$, respectively.[195,196] Along the same lines, mahanimbine (CS 1.112) inhibited the enzymatic activity of AChE.[197] These results converge with those obtained by Liu et al.[177] whereby a series of carbazole alkaloids from *Clausena lansium* (Lour.)

■ **CS 1.112** Mahanimbine.

■ **CS 1.113** Claulansine A.

■ **CS 1.114** Claulansine H.

Skeels (Rutaceae Juss.) including claulansine A (CS 1.113), claulansine H (CS 1.114), claulansine I (CS 1.115), claulansine J (CS 1.116), and murrayanine (CS 1.117) protected rat pheochromocytoma (PC12) cells against β-amyloid (Aβ_{25-35}) peptide insults.[176] Furthermore, claulansine F (CS 1.118), methyl 6-methoxycarbazole-3-carboxylate (CS 1.119), and clausine I (CS 1.120) protected PC12 cells against sodium nitroprusside oxidative insults.[176] Another antioxidant carbazole alkaloid is clausine Z (CS 1.121) from *Clausena excavata* Burm. F (Rutaceae Juss.), which protected

■ **CS 1.115** Claulansine I.

■ **CS 1.116** Claulansine J.

■ **CS 1.117** Murrayanine.

■ **CS 1.118** Claulansine.

■ **CS 1.119** Methyl 6-methoxycarbazole-3-carboxylate.

■ **CS 1.120** Clausine I.

cerebellar granule neurons *in vitro* against oxidative insults and inhibited the enzymatic activity of cyclin-dependent kinase 5 (CDK5).[198] In neurophysiological conditions, CDK5 is activated by calcium-dependent calpain cleavage of p35 into p25 and induces apoptosis by inactivating ERK1/2.[199] In addition, CDK5 is involved in the phosphorylation of microtubule-associated protein tau, mediates β-amyloid (Aβ_{1-42}) peptide insults, and participates in the pathophysiology of ALS. Small molecules capable of inhibiting CDK5 are thus robust candidates for the development of neurodegenerative diseases. It is therefore reasonably conceivable that clausine Z and derivatives may have some therapeutic values against AD, PD, and ALS.

■ **CS 1.121** Clausine Z.

1.2.6 *Melodinus fusiformis* **Champ. ex Benth.**

History The plant was first described by John George Champion in *Hooker's Journal of Botany and Kew Garden Miscellany* published in 1852.

Synonyms *Melodinus edulis* H. Lév., *Melodinus esquirolii* H. Lév., *Melodinus flavus* H. Lév., *Melodinus seguinii* H. Lév., *Melodinus wrightioides* Hand.-Mazz.

Family Apocynaceae Juss., 1789

Common Name Jian shan chen (Chinese)

Habitat and Description This woody climber grows in the forests of China to a length of 10 m. The stems are terete, hairy at apex, and produce a white latex. The leaves are opposite, simple, and exstipulate. The petiole is 0.5 cm long. The leaf blade is oblong, 5 cm–10 cm × 1 cm–5 cm, rounded at the base, acuminate at the apex, and presents 13–15 pairs of secondary nerves. The inflorescence is a terminal cyme, which is 5 cm long. The calyx consists of 5 sepals, which are 0.5 cm long. The corolla is tubular, white, 2 cm long, and produces 5 contorted lobes, which are 1.5 cm × 0.5 cm. The androecium includes minute 5 stamens, which are attached to the tube near the base. The fruit is a poisonous berry that is elongated, orange, and 3 cm–5 cm × 2.5 cm (Figure 1.8).

Medicinal Uses In China, the plant is used to treat rheumatism.

■ **FIGURE 1.8** *Melodinus fusiformis* Champ. ex Benth.

Phytopharmacology The plant abounds with monoterpenoid indole alkaloids, including the plumeran indole alkaloids venalstonine, venalstonidine, (+)-voaphylline, tabersonine,[200] 11-hydroxytabersonine, and melomorsine I[201]; quinoline alkaloids such as scandine, meloscandonine, 10-hydroxyscandine, and 19-epi-meloscandonine; the triterpenes 11,12-dehydroursolic acid lactone, ursolic acid, and oleanolic lactone[201]; the lignans (+)-pinoresinol, (+)-1-hydroxysyringaresinol, syringaresinol, and (+)-fraxiresinol; and the flavonoid astragalin.[201]

Proposed Research Pharmacological study of tabersonine and plumeran congeners for the treatment of neurodegenerative diseases.

Rationale Members of the family Apocynaceae Juss., Loganiaceae R. Br. ex Mart., and Rubiaceae Juss. conceive a bewildering array of monoterpenoid indole alkaloids formed by the fusion of tryptamine to a monoterpene. The monoterpene folding allows the classification of monoterpenoid indole alkaloids into several groups, including the vincosans, corynantheans, strychnans, ibogans, aspidospermatans, plumerans, and eburnans.[202] The vincosan strictosidinic acid (CS 1.122) isolated from *Psychotria myriantha* Müll. Arg. (family Rubiaceae Juss.) abated the level of hippocampal 3,4-hydroxyindoleacetic acid (DOPAC) in rodents at a dose of 10 mg/kg via inhibition of monoamine oxidase A (MAO-A) with an IC_{50} value equal to 150 μg/mL and lowered serotonin synthesis,[203] suggesting its development as a lead for the treatment of PD. Another vincosan alkaloid of neurological interest is desoxycordifoline (CS 1.123) from *Chimarrhis turbinata* DC (family Rubiaceae Juss.), which inhibited the enzymatic activity of AChE at a dose of 0.1 μM.[204]

The corynanthean alkaloids raubasine (CS 1.124) and corynanthine (CS 1.125) from *Catharanthus roseus* (L.) G. Don (family Apocynaceae Juss.) have the interesting ability to inhibit adrenergic receptors subtype $\alpha 1$,[205]

■ **CS 1.122** Strictosidinic acid.

■ **CS 1.123** Desoxycordifoline.

■ **CS 1.124** Raubasine.

■ **CS 1.125** Corynanthine.

resulting in increased cerebrovascular perfusion and mitochondrial respiration,[206] which declines with aging and stroke. In fact, raubasine (ajmalicine) was formulated with the synthetic piperazine almitrine into a drug used in the 1980s to enhance oxygen supply to the brain for post-stroke rehabilitation.[207] Likewise, yohimbine (CS 1.126) from *Rauvolfia serpentina* (L.) Benth. ex Kurz (family Apocynaceae) is an adrenoceptor subtype $\alpha2$ antagonist which mitigated Parkinson's-like tremor and rigidity at doses of 0.6 and 0.2 mg/kg, respectively, in rodents poisoned with reserpine.[208] Considering evidence that Parkinson's patients receiving 20 mg of the adrenoceptor subtype $\alpha2$ antagonist idazoxan showed 40% reduction

■ **CS 1.126** Yohimbine.

■ **CS 1.127** Hirsutine.

■ **CS 1.128** Hirsuteine.

in dyskinesia,[209] one can suggest that adrenoceptor subtype $\alpha2$ antagonist corythanean alkaloids might be of value for the treatment of PD. Other vasodilatory corynanthean alkaloids are hirsutine (CS 1.127) and hirsuteine (CS 1.128) from *Uncaria sinensis* (Oliv.) Havil. (family Rubiaceae Juss.), which protected neurons against glutamate insults at doses of 3×10^{-4} M by blocking the influx of Ca^{2+}.[210]

Note that 10-methoxy-*N*-methylpericyclivine (CS 1.129), akuammidine (CS 1.130), and yohimbine from *Haplophyton crooksii* (L.D. Benson) (family Apocynaceae Juss.) inhibited the enzymatic activity of AChE with IC_{50} values equal to 1.3×10^{-4} M, 1.8×10^{-4} M, and

■ **CS 1.129** 10-methoxy-*N*-methylpericyclivine.

■ **CS 1.130** Akuammidine.

4.3×10^{-4} M, respectively.[211] Likewise, serpentine (CS 1.131), geissoschizine methyl ether (CS 1.132), and turbinatine (CS 1.133) from *Catharanthus roseus* (L.) G. Don (family Apocynaceae Juss.), *Uncaria rhynchophylla* (Miq.) Miq. ex Havil. (family Rubiaceae Juss.), and *Chimarrhis turbinata* DC (Family Rubiaceae Juss.), respectively, inhibited the enzymatic activity of AChE with IC_{50} values equal to $0.7\,\mu$M,[212] $3.7\,\mu$g/mL,[213] and $1.8\,\mu$M,[204] respectively. In fact, geissoschizine methyl ether from *Uncaria sinensis* (Oliv.) Havil. (family Rubiaceae Juss.) is a vasodilatator via Ca^{2+} channel blockade[214] and binds to serotoninergic receptor subtype 2A (5-HT_{2A}), serotoninergic receptor subtype 2C (5-HT_{2c}), and serotoninergic receptor subtype 1A (5-HT_{1A}) with K_i values equal to $1.4\,\mu$M, $0.9\,\mu$M, and $0.8\,\mu$M, respectively.[215] Of particular interest is the fact that geissoschizine methyl ether is a 5-HT_{1A} agonist and a 5-$HT_{2A/2C}$ antagonist.[215] In this context, it is important to mention that 5-HT_{1A} agonists at low dose improve but at high dose impair learning, whereas 5-HT_{1A} antagonists enhance learning and memory.[216] Furthermore, 5-$HT_{2A/2C}$ antagonists such as ritanserin improve cognition.[217] Additional

■ **CS 1.131** Serpentine.

■ **CS 1.132** Geissoschizine methyl ether.

■ **CS 1.133** Turbinatine.

support to the contention that the corythanean framework may favor neuroprotection via serotoninergic receptors is provided by the finding that alstonine (CS 1.134) from *Picralima nitida* (Stapf) T. Durand & H. Durand (family Apocynaceae Juss.) protected rodents at a dose of 1 mg/kg against NMDA receptor antagonist MK801-induced dementia.[218] In the same experiment, the analeptic effects of alstonine were abated by the aforementioned serotonin antagonist ritanserin, implying the involvement of 5-HT$_{2A/2C}$ receptors.[218]

In neurophysiological conditions, the stimulation of 5-HT$_{2A/2C}$ expressed by prefrontal cortex pyramidal neurons results in the activation of

■ **CS 1.134** Alstonine.

phospholipase C (PLC) via protein guanosine triphosphate binding protein Gαq; conversion of phosphatidylinositol 4,5-bisphosphate (PIP2) into diacylglycerol (DAG); and activation of PKC, which phosphorylates Na^+ channel and ionotropic glutamate receptor NMDA, resulting in Na^+ current blockade[219] and NMDA receptor activation.[220]

A similar type of NMDA receptor activation was observed with the stimulation of opioid receptor subtype μ in spinal trigeminal neurons of rodents by D-Ala^2-MePhe^4-Gly-ol^5-enkephalin.[221] In fact, *Picralima nitida* (Stapf) T. Durand & H. Durand (family Apocynaceae Juss.) produces the corynanthean alkaloids akuammidine and akuammine (CS 1.135), which are opioid receptor subtype μ agonists and antagonists, respectively,[222] implying that the hydroxyl group at C10 of akuammine is crucial for opioid activity. Therefore, although highly speculative, it is not difficult to imagine that corynanthean alkaloids with opioid receptor subtype μ agonist properties may be useful to enhance NMDA clearance, which is impaired in ALS.[223] A notable example of corynanthean opioid receptor subtype μ agonist is mitragynine (CS 1.136) isolated from *Mitragyna speciosa* (Korth.) Havil. (family Rubiaceae Juss.),[224] the leaves of which are used in Malaysia and Thailand as a garden substitute for opium. Indeed, at a dose of $3\,\mu M$, mytragynine inhibited the contraction of guinea pig ileum induced by electrical stimulation.[224] This relaxant effect was abolished by the opioid receptor subtype μ antagonist naloxone.[224]

Data received by analgesic study in rodents provided further evidence supporting the notion that mytragynine is an opioid receptor subtype μ and opioid receptor subtype δ agonist.[225] Likewise, 7-hydroxymitragynine (CS 1.137) from the same plant at a dose of $5\,mg/kg$ protected rodents against both tail-flick and hot-plate-induced algesia, which was reversed by naloxone and exhibited opioid receptor subtype μ agonistic properties superior to that of mitragynine,[226] implying that a hydroxyl in C7 is beneficial for opioid activity. Likewise, the opioid receptor subtype μ agonist morphine

■ **CS 1.135** Akuammine.

■ **CS 1.136** Mitragynine.

■ **CS 1.137** 7-hydroxymitragynine.

and the opioid receptor subtype κ agonist spiradoline mitigated the secretion of glutamate by cerebrocortical slices of rodents exposed to K^+.[227] Note that the opioid receptor subtype μ agonist morphine from *Papaver somniferum* L. (family Papaveraceae Juss.) at a dose of $1\,\mu M$ inhibited by 70% the apoptosis of human SH-SY5Y neuroblastoma cells deprived of serum via protein kinase B (Akt) phosphorylation.[228] The neuroprotective and probably neuritogenic property of opioid agonists was further shown in astrocytes where the stimulation of opioid receptor subtype μ resulted in a transient activation of ERK1/2 via release of calmoduline (CaM), which activated PLC, which produced diacylglycerol (DAG), which stimulated PKC.[229] Furthermore, the stimulation of opioid receptor subtype κ evoked phosphoinositide 3-kinase (PI3K), activation of ERK1/2, and CREB activation.[229] If one considers evidence that the stimulation of epidermal growth factor receptor by EGF activates ERK1/2, Akt, and therefore survival, the present results suggest that corynanthean may very well be of therapeutic interest against neurodegeneration. Another compelling neurological feature of opioid receptor subtype μ agonists is their ability to modulate cyclin-dependent kinase-5 (CDK5). Indeed, rodents exposed chronically to morphine developed reduced levels of CDK5,[230] which is activated by β-amyloid ($A\beta_{1-42}$) peptide and accounts for tau hyperphosphorylation and formation of neurofibrillary tangles in hippocampal neurons.[231,232] CDK5 activated by the protein p35 accounts for the survival of neurons by regulating the phosphorylation of cytoskeletal neurofilaments, and its malfunction is linked to neurodegenerative processes leading to ALS, PD, and AD.[233]

Several lines of experimental evidence point to the fact that opioid receptor subtype δ antagonists are neuroprotective. Teng et al.[234] observed that HEK293T cells exposed to the δ-opioid agonist [D-Ala2 , D-Leu5]-enkephalin had increased β-secretase and γ-secretase activities by 143% and 156%, respectively.[234] In the same experiment, activated opioid receptor subtype δ aggregated with β-secretase and presenilin-1 to form a complex which underwent endocytosis into lysosomes that participated in the catabolism of amyloid precursor protein (APP) into insoluble $A\beta_{1-42}$ peptide,[234] raising the hypothesis that opioid receptor subtype δ antagonists may prevent the formation of plaques.[234] In effect, lower occurrence of hippocampal amyloid plaques and improved cognitive function were shown in APP/PS double-transgenic mice treated with the opioid receptor subtype δ antagonist naltrindole at a dose of 5 mg/kg/day.[234] The neuroprotective effect of δ-opioid receptor antagonists was further substantiated by the fact that SNC80 at a dose of 10 mg/kg protected rodents against reserpine-induced hypokynesia.[235] Likewise, 10 mg/kg of SNC80 protected rodents against 1-methyl-4-phenyl-1,2,3,6-tetrahydropyridine (MPTP)-induced parkinsonism.[235] It is therefore reasonable to contemplate corynanthean alkaloids such as pseudoakuammigine (CS 1.138), which binds to opioid receptor subtype δ with a Ki value equal to 0.9 μM, as a conceptually new chemical framework that might contribute to the development of neuroprotective leads.

There are reports that corynanthean alkaloids interact with the glycine receptors. For instance, akuammiline derivatives

■ **CS 1.138** Pseudoakuammigine.

■ **CS 1.139** *N*-demethyl-3-epi-dihydrocorymine.

N-demethyl-3-epi-dihydrocorymine (CS 1.139) from *Alstonia glauce-scens* (Wall. ex G. Don) Monach. (family Apocynaceae Juss.) and defor-mylcorymine (CS 1.140), corymine (CS 1.141), and pleiocarpamine (CS 1.142) from *Hunteria zeylanica* (Retz.) Gardner ex Thwaites (family Apocynaceae Juss.) inhibited glycine receptor Cl$^-$ channels with IC$_{50}$ values equal to 37 μM, 55 μM, 12 μM, and >1 mM, respectively.[236] However, agents blocking glycine receptors are dreadful poisons, as shown especially with the strychnan alkaloid strychnine (CS 1.143) from *Strychnos nux-vomica* L. (family Loganiaceae R. Br. ex Mart.), which inhibits glycine receptors with an IC$_{50}$ value equal to 0.05 μM,[236] resulting in quick motoneuron hyperactivity, muscular spasms, tetanus, epistotonos, *risus sardonicus*, and death by suffocation. Oddly enough, the glycine receptor antagonist GV150526 protected rodents against post-ischemic insults at a dose of 3 mg/kg by reducing the infarct volume by 83%, 65%, and 58% at 1 h, 3 h, and 6 h after single bolus administration of the drug at a dose of 3 mg/kg.[237] In this context, it is interesting to note that lower glycinergic activity in anterior horn due to lower densities of glycine receptors that results in motoneuron hyperactivity may very well contribute to the pathophysiology of ALS.[238] Therefore, one could draw an inference that glycine receptor agonists may be useful in delaying the progression of ALS, and such agonists could be very well derived from corynanthean alkaloids.

■ **CS 1.140** Deformylcorymine.

■ **CS 1.141** Corymine.

■ **CS 1.142** Pleiocarpamine.

■ **CS 1.143** Strychnine.

■ **CS 1.144** Ibogaine.

Regarding the development of neuroprotective drugs, Sershen et al.[244] provided evidence that the release of dopamine by striata by pretreating rodents with 40 mg/kg of ibogaine (CS 1.144) was decreased via κ- opioid receptors. The ibogan alkaloid ibogaine from *Tabernanthe iboga* Baill. (family Apocynaceae Juss.) used by African shamans to induce hallucination has drawn considerable attention by virtue of its compelling and yet undelineated ability to abrogate heroin and cocaine addiction.[239] Ibogaine underwent preliminary clinical trials,[239–241] which were eventually abandoned.[242,243] The use of ibogaine is illegal in several countries, but one might argue that ibogaine and congener should undergo proper clinical trials as their anti-addictive effects are potent. Furthermore, understanding the neuropharmacological events evoked by ibogaine may very well contribute to opioid receptors.[244] Considering the fact that opioid receptors subtype κ are located presynaptically on striatal dopamine terminals and modulate the release of dopamine,[245] one could reasonably frame the hypothesis that ibogaine inhibits the dopaminergic transmission imposed by cocaine by stimulating opioid receptor subtype κ receptors. Furthermore, ibogaine is an NMDA receptor antagonist with a Ki value of 1.1 μM and occupied opioid receptor subtype δ with a Ki value of 29.8 μM, whereas the ibogans coronaridine (CS 1.145) and tabernanthine (CS 1.146) interacted with NMDA receptors with Ki values equal to 6.2 μM and 10.5 μM, respectively.[246] Ibogaine and tabernanthine bind to sigma-1 (σ1) and sigma-2 (σ2) receptors with Ki values equal to 8554 nM, 2872 nM, and 201 nM, 194 nM, respectively.[247] Sigma-1 (σ1) endoplasmic reticulum receptors are of particular interest based on the fact that a cytoplasmic Ca^{2+} increase in neurons results in σ1-Bip dissociation and the release of σ1 in the cytoplasm,[248] which contravenes β-amyloid (A$β_{25-35}$) peptide-induced apoptosis.[249,250] In fact, the sigma-1 (σ1) receptor agonist PRE-084 at a dose of 0.1 μg/g abrogated neuronal insults induced by the glutamatergic agonist ibotenate by inhibition of microglial activation and protection of neurons against glutamate insults with inhibition of caspase 3.[251] It therefore seems reasonable to consider ibogan alkaloids as a vast source of untapped sigma-1 (σ1) agonists awaiting further studies for the treatment of AD and other neurodegenerative conditions.

■ **CS 1.145** Coronaridine.

■ **CS 1.146** Tabernanthine.

■ **CS 1.147** 18-methylaminocoronaridine.

Increasing evidence supports the notion that ibogans are nicotinic receptor antagonists, as shown with ibogaine, which antagonized $\alpha3\beta4$ nicotinic receptors with a K_d value equal to $0.4\,\mu M$,[252] as well as $\alpha1\beta1\gamma\delta$ nicotinic receptors with an IC_{50} value equal to $17\,\mu M$[253] and inhibited the enzymatic activity of AChE with an IC_{50} value equal to $520\,\mu M$.[254] In addition, $\alpha1\beta1\gamma\delta$ nicotinic receptors from human muscle embryonics were inhibited by the ibogan alkaloids 18-methylaminocoronaridine (CS 1.147), catharanthine (CS 1.148), albifloranine (CS 1.149), and 18-methoxycoronaridine (CS 1.150) with IC_{50} values equal to $5.9\,\mu M$, $20\,\mu M$, $46\,\mu M$, and $6.8\,\mu M$, respectively[253]; 18-methoxycoronaridine antagonized the $\alpha3\beta4$ nicotinic receptors with a Ki value equal to $0.7\,\mu M$.[255] One therefore might argue that the ibogan framework could be used as a starting point for the synthesis of nicotinic agonists. Furthermore, ibogan alkaloids have the tendency to inhibit the enzymatic activity of AChE. For instance, coronaridine and voacangine (CS 1.151) from *Tabernaemontana australis* Müll. Arg. (family Apocynaceae Juss.) inhibited the enzymatic activity of AChE at a dose

■ **CS 1.148** Catharanthine.

■ **CS 1.149** Albifloranine.

■ **CS 1.150** 18-methoxycoronaridine.

■ **CS 1.151** Voacangine.

of 0.01 mM.[256] Furthermore, the ibogan alkaloids coronaridine, 10-hydrox-ycoronaridine (CS 1.152), and voacangine from *Ervatamia hainanensis* Tsiang (family Apocynaceae Juss.) inhibited the enzymatic activity of AChE with IC_{50} values equal to 8.6 μM, 20.9 μM, and 4.4 μM, respectively.[257] Likewise, 19,20-dihydrotabernamine (CS 1.153) and 19,20-dihydroervahanine (CS 1.154) from *Tabernaemontana divaricata* (L.) R. Br. ex Roem. & Schult. (family Apocynaceae Juss.) potently inhibited the enzymatic activity of AChE with IC_{50} values equal to 227 nM and 71 nM.[258]

Another compelling feature of ibogan alkaloids is their ability to antagonize the cannabinoid (CB1) receptor. For instance voacamine (CS 1.155),

■ **CS 1.152** 10-hydroxycoronaridine.

■ **CS 1.153** 19,20-dihydrotabernamine.

■ **CS 1.154** 19,20-dihydroervahanine A.

■ **CS 1.155** Voacamine.

■ **CS 1.156** 3,6-oxidovoacangine.

3,6-oxidovoacangine (CS 1.156), and 5-hydroxy-3,6-oxidovoacangine (CS 1.157) from *Voacanga africana* Stapf (family Apocynaceae Juss.) antagonized the CB1 receptor with IC_{50} values of 0.04 μM, 0.1 μM, and 0.1 μM, respectively, whereas voacangine and tabersonine were inactive,[259] implying that the C3–C6 ether moiety is favorable for activity. The activation of the CB1 receptor induces amnesia due to acetylcholine depletion.[259,260] In fact, the CB1 receptor antagonist SR141716A at a dose of 1 mg/kg protected rodents against dementia induced by 16 nmol of β-amyloid (Aβ_{25-35}) peptide or 800 pmol of β-amyloid (Aβ_{1-42}) peptide.[261] Thus, it can be hypothesized that oxidovoacangin and congeners, by antagonizing the CB1 receptor, may be able to provide some leads for the treatment of AD. In addition, ibogan alkaloids voacangine, 3-oxovoacangine (CS 1.158), voacristine (CS 1.159), and (7α)-voacangine hydroxyindolenine (CS 1.160) from *Voacanga africana* Stapf (family Apocynaceae Juss.) and coronaridine and isovoacangine (CS 1.161) isolated from *Tabernaemontana subglobosa* Merr. (family Apocynaceae Juss.) at a dose of 10 μM displayed

■ **CS 1.157** 5-hydroxy-3,6-oxidovoacangine.

■ **CS 1.158** 3-oxovoacangine.

■ **CS 1.159** Voacristine.

■ **CS 1.160** (7α)-voacangine hydroxyindolenine.

■ **CS 1.161** Isovoacangine.

■ **CS 1.162** Heyneanine.

transient receptor potential vanilloid type 1 (TRPV1) receptor antagonist activities *in vitro*.[262] TRPV1 receptors are expressed by dopaminergic neurons and microglia in the *substantia nigra*, and their stimulation by capsaicin induced Ca^{2+} influx, mitochondrial insults, release of cytochrome c, and apoptosis.[263,264] Therefore, ibogans may very well be of value for the development of agents for the treatment of PD.

Note that coronaridine from *Tabernaemontana heyneana* Wall. (family Apocynaceae Juss.) displayed estrogenic activity in rodents.[265] In line with the aforementioned evidence, the ibogans heyneanine (CS 1.162) and voacristine at a dose of 25 mg/kg displayed anti-implantation activity in rodents together with uterus hypertrophy.[266] Furthermore, 10-hydroxy-coronaridine from *Tabernaemontana penduliflora* K. Schum (family Apocynaceae Juss.) displayed estrogenic activity *in vitro* as shown by its ability to bind to estrogen receptors and boost the multiplication of T47D cells by 183%.[267] Given the neuroprotective effect of estrogenic agents, one could draw an inference that ibogans may exert neuroprotective effects not only via conventional receptors but also via the activation of neuronal estrogen receptors α, and activation of Akt and ERK1/2[268] and the production of transforming growth factor-β (TGF-β) by astrocytes.[269]

Haplophyton crooksii (L.D. Benson) L.D. Benson (family Apocynaceae Juss.) produces the aspidospermatan tubotaiwine (CS 1.163) and the strychnan 16-decarbomethoxyvinervine (CS 1.164), which inhibited the enzymatic activity of AChE with IC_{50} values equal to 1×10^{-4} M Likewise, the strychnan geissospermine (CS 1.165) from *Geissospermum vellosii* Allemão (family Apocynaceae Juss.) inhibits AChEs with an IC_{50} value equal to 39.3 μg/mL.[270] The aspidospermans tubotaiwine and apparicine (CS 1.166) from *Tabernaemontana pachysiphon* Stapf (family Apocynaceae Juss.) displayed some affinities to opioid receptors subtype μ with Ki values equal to 1.6 μmol and 2.6 μmol, respectively; tubotaiwine is an opioid receptor subtype μ agonist.[271] Likewise, the strychnan alkaloid akuammicine (CS 1.167) isolated from *Picralima nitida* (Stapf) T. Durand & H. Durand (family Apocynaceae Juss.) is an opioid receptor

■ **CS 1.163** Tubotaiwine.

■ **CS 1.164** 16-decarbomethoxyvinervine.

■ **CS 1.165** Geissospermine.

■ **CS 1.166** Apparicine.

■ **CS 1.167** Akuammicine.

subtype κ agonist with a Ki value of 0.2 μM. Evidence pointing to the involvement of opioid receptors subtype κ in numerous neurophysiological processes including pain and mood as well as neurodegeneration has been recently accumulating.[272] Indeed, the opioid receptor subtype κ agonist dynorphin at a dose of 0.1 μM inhibited K^+-induced release of dopamine from striatal neurons by 30%, and at 1 μM mitigated the release of acetylcholine by 20%, and the opioid receptor subtype δ agonist [Leu^5]enkephalin at 0.3 μM abrogated the K^+-induced release of acetylcholine with an EC_{50} value equal to 0.1 μM,[273] supporting the concept that opioid receptor subtype κ and opioid receptor subtype δ antagonists may very well be useful for the treatment of neurodegenerative diseases. In addition, stimulation of opioid receptors subtype κ by dynorphin A in the hippocampal region and *dentate gyrus* decreased glutamate release and therefore the excitatory corticostriatal pathway observed in Parkinson's disease.[274] This concept is further supported by the fact that the release of glutamate by synaptosomes from rodent *striatum* exposed to 4-aminopyridine was nullified by the opioid receptors subtype κ agonist enadoline at a dose of 300 μM.[275] In fact, the activation of opioid receptors subtype κ agonist inhibits adenylate cyclase (AC) via guanosine triphosphate binding protein $G_{i/o}$ protein, hence depletion of cAMP, inhibition of PKA, blockade of Ca^{2+} influx, and stimulation of K^+ efflux,[276] which results in the inhibition of neurotransmitter release.[277] Another interesting fact is that the opioid receptors subtype κ agonist U69593 induced the proliferation of C6 glioma cells via guanosine triphosphate binding protein $G_{i/o}$ protein activation, influx of Ca^{2+} by L-type Ca^{2+} channels opening, followed by activation of PLC, hence an increase in inositol 1,4,5-trisphosphate, induction of PKC, and phosphorylation of ERK1/2,[278,279] raising the captivating possibility that guanosine triphosphate binding protein $G_{i/o}$ protein receptor agonist may favor the growth and survival of neurons.

Little attention has been focused on the neuroprotective potencies of plumerans although echitovenidine (CS 1.168) from *Alstonia venenata* R. Br. (family Apocynaceae Juss.) inhibited the enzymatic activity of MAO-B by 47% at a dose of 3×10^{-4} M,[280] and 12-methoxy-1-methyl-aspidospermidine (CS 1.169) from *Geissospermum vellosii* Allemão (family Apocynaceae Juss.) incurred analgesic effects in rodents via the stimulation of 5-HT_{1A}.[281] Biopsies from brains of Alzheimer's patients showed that the degeneration of cortical pyramidal neurons induces a collapse in glutamatergic neurotransmission, which could be corrected by using a 5-HT_{1A} antagonist,[282] and plumerans may very well be of value in the treatment of dementia. Furthermore, the plumeran tabersonine (CS 1.170) binds to adrenergic receptors subtype α1 and α 2 receptors with *K*i values equal to 5.1 μM and 4.2 μM, respectively,[283] and has been used for the synthesis of vincamine (CS 1.171) and vinpocetine (CS 1.172).

The eburnan vincamine from *Vinca minor* L. (family Apocynaceae Juss.) displayed some levels of clinical efficacy against dementia,[284] where it enhanced blood perfusion, oxygen consumption, and glucose metabolism,[285] hence its use for the treatment of ischemic stroke and cognitive dysfunction in some countries although it does not treat AD or PD. Likewise, the vincamine derivative vinpocetine has been proposed for the treatment of cerebrovascular disorders but in clinical trials was unable

■ **CS 1.168** Echitovenidine.

to fight cognitive impairment and dementia.[286] Vinpocetine inhibited the enzymatic activity of phosphodiesterase (PDE), which resulted in an increase of cAMP and cyclic guanosine monophosphate (cGMP) with an IC_{50} value equal to 5 2 μmol/L.[287] The resulting increase in cAMP stimulates the enzymatic activity of PKA, which commands the activation of CREB and therefore the synthesis of anti-apoptotic proteins as well as neuritogenesis. Similarly, an increase of cGMP activates ERK1/2 and therefore the phosphorylation of CREB.[288] Furthermore, vinpocetine blocked voltage-gated Na^+ channels with an IC_{50} value equal to 0.3 μM[289] and inhibited the increase of free cytoplasmic Ca^{2+} concentration of neurons exposed to veratridine.[290] Furthermore, the secretion of dopamine and acetylcholine from striatal slices exposed to NMDA was abated by 40 μM of vinpocetine, suggesting vinpocetine to be a mild NMDA receptor antagonist.[291] Another vincamine derivative of neurological interest is vindeburnol (CS 1.173), which at a dose of 20 mg/kg protected rodents against myelin oligodendrocyte glycoprotein (MOG)$_{35-55}$-induced paralysis by counteracting demyelination, astrocyte activation in the *cerebellum*, and induction of noradrenergic transcription factor (Mash1), noradrenergic transporter (NET1), and dopamine decarboxylase and tyrosine hydroxylase (TH) in the *locus coeruleus*[292] and may therefore be valuable for the treatment of multiple sclerosis (MS) and amyotrophic lateral sclerosis (ALS).

■ **CS 1.169** 12-methoxy-1-methyl-aspidospermidine.

■ **CS 1.170** Tabersonine.

■ **CS 1.171** Vincamine.

■ **CS 1.172** Vinpocetine.

■ **CS 1.173** Vindeburnol.

The unusual monoterpenoid indole alkaloid psychollatine (CS 1.174) isolated from *Psychotria umbellata* Thonn. (family Rubiaceae Juss.) incurred anxiolytic effects in rodents subjected to the hole-board model at a dose of 15 mg/kg, which was nullified by 5-HT$_{2A/2C}$ antagonist ritanserin.[293] Furthermore, psychollatine 100 mg/kg compromised the cognitive function of rodents,[293] further confirming the contention that 5-HT$_{2A/2C}$ agonists induce dementia. Notwithstanding these neurological limitations, psychollatine at a dose of 100 mg/kg protected rodents against NMDA-induced seizures, and NMDA receptor antagonist MK801-induced hyperlocomotion, and at 10 mg/kg mitigated amphetamine poisoning and nullified the climbing behavior induced by apomorphine, supporting the notion that psychollatine is an NMDA receptor antagonist.[294] The loss of motor neurons in ALS results from intense glutamate activation of NMDA, followed by massive influx of Ca^{2+}, resulting in the activation of calpain, mitogen-activated protein kinase (MAPK) p38, and JNK; NO generation; mitochondrial insults; caspase 3 activation; endoplasmic reticulum (ER) stress; burst in ROS; and ultimately apoptosis, as the antioxidant capacities of SOD are weakened.[295] Note that 250 μg/mL of psychollatine protected SOD-deficient yeasts against paraquat, and H_2O_2[296] and psychollatine and derivatives may be viewed as an unusual chemical template that might very well contribute to the development of drugs for the treatment of AD or ALS.

The oxindole alkaloids rhynchophylline (CS 1.175), isorhynchophylline (CS 1.176), and isocorynoxeine (CS 1.177) from *Uncaria sinensis* (Oliv.) Havil. (family Rubiaceae Juss.) protected neurons against glutamate insults at doses of 3×10^{-4} M with inhibition of Ca^{2+} influx.[210] One should recall that small molecules that are able to block excessive influx of Ca^{2+} into neurons which results from glutamate receptor stimulation are of tremendous neurological interest. Rhynchophylline and isorhynchophylline inhibited muscarinic receptor subtype 1 (M1)-mediated Ca^{2+} cellular influx in oocytes with IC$_{50}$ values equal to 84.5 nM and 2.6 μM, respectively.[213] As mentioned

■ **CS 1.174** Psychollatine.

■ **CS 1.175** Rhynchophylline.

■ **CS 1.176** Isorhynchophylline.

■ **CS 1.177** Isocorynoxeine.

CS 1.178 Corynoxeine.

previously, a high concentration of Ca^{2+} in the cytoplasm of neurons precipitates the generation of NO via the stimulation of neuronal nitric oxide synthase (nNOS),[297] and a Ca^{2+} influx blockade may very well account for the inhibition of NO by microglial cells exposed to LPS by corynoxeine (CS 1.178), isocorynoxeine, rhynchophylline, and isorhynchophylline from *Uncaria rhynchophylla* (Miq.) Jacks. (family Rubiaceae Juss.) with IC_{50} values equal to 15.7 μM, 13.7 μM, 18.5 μM, and 19 μM, respectively.[298]

Another source of NO is iNOS, the levels of which were abated by rhynchophylline and isorhynchophylline at a dose of 30 μM in microglial cells challenged with LPS.[299] In the same experiment, rhynchophylline and isorhynchophylline hindered the degradation of nuclear factor of kappa light polypeptide gene enhancer in B-cells inhibitor α (IκBα), hence blockade of NF-κB and fall in TNF-α production with IC_{50} values equal to 12.1 μM and 2.3 μM, respectively, and inhibition of IL-1β with IC_{50} values of 6.1 μM and 3.3 μM, respectively.[299] Furthermore, isorhynchophylline at a dose of 50 μM abrogated β-amyloid (Aβ_{25-35}) peptide-induced PC12 cell apoptosis with a reduction of ROS and malondialdehyde (MDA), an increase in glutathione (GSH), a decrease of pro-apoptotic Bax, protection of mitochondrial integrity, and inhibition of caspase 3.[300,301] The oxindoles pteropodine (CS 1.179) and isopteropodine (CS 1.180) from *Uncaria tomentosa* (Willd. ex Roem. & Schult.) DC. (family Rubiaceae Juss.) boosted the effects of acetylcholine and serotonin toward muscarinic receptors subtype 1 (M_1) and 5 serotoninergic receptor subtype 2 (5-HT$_2$) expressed in oocytes with IC_{50} values equal to 9.5 μM, 9.9 μM, 13.5 μM, and 14.5 μM, respectively.[302] In this context, it is of interest to note that agonists of 5 serotoninergic receptor subtype 2 (5-HT$_2$) improve cognition in rodents[303] by enhancing cholinergic function.[304,305] Further study on pteropodine and isopteropodine and congeners may result in the development of drugs for the treatment of AD. Note that the oxindole rupicoline (CS 1.181) from *Tabernaemontana australis* (Müell. Arg) Miers (family Apocynaceae Juss.) inhibited the enzymatic activity of AChE.[256]

■ **CS 1.179** Pteropodine.

■ **CS 1.180** Isopteropodine.

■ **CS 1.181** Rupicoline.

REFERENCES

[122] Abe F, Yamauchi T. Megastigmanes and flavonoids from the leaves of *Scorodocarpus borneensis*. Phytochemistry 1993;33(6):1499–501.

[123] Lim H, Kubota K, Kobayashi A, Sugawara F. Sulfur-containing compounds from *Scorodocarpus borneens*is and their antimicrobial activity. Phytochemistry 1998;48(5):787–90.

[124] Wiart C, Martin MT, Awang K, Hue N, Serani L, Laprévote O, et al. Sesquiterpenes and alkaloids from *Scorodocarpus borneensis*. Phytochemistry 2001;58(4):653–6.

[125] Ubhi K, Inglis C, Mante M, Patrick C, Adame A, Spencer B, et al. Fluoxetine ameliorates behavioral and neuropathological deficits in a transgenic model mouse of α-synucleinopathy. Exp Neurol 2012;234(2):405–16.

[126] Nibuya M, Nestler EJ, Duman RS. Chronic antidepressant administration increases the expression of cAMP response element binding protein (CREB) in rat hippocampus. J Neurosci 1996;16:2365–72.

[127] Grote HE, Bull ND, Howard ML, Van Dellen A, Blakemore C, Bartlett PF, et al. Cognitive disorders and neurogenesis deficits in Huntington's disease mice are rescued by fluoxetine. Eur J Neurosci 2005;22(8):2081–8.

[128] Chung YC, Kim SR, Park JY, Chung ES, Park KW, Won SY, et al. Fluoxetine prevents MPTP-induced loss of dopaminergic neurons by inhibiting microglial activation. Neuropharmacology 2011;60(6):963–74.

[129] Li B, Zhang S, Li M, Hertz L, Peng L. Serotonin increases ERK1/2 phosphorylation in astrocytes by stimulation of 5-HT(2B) and 5-HT(2C) receptors. Neurochem Int 2010;57(4):432–9.

[130] Mercier G, Lennon AM, Renouf B, Dessouroux A, Ramauge M, Courtin F, et al. MAP kinase activation by fluoxetine and its relation to gene expression in cultured rat astrocytes. J Mol Neurosci 2004;4:207–16.

[131] Saito M, Hori M, Obara Y, Ohizumi Y, Ohkubo S, Nakahata N. Neurotrophic factor production in human astrocytoma cells by 2,5,6-tribromogramine via activation of epsilon isoform of protein kinase C. Eur J Pharm Sci 2006;28(4):263–71.

[132] Feng Z, Zhang JT. Protective effect of melatonin on beta-amyloid-induced apoptosis in rat astroglioma c6 cells and its mechanism. Free Radical Biol Med 2004;37(11):1790–801.

[133] Schmidt F, Douaron GL, Champy P, Amar M, Séon-Méniel B, Raisman-Vozari R, et al. Tryptamine-derived alkaloids from Annonaceae exerting neurotrophin-like properties on primary dopaminergic neurons. Bioorg Med Chem 2010;18(14):5103–13.

[134] Štolc S. Indole derivatives as neuroprotectants. Life Sci 1999;65(18–19):1943–50.

[135] Coowar D, Bouissac J, Hanbali M, Paschaki M, Mohier E, Luu B. Effects of indole fatty alcohols on the differentiation of neural stem cell derived neurospheres. J Med Chem 2004;47(25):6270–82.

[136] Pierfelice T, Alberi L, Gaiano N. Notch in the vertebrate nervous system: an old dog with new tricks. Neuron 2011;69(5):840–55.

[137] Thomsen T, Kaden B, Fischer JP, Bickel U, Barz H, Gusztony G, et al. Inhibition of acetylcholinesterase activity in human brain tissue and erythrocytes by galanthamine, physostigmine and tacrine. Eur J Clin Chem Clin Biochem 1991;29(8):487–92.

[138] Robinson B, Moorcroft D. Alkaloids of *Physostigma venenosum*. Part IX. The absolute configuration of geneserine an application of the nuclear Overhauser effect. J Chem Soc: Organic 1970;15:2077–8.

[139] Windisch M, Hutter-Paier B, Jerkovic L, Imbimbo B, Villetti G. The protective effect of ganstigmine against amyloid beta 25–35 neurotoxicity on chicken cortical neurons is independent from the cholinesterase inhibition. Neurosci Lett 2003;341(3):181–4.

[140] Hoessel R, Leclerc S, Endicott JA, Nobel ME, Lawrie A, Tunnah P, et al. Indirubin, the active constituent of a Chinese antileukaemia medicine, inhibits cyclin-dependent kinases. Nat Cell Biol 1999;1:60–7.

[141] Leclerc S, Garnier M, Hoessel R, Marko D, Bibb JA, Snyder GL, et al. Indirubins inhibit glycogen synthase kinase-3 beta and CDK5/P25, two protein kinases involved in abnormal tau phosphorylation in Alzheimer's disease. A property common to most cyclin-dependent kinase inhibitors? J Biol Chem 2001;276(1):251–60.

[142] Xie Y, Liu Y, Ma C, Yuan Z, Wang W, Zhu Z, et al. Indirubin-3′-oxime inhibits c-Jun NH2-terminal kinase: anti-apoptotic effect in cerebellar granule neurons. Neurosci Lett 2004;367(3):355–9.

[143] Coffey ET, Smiciene G, Hongisto V, Cao J, Brecht S, Herdegen T, et al. c-Jun N-terminal protein kinase (JNK) 2/3 is specifically activated by stress, mediating c-Jun activation, in the presence of constitutive JNK1 activity in cerebellar neurons. J Neurosci 2002;22:4335–45.

[144] Stoit AR, den Hartog AP, Mons H, van Schaik S, Barkhuijsen N, Stroomer C, et al. 7-Azaindole derivatives as potential partial nicotinic agonists. Bioorg Med Chem Lett 2008;18(1):188–93.

[145] Sahakian B, Jones G, Levy R, Gray J, Warburton D. The effects of nicotine on attention, information processing, and short-term memory in patients with dementia of the Alzheimer type. Br J Psychiatry 1989;154:797–800.

[146] Wang CH, Zheng XY, Zhang ZJ, Chou GX, Wu T, Cheng XM, et al. Acetylcholinesterase inhibitive activity-guided isolation of two new alkaloids from seeds of *Peganum nigellastrum* Bunge by an *in vitro* TLC-bioautographic assay. Arch Pharm Res 2009;32(9):1245–51.

[147] Herraiz T, Guillén H, Galisteo J. *N*-methyltetrahydro-beta-carboline analogs of 1-methyl-4-phenyl-1,2,3,6-tetrahydropyridine (MPTP) neurotoxin are oxidized to neurotoxic beta-carbolinium cations by heme peroxidases. Biochem Biophys Res Commun 2007;356(1):118–23.

[148] Gruss M, Appenroth D, Flubacher A, Enzensperger C, Bock J, Fleck C, et al. 9-Methyl-β-carboline-induced cognitive enhancement is associated with elevated hippocampal dopamine levels and dendritic and synaptic proliferation. J Neurochem 2012;121(6):924–31.

[149] Wernicke C, Hellmann J, Zieba B, Kuter K, Ossowska K, Frenzel M, et al. 9-Methyl-beta-carboline has restorative effects in an animal model of Parkinson's disease. Pharmacol Rep 2010;62(1):35–53.

[150] Polanski W, Enzensperger C, Reichmann H, Gille G. The exceptional properties of 9-methyl-beta-carboline: stimulation, protection and regeneration of dopaminergic neurons coupled with anti-inflammatory effects. J Neurochem 2010;113(6):1659–75.

[151] Sang WY, Ju SK, Sam SK, Kun HS, Hyeun WC, Hyun PK, et al. Constituents of the fruits and leaves of Euodia daniellii. Arch Pharm Res 2002;25(6):824–30.

[152] Park CH, Kim S–H, Choi W, Lee Y–J, Kim JS, Kang SS, et al. Novel anticholinesterase and antiamnesic activities of dehydroevodiamine, a constituent of Evodia rutaecarpa. Planta Med 1996;62(5):405–9.

[153] Wang H-H, Chou C-J, Liao J-F, Chen C-F. Dehydroevodiamine attenuates beta-amyloid peptide-induced amnesia in mice. Eur J Pharmacol 2001;413(2–3):221–5.

[154] Lim DK, Lee YB, Kim HS. Effects of dehydroevodiamine exposure on glutamate release and uptake in the cultured cerebellar cells. Neurochem Res 2004;29(2):407–11.

[155] Fang J, Liu R, Tian Q, Hong XP, Wang SH, Cao FY, et al. Dehydroevodiamine attenuates calyculin A-induced tau hyperphosphorylation in rat brain slices. Acta Pharmacol Sin 2007;28(11):1717–23.

[156] Brown GC. Nitric oxide and neuronal death. Nitric Oxide 2010;23:153–65.

[157] Saez TE, Pehar M, Vargas M, Barbeito L, Maccioni RB. Astrocytic nitric oxide triggers tau hyperphosphorylation in hippocampal neurons. In Vivo 2004;18(3):275–80.

[158] Ko HC, Wang YH, Liou KT, Chen CM, Chen CH, Wang WY, et al. Anti-inflammatory effects and mechanisms of the ethanol extract of *Evodia rutaecarpa* and its bioactive components on neutrophils and microglial cells. Eur J Pharmacol 2007;555:211–27.

[159] Decker M. Novel inhibitors of acetyl-and butyrylcholinesterase derived from the alkaloids dehydroevodiamine and rutaecarpine. Eur J Med Chem 2005;40(3):305–13.

[160] Wang B, Mai YC, Li Y, Hou JQ, Huang SL, Ou TM, et al. Synthesis and evaluation of novel rutaecarpine derivatives and related alkaloids derivatives as selective acetylcholinesterase inhibitors. Eur J Med Chem 2010;45(4):1415–23.

[161] Geissler T, Brandt W, Porzel A, Schlenzig D, Kehlen A, Wessjohann L, et al. Acetylcholinesterase inhibitors from the toadstool *Cortinarius infractus*. Bioorg Med Chem 2010;18(6):2173–7.

[162] Horakova L, Lukovic L, Stolc S. Effect of stobadine and vitamin E on the ischemic reperfused brain tissue. Pharmazie 1990;45:223–4.

[163] Stasko A, Ondrias K, Misk V, Szocsova H, Gergel D. Stobadine: a novel scavenger of free radicals. Chem Pap 1990;44:493–500.

[164] Racay P, Kaplan P, Lehotsky J, Mezesova V. Rabbit brain endoplasmic reticulum membranes as target for free radicals. Changes in Ca^{2+}-transport and protection by stobadine. Biochem Mol Biol Int 1995;36:569–77.

[165] Lehotský J, Kaplán P, Račay P, Matejovičová M, Drgová A, Mézešová V. Membrane ion transport systems during oxidative stress in rodent brain: protective effect of stobadine and other antioxidants. Life Sci 1999;65(18–19):1951–8.

[166] Horakova L, Juranek I, Boknkova B. Antioxidative effect of stobadine on lipid peroxidation of rat brain mitochondria. Biologia 1990;45:313–18.

[167] Bachurin S, Bukatina E, Lermontova N, Tkachenko S, Afanasiev A, Grigoriev V, et al. Antihistamine agent dimebon as a novel neuroprotector and a cognition enhancer. Ann N Y Acad Sci 2001;939:425–35.

[168] Sabbagh MN, Shill HA. Latrepirdine, a potential novel treatment for Alzheimer's disease and Huntington's chorea. Curr Opin Investig Drugs 2010;11(1):80–91.

[169] Miller G. Pharmacology. The puzzling rise and fall of a dark-horse Alzheimer's drug. Science 2010;327:1309.

[170] Yang MH, Cao YH, Li WX, Yang YQ, Chen YY, Huang L. Isolation and structural elucidation of clausenamide from the leaves of *Clausena lansium* (Lour.) Skeels. Acta Pharmacol Sin 1987;22(1):33–40.

[171] Lakshmi V, Raj K, Kapil RS. A triterpene alcohol, lansiol, from *Clausena lansium*. Phytochemistry 1989;28(3):943–5.

[172] Khan NU, Naqvi SWI, Ishratullah K. Wampetin, a furocoumarin from *Clausena wampi*. Phytochemistry 1983;22(11):2624–5.

[173] Ito C, Katsuno S, Furukawa H. Structures of lansiumarin-A, -B, -C, three new furocoumarins from *Clausena lansium*. Chem Pharm Bull 1998;46(2):341–3.

[174] Wen-Shyong L, McChesney JD, El-Feraly FS. Carbazole alkaloids from *Clausena lansium*. Phytochemistry 1991;30(1):343–6.

[175] Maneerat W, Laphookhieo S. Antitumoral alkaloids from *Clausena lansium*. Heterocycles 2010;81(5):1261–9.

[176] Maneerat W, Ritthiwigrom T, Cheenpracha S, Laphookhieo S. Carbazole alkaloids and coumarins from *Clausena lansium* roots. Phytochem Lett 2012;5(1):26–8.

[177] Liu H, Li CJ, Yang JZ, Ning N, Si YK, Li L, et al. Carbazole alkaloids from the stems of *Clausena lansium*. J Nat Prod 2012;75(4):677–82.

[178] Tang YZ, Liu ZQ. Free-radical-scavenging effect of carbazole derivatives on AAPH-induced hemolysis of human erythrocytes. Bioorg Med Chem 2007;15(5):1903–13.

[179] Lysko PG, Lysko KA, Yue TL, Webb CL, Gu JL, Feuerstein G, et al. Neuroprotective effects of carvedilol, a new antihypertensive agent, in cultured rat cerebellar neurons and in gerbil global brain ischemia. Stroke 1992;23(11):1630–6.

[180] Lysko PG, Lysko KA, Webb CL, Feuerstein G. Neuroprotective effects of carvedilol, a new antihypertensive, at the *N*-Methyl-D-aspartate receptor. Neurosci Lett 1992;148(1–2):34–8.

[181] Lysko PG, Webb CL, Feuerstein G. Protective effects of carvedilol, a new antihypertensive, as a Na^+ channel modulator and glutamate transport inhibitor. Neurosci Lett 1994;171(1–2):77–80.

[182] Ulbricht W. Effects of veratridine on sodium currents and fluxes. Rev Physiol Biochem Pharmacol 1998;133:1–54.

[183] Kumar A, Prakash A, Dogra S. Neuroprotective effect of carvedilol against aluminium induced toxicity: possible behavioral and biochemical alterations in rats. Pharmacol Rep 2011;63(4):915–23.

[184] Manukhina EB, Downey HF, Mallet RT. Role of nitric oxide in cardiovascular adaptation to intermittent hypoxia. Exp Biol Med 2006;231(4):343–65.

[185] Kotoda N, Shin-Ya K, Furihata K, Hayakawa Y, Seto H. Isolation and structure elucidation of novel neuronal cell protecting substances, carbazomadurins A and B produced by *Actinomadura madurae*. J Antibiot 1997;50(9):770–2.

[186] Pieper AA, Xie S, Capota E, Estill SJ, Zhong J, Long JM, et al. Discovery of a proneurogenic, neuroprotective chemical. Cell 2010;142(1):39–51.

[187] Hara H, Onodera H, Yoshidomi M, Matsuda Y, Kogure K. Staurosporine, a novel protein kinase C inhibitor, prevents postischemic neuronal damage in the gerbil and rat. J Cereb Blood Flow Metab 1990;10(5):646–53.

[188] Goodman Y, Mattson MP. Staurosporine and K-252 compounds protect hippocampal neurons against amyloid beta-peptide toxicity and oxidative injury. Brain Res 1994;650(1):170–4.

[189] Abe K, Takayanagi M, Saito H. Neurotrophic effects of epidermal growth factor on cultured brain neurons are blocked by protein kinase inhibitors. Jpn J Pharmacol 1992;59(2):259–61.

[190] Maroney AC, Glicksman MA, Basma AN, Walton KM, Knight Jr. E, Murphy CA, et al. Motoneuron apoptosis is blocked by CEP-1347 (KT 7515), a novel inhibitor of the JNK signaling pathway. J Neurosci 1998;18(1):104–11.

[191] Waldmeier P, Bozyczko-Coyne D, Williams M, Vaught JL. Recent clinical failures in Parkinson's disease with apoptosis inhibitors underline the need for a paradigm shift in drug discovery for neurodegenerative diseases. Biochem Pharmacol 2006;72(10):1197–206.

[192] Vasudevan M, Parle M. Antiamnesic potential of *Murraya koenigii* leaves. Phytother Res 2009;23:308–16.

[193] Mani V, Ramasamy K, Ahmad A, Parle M, Shah SAA, Majeed ABA. Protective effects of total alkaloidal extract from *Murraya koenigii* leaves on experimentally induced dementia. Food Chem Toxicol 2012;50(3–4):1036–44.

[194] Mani V, Ramasamy K, Ahmad A, Wahab SN, Jaafar SM, Kek TL, et al. Effects of the total alkaloidal extract of *Murraya koenigii* leaf on oxidative stress and cholinergic transmission in aged mice. Phytother Res 2013;27(1):46–53.

[195] Tachibana Y, Kikuzaki H, Lajis NH, Nakatani N. Antioxidative activity of carbazoles from *Murraya koenigii* leaves. J Agric Food Chem 2001;49(11):5589–94.

[196] Tachibana Y, Kikuzaki H, Lajis NH, Nakatani N. Comparison of antioxidative properties of carbazole alkaloids from *Murraya koenigii* leaves. J Agric Food Chem 2003;51(22):6461–7.

[197] Kumar NS, Mukherjee PK, Bhadra S, Saha BP, Pal BC. Acetylcholinesterase inhibitory potential of a carbazole alkaloid, mahanimbine, from *Murraya koenigii*. Phytother Res 2010;24:629–31.

[198] Porterat O, Puder C, Bolek W, Wagner K, Ke C, Ye Y, et al. Clausine Z, a new carbazole alkaloid from *Clausena excavata* with inhibitory activity on CDK5. Pharmazie 2005;60(8):637–9.

[199] Weishaupt JH, Neusch C, Bähr M. Cyclin-dependent kinase 5 (CDK5) and neuronal cell death. Cell Tissue Res 2003;312(1):1–8.

[200] Cai XH, Jiang H, Li Y, Cheng GG, Liu YP, Feng T, et al. Cytotoxic indole alkaloids from *Melodinus fusiformis* and *M. morsei*. Chin J Nat Med 2011;9(4):259–63.

[201] Wang DW, Luo XD, Jiang B. Chemical constituents in twigs and leaves of *Melodinus fusiformis*. Chin Trad Herbal Drugs 2012;43(4):653–7.

[202] Ziegler J, Facchini PJ. Alkaloid biosynthesis: metabolism and trafficking. Annu Rev Plant Biol 2008;59:735–69.

[203] Farias FM, Passos CS, Arbo MD, Barros DM, Gottfried C, Steffen VM, et al. Strictosidinic acid, isolated from *Psychotria myriantha* Mull. Arg. (Rubiaceae), decreases serotonin levels in rat hippocampus. Fitoterapia 2012;83(6):1138–43.

[204] Cardoso CL, Castro-Gamboa I, Siqueira Silva DH, Furlan M, Epifanio RDA, Da Cunha Pinto A, et al. Indole glucoalkaloids from *Chimarrhis turbinata* and their evaluation as antioxidant agents and acetylcholinesterase inhibitors. J Nat Prod 2004;67(11):1882–5.

[205] Demichel P, Gomond P, Roquebert J. Pre-and postsynaptic alpha-adrenoceptor blocking activity of raubasine in the rat vas deferens. Br J Pharmacol 1981;74(4):739–45.

[206] Nowicki JP, MacKenzie ET, Spinnewyn B. Effects of agents used in the pharmacotherapy of cerebrovascular disease on the oxygen consumption of isolated cerebral mitochondria. J Cereb Blood Flow Metab 1982;2(1):33–40.

[207] Li S, Long J, Ma Z, Xu Z, Li J, Zhang Z. Assessment of the therapeutic activity of a combination of almitrine and raubasine on functional rehabilitation following ischaemic stroke. Curr Med Res Opin 2004;20(3):409–15.

[208] Colpaert FC. Pharmacological characteristics of tremor, rigidity and hypokinesia induced by reserpine in rat. Neuropharmacology 1987;26(9):1431–40.

[209] Rascol O, Arnulf I, Peyro-Saint Paul H, Brefel-Courbon C, Vidailhet M, Thalamas C, et al. Idazoxan, an alpha-2 antagonist, and L-DOPA-induced dyskinesias in patients with Parkinson's disease. Mov Disord 2001;16(4):708–13.

[210]   Shimada Y, Goto H, Itoh T, Sakakibara I, Kubo M, Sasaki H, et al. Evaluation of the protective effects of alkaloids isolated from the hooks and stems of *Uncaria sinensis* on glutamate-induced neuronal death in cultured cerebellar granule cells from rats. J Pharm Pharmacol 1999;51:715–22.

[211]   Mroue MA, Euler KL, Ghuman MA, Alam M. Indole alkaloids of *Haplophyton crooksii*. J Nat Prod 1996;59(9):890–3.

[212]   Pereira DM, Ferreres F, Oliveira JMA, Gaspar L, Faria J, Valentão P, et al. Pharmacological effects of *Catharanthus roseus* root alkaloids in acetyl-cholinesterase inhibition and cholinergic neurotransmission. Phytomedicine 2010;17(8–9):646–52.

[213]   Yang ZD, Duan DZ, Du J, Yang MJ, Li S, Yao XJ. Geissoschizine methyl ether, a corynanthean-type indole alkaloid from *Uncaria rhynchophylla* as a potential acetylcholinesterase inhibitor. Nat Prod Res 2011;26(1):22–8.

[214]   Yuzurihara M, Ikarashi Y, Goto K, Sakakibara I, Hayakawa T, Sasaki H. Geissoschizine methyl ether, an indole alkaloid extracted from *Uncariae Ramulus et Uncus*, is a potent vasorelaxant of isolated rat aorta. Eur J Pharmacol 2002;444(3):183–9.

[215]   Pengsuparp T, Indra B, Nakagawasai O, Tadano T, Mimaki Y, Sashida Y, et al. Pharmacological studies of geissoschizine methyl ether, isolated from *Uncaria sinensis* Oliv., in the central nervous system. Eur J Pharmacol 2001;425(3):211–18.

[216]   King MV, Marsden CA, Fone KCF. A role for the 5-HT$_{1A}$, 5-HT$_4$ and 5-HT$_6$ receptors in learning and memory. Trends Pharmacol Sci 2008;29(9):482–92.

[217]   van Laar M, Volkerts E, Verbaten M. Subchronic effects of the GABA-agonist lorazepam and the 5-HT2A/2C antagonist ritanserin on driving performance, slow wave sleep and daytime sleepiness in healthy volunteers. Psychopharmacology 2001;154:189–97.

[218]   Linck VM, Bessa MM, Herrmann AP, Iwu MM, Okunji CO, Elisabetsky E. 5-HT2A/C receptors mediate the antipsychotic-like effects of alstonine. Prog Neuropsychopharmacol Biol Psychiatry 2012;36:29–33.

[219]   Carr DB, Cooper DC, Ulrich SL, Spruston N, Surmeier DJ. Serotonin receptor activation inhibits sodium current and dendritic excitability in prefrontal cortex via a protein kinase C-dependent mechanism. J Neurosci 2002;22(16):6846–55.

[220]   Grosshans DR, Browning MD. Protein kinase C activation induces tyrosine phosphorylation of the NR2A and NR2B subunits of the NMDA receptor. J Neurochem 2001;76:737–44.

[221]   Chen L, Huang LY. Sustained potentiation of NMDA receptor-mediated glutamate responses through activation of protein kinase C by a mu opioid. Neuron 1991;7(2):319–26.

[222]   Menzies JR, Paterson SJ, Duwiejua M, Corbett AD. Opioid activity of alkaloids extracted from *Picralima nitida* (fam. Apocynaceae). Eur J Pharmacol 1998;350(1):101–8.

[223]   Rothstein JD, Martin LJ, Kuncl RW. Decreased glutamate transport by the brain and spinal cord in amyotrophic lateral sclerosis. N Engl J Med 1992;22:1464–8.

[224]   Watanabe K, Yano S, Horie S, Yamamoto LT. Inhibitory effect of mitragynine, an alkaloid with analgesic effect from Thai medicinal plant *Mitragyna speciosa*, on electrically stimulated contraction of isolated guinea-pig ileum through the opioid receptor. Life Sci 1997;60(12):933–42.

[225] Thongpradichote S, Matsumoto K, Tohda M, Takayama H, Aimi N, Sakai SI, et al. Identification of opioid receptor subtypes in antinociceptive actions of supraspinally-administered mitragynine in mice. Life Sci 1998;62(16):1371–8.

[226] Matsumoto K, Horie S, Ishikawa H, Takayama H, Aimi N, Ponglux D, et al. Antinociceptive effect of 7-hydroxymitragynine in mice: discovery of an orally active opioid analgesic from the Thai medicinal herb *Mitragyna speciosa*. Life Sci 2004;74(17):2143–55.

[227] Nicol B, Rowbotham DJ, Lambert DG. Mu- and kappa-opioids inhibit K^+ evoked glutamate release from rat cerebrocortical slices. Neurosci Lett 1996;218(2):79–82.

[228] Iglesias M, Segura MF, Comella JX, Olmos G. Mu-opioid receptor activation prevents apoptosis following serum withdrawal in differentiated SH-SY5Y cells and cortical neurons via phosphatidylinositol 3-kinase. Neuropharmacology 2003;44:482–92.

[229] Belcheva MM, Clark AL, Haas PD, Serna JS, Hahn JW, Kiss A, et al. Mu and kappa opioid receptors activate ERK1/2/MAPK via different protein kinase C isoforms and secondary messengers in astrocytes. J Biol Chem 2005;280(30):27662–9.

[230] Ferrer-Alcón M, La Harpe R, Guimón J, García-Sevilla JA. Downregulation of neuronal cdk5/p35 in opioid addicts and opiate-treated rats: relation to neurofilament phosphorylation. Neuropsychopharmacology 2003;28(5):947–55.

[231] Alvarez A, Toro R, Cáceres A, Maccioni RB. Inhibition of tau phosphorylating protein kinase cdk5 prevents beta-amyloid-induced neuronal death. FEBS Lett 1999;459(3):421–6.

[232] Jaquet PE, Ferrer-Alcón M, Ventayol P, Guimón J, García-Sevilla JA. Acute and chronic effects of morphine and naloxone on the phosphorylation of neurofilament-H proteins in the rat brain. Neurosci Lett 2001;304:37–40.

[233] Maccioni RB, Otth C, Concha H, Munoz JP. The protein kinase Cdk5. Structural aspects, roles in neurogenesis and involvement in Alzheimer's pathology. Eur J Biochem 2001;268:1518–27.

[234] Teng L, Zhao J, Wang F, Ma L, Pei G. A GPCR/secretase complex regulates beta- and gamma-secretase specificity for Abeta production and contributes to AD pathogenesis. Cell Res 2010;20:138–53.

[235] Hille CJ, Fox SH, Maneuf YP, Crossman AR, Brotchie JM. Antiparkinsonian action of a delta opioid agonist in rodent and primate models of Parkinson's disease. Exp Neurol 2001;172(1):189–98.

[236] Leewanich P, Tohda M, Matsumoto K, Subhadhirasakul S, Takayama H, Aimi N, et al. Inhibitory effects of corymine-related compounds on glycine receptors expressed in *Xenopus oocytes*. Jpn J Pharmacol 1998;77(2):169–72.

[237] Bordi F, Pietra C, Ziviani L, Reggiani A. The glycine antagonist GV150526 protects somatosensory evoked potentials and reduces the infarct area in the MCAo model of focal ischemia in the rat. Exp Neurol 1997;145:425–33.

[238] Chang Q, Martin LJ. Glycine receptor channels in spinal motoneurons are abnormal in a transgenic mouse model of amyotrophic lateral sclerosis. J Neurosci 2011;31(8):2815–27.

[239] Alper KR, Lotsof HS, Kaplan CD. The ibogaine medical subculture. J Ethnopharmacol 2008;115(1):9–24.

[240] Mash DC, Kovera CA, Pablo J, Tyndale RF, Ervin FD, Williams IC, et al. Ibogaine: complex pharmacokinetics, concerns for safety, and preliminary efficacy measures. Ann N Y Acad Sci 2000;914:394–401.

[241] Mash DC, Kovera CA, Buck BE, Norenberg MD, Shapshak P, Hearn WL. Medication development of ibogaine as a pharmacotherapy for drug dependence. Ann N Y Acad Sci 1998;844:274–92.

[242] O'Hearn E, Long DB, Molliver ME. Ibogaine induces glial activation in parasagittal zones of the cerebellum. Neuro Report 1993;4:299–302.

[243] Molinari HH, Maisonneuve IM, Glick SD. Dose dependence of ibogaine neurotoxicity. Soc Neurosci Abstr 1994;20:1236.

[244] Sershen H, Hashim A, Lajtha A. The effect of ibogaine on kappa-opioid- and 5-HT3-induced changes in stimulation-evoked dopamine release *in vitro* from striatum of C57BL/6By mice. Brain Res Bull 1995;36(6):587–91.

[245] Jackisch R, Hotz H, Heming G. No evidence for presynaptic opioid receptors on cholinergic, but presence of kappa-receptors on dopaminergic neurons in the rabbit caudate nucleus: involvement of endogenous opioids. Naunyn-Schmiedeberg's Arch Pharmacol 1993;348:234–41.

[246] Layer RT, Skolnick P, Bertha CM, Bandarage UK, Kuehne ME, Popik P. Structurally modified ibogaine analogs exhibit differing affinities for NMDA receptors. Eur J Pharmacol 1996;309(2):159–65.

[247] Bowen WD, Vilner BJ, Williams W, Bertha CM, Kuehne ME, Jacobson AE. Ibogaine and its congeners are sigma 2 receptor-selective ligands with moderate affinity. Eur J Pharmacol 1995;279(1):1–3.

[248] Hayashi T, Su TP. Sigma-1 receptor chaperones at the ER-mitochondrion interface regulate Ca(2+) signaling and cell survival. Cell 2007;131(3):596–610.

[249] Mauricer T, Lockhart BP. Neuroprotective and anti-amnesic potentials of s (sigma) receptor ligands. Prog Neuropsychopharmacol Biol Psychiatry 1997;21:69–102.

[250] Marrazzo A, Caraci F, Salinaro ET, Su TP, Copani A, Ronsisvalle G. Neuroprotective effects of sigma-1 receptor agonists against beta-amyloid-induced toxicity. Neuroreport 2005;16:1223–6.

[251] Griesmaier E, Posod A, Gross M, Neubauer V, Wegleiter K, Hermann M, et al. Neuroprotective effects of the sigma-1 receptor ligand PRE-084 against excitotoxic perinatal brain injury in newborn mice. Exp Neurol 2012;237(2):388–95.

[252] Arias H,R, Rosenberg A, Targowska-Duda KM, Feuerbach D, Yuan XJ, Jozwiak K, et al. Interaction of ibogaine with human alpha3beta4-nicotinic acetylcholine receptors in different conformational states. Int J Biochem Cell Biol 2010;42(9):1525–35.

[253] Arias HR, Feuerbach D, Targowska-Duda KM, Jozwiak K. Structure-activity relationship of ibogaine analogs interacting with nicotinic acetylcholine receptors in different conformational states. Int J Biochem Cell Biol 2011;43(9):1330–9.

[254] Alper K, Reith MEA, Sershen H. Ibogaine and the inhibition of acetylcholinesterase. J Ethnopharmacol 2012;139(3):879–82.

[255] Kuehne ME, He L, Jokiel PA, Pace CJ, Fleck MW, Maisonneuve IM, et al. Synthesis and biological evaluation of 18-methoxycoronaridine congeners. Potential antiaddiction agents. J Med Chem 2003;46(13):2716–30.

[256] Andrade MT, Lima JA, Pinto AC, Rezende CM, Carvalho MP, Epifanio RA. Indole alkaloids from *Tabernaemontana australis* (Müell. Arg) Miers that inhibit acetylcholinesterase enzyme. Bioorg Med Chem 2005;13(12):4092–5.

[257] Zhan ZJ, Yu Q, Wang ZL, Shan WG. Indole alkaloids from *Ervatamia hainanensis* with potent acetylcholinesterase inhibition activities. Bioorg Med Chem Lett 2010;20(21):6185–7.

[258] Ingkaninan K, Changwijit K, Suwanborirux K. Vobasinyl-iboga bisindole alkaloids, potent acetylcholinesterase inhibitors from *Tabernaemontana divaricata* root. J Pharm Pharmacol 2006;58(6):847–52.

[259] Kitajima M, Iwai M, Kikura-Hanajiri R, Goda Y, Iida M, Yabushita H, et al. Discovery of indole alkaloids with cannabinoid CB1 receptor antagonistic activity. Bioorg Med Chem Lett 2011;21(7):1962–4.

[260] Lichtman AH, Dimen KR, Martin BR. Systemic or intrahippocampal cannabinoid administration impairs spatial memory in rats. Psychopharmacology 1995;119:282–90.

[261] Mazzola C, Micale V, Drago F. Amnesia induced by beta-amyloid fragments is counteracted by cannabinoid CB1 receptor blockade. Eur J Pharmacol 2003;477(3):219–25.

[262] Lo MW, Matsumoto K, Iwai M, Tashima K, Kitajima M, Horie S, et al. Inhibitory effect of Iboga-type indole alkaloids on capsaicin-induced contraction in isolated mouse rectum. J Nat Med 2011;65(1):157–65.

[263] Kim SR, Lee DY, Chung ES, Oh UT, Kim SU, Jin BK. Transient receptor potential vanilloid subtype 1 mediates cell death of mesencephalic dopaminergic neurons *in vivo* and *in vitro*. J Neurosci 2005;25(3):662–71.

[264] Kim SR, Kim SU, Oh U, Jin BK. Transient receptor potential vanilloid subtype 1 mediates microglial cell death *in vivo* and *in vitro* via Ca^{2+}-mediated mitochondrial damage and cytochrome c release. J Immunol 2006;177(7):4322–9.

[265] Mehrotra PK, Kamboj VP. Hormonal profile of coronaridine hydrochloride—an antifertility agent of plant origin. Planta Med 1978;33(4):345–9.

[266] Srivastava S, Singh MM, Kulshreshtha DK. A new alkaloid and other anti-implantation principles from *Tabernaemontana heyneana*. Planta Med 2001;67(6):577–9.

[267] Masuda K, Akiyama T, Taki M, Takaishi S, Iijima Y, Yamazaki M, et al. Isolation of 10-hydroxycoronaridine from *Tabernaemontana penduliflora* and its estrogen-like activity. Planta Med 2000;66(2):169–71.

[268] Singh M. Ovarian hormones elicit phosphorylation of Akt and extracellular signal regulated kinase in explants of the cerebral cortex. Endocrine 2001;14:407–15.

[269] Dhandapani KM, Wade FM, Mahesh VB, Brann DW. Astrocyte-derived transforming growth factor-β mediates the neuroprotective effects of 17 β-estrogen: involvement of nonclassical genomic signaling pathways. Endocrinology 2005;146:2749–59.

[270] Lima JA, Costa RS, Epifânio RA, Castro NG, Rocha MS, Pinto AC. *Geissospermum vellosii* stembark. Anticholinesterase activity and improvement of scopolamine-induced memory deficits. Pharmacol Biochem Behav 2009;92(3):508–13.

[271] Ingkaninan K, Ijzerman AP, Taesotikul T, Verpoorte R. Isolation of opioid-active compounds from *Tabernaemontana pachysiphon* leaves. J Pharm Pharmacol 1999;51(12):1441–6.

[272] Cai Z, Ratka A. Opioid system and Alzheimer's disease. Neuromolecular Med 2012;14(2):91–111.

[273] Mulder AH, Wardeh G, Hogenboom F, Frankhuyzen AL. Kappa- and delta-opioid receptor agonists differentially inhibit striatal dopamine and acetylcholine release. Nature 1984;308(5956):278–80.

[274] Wagner JJ, Caudle RM, Chavkin C. Kappa-opioids decrease excitatory transmission in the *dentate gyrus* of the guinea pig *hippocampus*. J Neurosci 1992;12(1):132–41.

[275] Hill MP, Brotchie JM. Modulation of glutamate release by a kappa-opioid receptor agonist in rodent and primate *striatum*. Eur J Pharmacol 1995;281(1):1–2.

[276] Standifer KM, Pasternak GW. G proteins and opioid receptor-mediated signalling. Cell Signal 1997;9(3–4):237–48.

[277] Williams JT, Christie MJ, Manzoni O. Cellular and synaptic adaptations mediating opioid dependence. Physiol Rev 2001;81(1):299–343.

[278] Bohn LM, Belcheva MM, Coscia CJ. Mitogenic signaling via endogenous kappa-opioid receptors in C6 glioma cells: evidence for the involvement of protein kinase C and the mitogen-activated protein kinase signaling cascade. J Neurochem 2000;74(2):564–73.

[279] Heidbreder CA, Goldberg SR, Shippenberg TS. The kappa-opioid receptor agonist U-69593 attenuates cocaine-induced behavioral sensitization in the rat. Brain Res 1993;616:335–8.

[280] Bhattacharya SK, Ray AB, Guha SR. Psychopharmacological studies on echitovenidine. Pharmacol Res Commun 1976;8(2):159–66.

[281] Werner JA, Oliveira SM, Martins DF, Mazzardo L, Dias JF, Lordello AL, et al. Evidence for a role of 5-HT(1A) receptor on antinociceptive action from Geissospermum vellosii. J Ethnopharmacol 2009;125(1):163–9.

[282] Fletcher A, Cliffe IA, Dourish CT. Silent 5-HT1A receptor antagonists: utility as research tools and therapeutic agents. Trends Pharmacol Sci 1993;14(12):41–8.

[283] Lewin G, Schaeffer C, Dacque C. Perhydrogenation of tabersonine, an aspidosperma indole alkaloid. J Nat Prod 1997;60:419–20.

[284] Fischhof PK, Möslinger-Gehmayr R, Herrmann WM, Friedmann A, Rubmann DL. Therapeutic efficacy of vincamine in dementia. Neuropsychobiology 1996;34(1):29–35.

[285] Tesseris J, Roggen G, Caracalos A, Triandafillou D. Effects of vincamin on cerebral metabolism. Eur Neurol 1975;13:195–202.

[286] Szatmari SZ, Whitehouse PJ. Vinpocetine for cognitive impairment and dementia. Cochrane Database Syst Rev 2003;1:CD003119.

[287] Hagiwara M, Endo T, Hidaka H. Effects of vinpocetine on cyclic nucleotide metabolism in vascular smooth muscle. Biochem Pharmacol 1984;33(3):453–7.

[288] Beavo JA. Cyclic nucleotide phosphodiesterases: functional implications of multiple isoforms. Physiol Rev 1995;75:725–48.

[289] Erdo SL, Molnar P, Lakics V, Bence J, Tomoskozi Z. Vincamine and vincanol are potent blockers of voltage-gated Na^+ channels. Eur J Pharmacol 1996;314:69–73.

[290] Zelles T, Franklin L, Koncz I, Lendvai B, Zsilla G. The nootropic drug vinpo-
       cetine inhibits veratridine-induced $[Ca^{2+}]i$ increase in rat hippocampal CA1
       pyramidal cells. Neurochem Res 2001;26(8–9):1095–100.

[291] Kiss B, Cai NS, Erdo SL. Vinpocetine preferentially antagonizes quisqualate/
       AMPA receptor responses: evidence from release and ligand binding studies. Eur
       J Pharmacol 1991;209:109–12.

[292] Polak PE, Kalinin S, Braun D, Sharp A, Lin SX, Feinstein DL. The vincamine
       derivative vindeburnol provides benefit in a mouse model of multiple sclerosis:
       effects on the *locus coerulus*. J Neurochem 2012;121(2):206–16.

[293] Both FL, Meneghini L, Kerber VA, Henriques AT, Elisabetsky E.
       Psychopharmacological profile of the alkaloid psychollatine as a 5HT2A/C sero-
       tonin modulator. J Nat Prod 2005;68(3):374–80.

[294] Both FL, Meneghini L, Kerber VA, Henriques AT, Elisabetsky E. Role of gluta-
       mate and dopamine receptors in the psychopharmacological profile of the indole
       alkaloid psychollatine. J Nat Prod 2006;69(3):342–5.

[295] Pasinelli P, Brown RH. Molecular biology of amyotrophic lateral sclerosis:
       insights from genetics. Nat Rev Neurosci 2006;7(9):710–23.

[296] Fragoso V, do Nascimento NC, Moura DJ, Silva AC, Richter MF, Saffi J, et al.
       Antioxidant and antimutagenic properties of the monoterpene indole alkaloid
       psychollatine and the crude foliar extract of *Psychotria umbellata* Vell. Toxicol
       in Vitro 2008;22(3):559–66.

[297] Nakamura T, Lipton SA. Redox modulation by S-nitrosylation contributes to
       protein misfolding, mitochondrial dynamics, and neuronal synaptic damage in
       neurodegenerative diseases. Cell Death Differ 2011;18:1478–86.

[298] Yuan D, Ma B, Wu C, Yang J, Zhang L, Liu S, et al. Alkaloids from the leaves
       of *Uncaria rhynchophylla* and their inhibitory activity on NO production in lipo-
       polysaccharide-activated microglia. J Nat Prod 2008;71(7):1271–4.

[299] Yuan D, Ma B, Yang JY, Xie YY, Wang L, Zhang LJ, et al. Anti-inflammatory
       effects of rhynchophylline and isorhynchophylline in mouse N9 microglial cells
       and the molecular mechanism. Int Immunopharmacol 2009;9(13–14):1549–54.

[300] Xian YF, Lin ZX, Mao QQ, Ip SP, Su ZR, Lai XP. Protective effect of isorhyn-
       chophylline against beta-amyloid-induced neurotoxicity in PC12 cells. Cell Mol
       Neurobiol 2011;32(3):353–60.

[301] Supnet C, Bezprozvanny I. The dysregulation of intracellular calcium in
       Alzheimer disease. Cell Calcium 2010;47(2):183–9.

[302] Kang TH, Matsumoto K, Tohda M, Murakami Y, Takayama H, Kitajima M, et al.
       Pteropodine and isopteropodine positively modulate the function of rat musca-
       rinic M(1) and 5-HT(2) receptors expressed in *Xenopus oocyte*. Eur J Pharmacol
       2000;444:39–45.

[303] Harvey JA. Serotonergic regulation of acetylcholine release in rat frontal cortex.
       Behav Brain Res 1996;73:47–50.

[304] Hirano H, Day J, Fibiger HC. Serotonergic regulation of acetylcholine release in
       rat frontal cortex. J Neurochem 1995;65:1139–45.

[305] Abdel-Fattah M, Matsumoto K, Tabata K, Takayama H, Kitajima M, Aimi N,
       et al. Effects of *Uncaria tomentosa* total alkaloid and its components on experi-
       mental amnesia in mice: elucidation using the passive avoidance test. J Pharm
       Pharmacol 2000;52:1553–61.

Isoquinoline Alkaloids

1.3.1 *Phellodendron amurense* **Rupr.**

History The plant was first described by Franz Josef Ruprecht in *Bulletin de la Classe Physico-Mathématique de l'Académie Impériale des Sciences de Saint-Pétersbourg* published in 1857.

Family Rutaceae Juss., 1789

Common Names Amur cork-tree, kihada (Japanese), huang bo (Chinese)

Habitat and Description This massive tree grows in the forests of China (Amur), Korea, and Japan. The plant is cultivated to ornament parks and streets in the USA, Canada, and Europe. The bark is deeply cracked, and the wood is intensely dark yellow. The stems are warty. The leaves are opposite, imparipinnate, and without stipules. The rachis is swollen at the base and bears 2–5 pairs of folioles and a terminal one. The folioles are 5 cm–12 cm × 2 cm–5 cm, asymmetrical, thin, glossy, elliptic, larger toward the apex of the leaf, with 5–7 pairs of inconspicuous secondary nerves. The inflorescence is a terminal panicle of little flowers. The perianth includes 5 sepals and 5 petals. The androecium includes 5 stamens inserted around a conspicuous disc. The gynoecium comprises 5 carpels. The fruit is a berry, which is 1 cm across, fragrant, darkish, globose, containing several minutely warty seeds (Figure 1.9).

Medicinal Uses Koreans use the plant to break fever. In China the plant is considered, like ginseng, as a panacea which is used to treat, among other things, diabetes, tuberculosis, dysentery, pneumonia, and meningitis.

Phytopharmacology The plant contains the flavonoids amurensin and phellamurin[306]; kaempferol; kaempferol-3-*O*-β-D-glucoside; kaempferol-3-*O*-β-D-galactoside; quercetin; quercetin-3-*O*-β-D-glucoside; quercetin-3-*O*-β-D-galactoside[307]; isoquinoline (CS 1.182); alkaloids including berberine, palmatine, magnoflorine, phellodendrine,[308]

■ **FIGURE 1.9** *Phellodendron amurense* Rupr.

■ **CS 1.182** Isoquinoline.

■ **CS 1.183** Noradrenaline.

oxyberberine, and oxypalmatine[309]; the indole alkaloid canthin-6-one[310]; 7-hydroxyrutaecarpine[311]; limonoids including limonin, aubacunone,[312] 12α-hydroxylimonin,[309] kihadalactones A and B[313]; the triterpenes niloticin, dihydroniloticin, niloticin acetate, piscidinol A, hispidol B, bourjotinolone A, and hispidone[313]; coumarines such as umbelliferon esculetin phellodenol A and F–H; scopoletin; demethylsuberosin; γ-fagarine[309]; and essential oil containing β-elemol and β-ocimene.[314]

Proposed Research Pharmacological study of berberine and congeners for the treatment of neurodegenerative diseases.

Rationale The most parsimonious explanation for the neuropharmacological properties of isoquinoline alkaloids is their ability to bind to dopaminergic and adrenergic receptors by virtue of their structural similitudes with dopamine and noradrenaline (CS 1.183). This concept was shown by Nimit et al.,[315] whereby series of benzylisoquinoline alkaloids were able to bind mainly to adrenergic receptor subtype β1, adrenergic receptor subtype

$\alpha 2$, and dopaminergic receptor subtype 2 (D_2), and to a lesser extent to serotoninergic receptors. For example, tetrahydropapaveroline (CS 1.184), 3'-methoxytetrahydropapaveroline (CS 1.185), 6-methoxytetrahydropapaveroline (CS 1.186), norreticuline (CS 1.187), and reticuline (CS 1.188) bind adrenergic receptor subtype $\beta 1$ with IC_{50} values equal to $0.3 \mu M$, $1 \mu M$, $2 \mu M$, $6 \mu M$, and $6.5 \mu M$.[315] Likewise, norreticuline and reticuline from *Papaver somniferum* L. (family Papaveraceae Juss.), 6-methoxytetrahydropapaveroline, and 3'-methoxytetrahydropapaveroline bind D_2 with IC_{50} values equal to $3 \mu M$, $4.5 \mu M$, $4.5 \mu M$, and $5 \mu M$, respectively.[315] Reticuline, 6-methoxytetrahydropapaveroline, and tetrahydropapaveroline bind to adrenergic receptor subtype $\alpha 2$ with IC_{50} values equal to $5 \mu M$, $8 \mu M$, and $8 \mu M$, respectively.[315] 6-Methoxytetrahydropapaveroline, reticuline, and norreticuline bind to serotoninergic receptors with IC_{50} values equal to $10 \mu M$. 3'-Methoxytetrahydropapaveroline binds to adrenergic receptors subtype $\alpha 1$ with an IC_{50} value equal to $5 \mu M$.[315]

Note that none of these isoquinoline alkaloids were able to bind to muscarinic receptors[315] simply because their frameworks differ from acetylcholine. However, dopaminergic receptors subtype 2 (D_2) are involved in learning via acetylcholine release in ventral *hippocampus*.[316] Furthermore, stephanine from *Stephania dielsiana* Y.C. Wu (family Menispermaceae Juss.) and xylopine from *Annona rugulosa* (Schltdl.) H. Rainer (family Annonaceae Juss.) are selective adrenergic receptor subtype $\alpha 1$ antagonists which inhibited *in vitro* muscle contractions imposed by phenylephrine with pA2 values of 6.7 and 6.6, respectively.[317] In addition, tetrahydroberberines bind to D_2. Thus, taking into account their unique selectivity for dopaminergic and adrenergic receptors, it is not difficult to imagine that isoquinolines may induce, enhance, or encumber the progression of Parkinson's disease (PD). Although the precise molecular mechanism underlying the pathophysiology of PD remains unknown, a massive body of evidence demonstrates the critical involvement of environmental toxins such as 11-methyl-4-phenyl-1,2,3,6-tetrahydropyridine (MPTP). Indeed, MPTP is converted by monoamine oxidase B (MAO-B) into 1-methyl-4-phenylpyridinium ion (MPP^+), which is selectively transported into dopaminergic neurons via the plasma membrane dopamine transporter (DAT).[318] Once in the cytoplasm, MPP^+ diffuses into mitochondria, where it blocks respiration, hence mitochondrial insult and apoptosis[318] of dopaminergic neurons in the *substantia nigra* and therefore Parkinsonism.[319,320] In this context, it is important to mention the isoquinoline scaffold shares some structural similarity with MPTP, implying that it may be metabolized into quinolinium ions which selectively impair selective mitochondrial respiration.[321] Therefore, one might logically argue that isoquinolines

■ **CS 1.184** Tetrahydropapaveroline.

■ **CS 1.185** 3'-methoxytetrahydropapaveroline.

■ **CS 1.186** 6-methoxytetrahydropapaveroline.

CS 1.187 (S)-norreticuline.

CS 1.188 (S)-reticuline.

CS 1.189 2-N-methyl-isoquinolimium.

CS 1.190 Norsalsolinol.

are environmental toxins which induce PD. The question was addressed by Storch et al.,[322] whereby over 20 different isoquinoline alkaloids were tested against non-neuronal human embryonic kidney (HEK-293) cells and mouse neuroblastoma cells (Neuro-2A) evidencing that most isoquinolines were not cytotoxic or neurotoxic.[322]

In addition, 1-methyl-4-phenylpyridinium ion (MPP$^+$), 2-N-methyl-isoquinolimium (CS 1.189), norsalsolinol (CS 1.190), 2-N-methyl-norsalsolinol (CS 1.191), and 2-N-methyl-salsolinol (CS 1.192) were predominantly cytotoxic against murine neuroblastoma (Neuro-2A) cells transfected with DAT with IC$_{50}$ values equal to 1.2 μM, 67.3 μM, 59 μM, 179.3 μM, and 416.9 μM, respectively, whereas papaverine (CS 1.193) and tetrahydropapaveroline had superior cytotoxic potencies against Neuro-2A cells with IC$_{50}$ values equal to 8.6 μM and 35.3 μM, respectively, and iso-quinoline, 1,2,3,4-tetrahydroisoquinoline (CS 1.194) and reticuline were inactive.[322] The toxicological significance of these data are quite clear in that they support the contention that isoquinolines are not systematically neurotoxic and in fact imply that isoquinoline, including benzyl isoquino-lines alkaloids, may be very well contemplated as a vast source of leads for the treatment of PD. This assertion is supported by the demonstration that the protoberberine alkaloid stepholidine (CS 1.195) from the *Stephania intermedia* H.S. Lo (family Menispermaceae Juss.) is first a serotoniner-gic receptor subtype 2 (5-HT$_2$) antagonist which relaxed preparations of thoracic aorta exposed to serotonin with a pA2 value equal to 9.7,[323] a dopaminergic receptor subtype 1 (D$_1$) agonist, and a D$_2$ antagonist.[324,325] Furthermore, l-stepholidine protected neurons against oxidative apopto-sis via activation of phosphoinositide 3-kinase (PI3K) and therefore pro-tein kinase B (Akt).[326] Along the same lines, the D$_1$ agonist dihydrexidine elicited antiparkinsonian effects in humans as efficiently as levodopa.[327] In addition, the stimulation of serotoninergic receptor subtype$_{1A}$ (5-HT$_{1A}$) increased the secretion of dopamine in the medial prefrontal cortex, and indeed, l-stepholidine given at a dose of 1 mg/kg imposed the release of dopamine in the ventral tegmental area (VTA) by 120% via D$_2$ blockade and 5-HT$_{1A}$ receptor stimulation.[328] However, stepholidine may not be beneficial in Alzheimer's disease because 5-HT$_{1A}$ agonists reduce cogni-tion and D$_2$ antagonists induce dementia in rodents due to acetylcholine depletion. The methoxylation of l-stepholidine at C10 provides l-isocor-ypalmine (CS 1.196) produced by *Corydalis yanhusuo* W.T. Wang ex Z.Y. Su & C.Y. Wu (family Papaveraceae Juss.), which exhibited binding affini-ties against D$_1$ with a Ki value equal to 83 nM.[329] The methoxylation of l-stepholidine at C2 and C10 yields l-tetrahydropalmatine (CS 1.197), which has the striking capability to abate the levels of cerebral serotonin, dopamine, and norepinephrine concentrations and to antagonize both pre- and postsynaptic dopaminergic receptors in the nigrostriatal pathway.[330–332] Note that the fusion of C2 and C3 hydroxyls on l-tetrahydropalmatine into a methylene dioxide produces tetrahydroberberine (CS 1.198), which blocks ATP-sensitive potassium (K⁺-ATP) channels at a dose of 100 μM.[333] The opening of potassium (K⁺-ATP) channels is downstream of mitochondrial asphyxia induced by 1-methyl-4-phenylpyridinium ion (MPP⁺) in dopaminergic neurons, which become paralyzed.[334,335] Thus, protoberberines, like l-tetrahydropalmatine, may abort dopaminergic apoptosis and may be considered as first-line leads for the treatment of PD.

The formation of a methylene dioxide at C2 and C3 of tetrahydropalmatine followed by removal of hydrogen atoms C8 and C13 yields berberine (CS 1.199), which is planar, cationic, neuroprotective, and neurotrophic and is produced by *Phellodendron amurense* Rupr. (family Rutaceae Juss.). Berberine, palmatine (CS 1.200), and coptisine (CS 1.201) at a dose of 5 μg/mL enhanced the neurite formation of rat pheochromocytoma (PC12) cells exposed to 10 ng/mL of nerve growth factor (NGF) by 30%, 5%, and 20%, respectively,[336] suggesting transient activation of extracellular signal-regulated kinase (ERK1/2). Berberine at a dose of 25 μM mitigated the apoptosis of hippocampal neurons deprived of oxygen and glucose with hypophosphorylation of anti-apoptotic protein (Bcl-2)[337] via probable induction of PI3K and consequently Akt.[338] Likewise, 2 μmol/L of berberine defended primary cortical neurons against β-amyloid (Aβ_{25-35}) peptide by 20%.[339] *In vivo* 20 mg/kg of berberine improved the cognition of rodents intoxicated with scopolamine by confronting nuclear factor kappa-light-chain-enhancer of activated B cells (NF-κB), enhancing the enzymatic activity of superoxide dismutase (SOD) and impairing acetylcholinesterase (AChE).[339] In accordance, motor 10 nM of berberine compelled the activation of PI3K, protein kinase in neuron cell-like NSC34 cells.[340] Lee et al.[341] synthesized a protoberberine derivative which inhibited by 95% the 2′,3′-*O*-(4-benzoyl-benzoyl)adenosine triphosphate (BzATP)-induced formation of pores in purinergic receptor (P2 × 7)-expressing human embryonic kidney (HEK293) cells at a dose of 10 μM. In the brain, the activation of P2 × 7 by ATP results in Ca²⁺ entry into microglia and astrocytes, and, hence, activation of calcineurin, G protein Ras, mitogen-activated protein kinase (MAPK) p38, ERK1/2, NF-κB, and expression of inducible nitric oxide synthetase (iNOS) and cyclo-oxygenase-2 (COX-2).[342,343] The clinical significance of these data are quite clear in that they support a possible role for berberine and synthetic derivatives for the treatment of Alzheimer's disease (AD), stroke, amyotrophic lateral sclerosis (ALS), and Parkinson's disease (PD).

■ **CS 1.191** 2-*N*-methyl-norsalsolinol.

■ **CS 1.192** 2-*N*-methyl-(R)-salsolinol.

■ **CS 1.193** Papaverine.

■ **CS 1.194** 1,2,3,4-tetrahydroisoquinoline.

■ **CS 1.195** Stepholidine.

■ **CS 1.196** (-)-isocorypalmine.

■ **CS 1.197** S-(-)-tetrahydropalmatine.

■ **CS 1.198** S-(-)-tetrahydroberberine.

■ **CS 1.199** Berberine.

■ **CS 1.200** Palmatine.

■ **CS 1.201** Coptisine.

Collective evidence suggests that the protoberberines, *id est*, palmatine, berberine, coptisine, groenlandicine, and jatrorrhizine are puissant AChE inhibitors on account of their charged quaternary amine and methoxyl moieties, which mimic acetylcholine. In effect, corynoxidine (CS 1.202), protopine (CS 1.203), palmatine, and berberine from *Corydalis speciosa* Maxim. (family Papaveraceae Juss.) hindered the enzymatic activity of AChE with IC_{50} values equal to $89\,\mu M$, $16.1\,\mu M$, $5.8\,\mu M$, and $3.3\,\mu M$, respectively.[344] Members of the *Stephania* Lour. (family Menispermaceae Juss.) shelter the protoberberine alkaloids stepharanine (CS 1.204),

■ **CS 1.202** Corynoxidine.

■ **CS 1.203** Protopine.

■ **CS 1.204** Stepharanine.

cyclanoline (CS 1.205), *N*-methyl stepholidine (CS 1.206), stepholidine, and corydalmine (CS 1.207), which inhibited the enzymatic activity of AChE with IC_{50} values equal to $14\,\mu$M, $9.2\,\mu$M, $31.3\,\mu$M, $>100\,\mu$M, and $100\,\mu$M, respectively, further implying that the occurrence of an iminium at 6a and unsaturation at C8 and C13 are beneficial for activity as well as planarity.[345] In the same experiment, palmatine, jatrorrhizine (CS 1.208), and berberine inhibited the enzymatic activity of AChE with IC_{50} values equal to $0.2\,\mu$M, $0.9\,\mu$M, and $0.5\,\mu$M, respectively, showing further the beneficial effect of quaternary nitrogen and methoxylation.[345] Corydaline (CS 1.209)

■ **CS 1.205** Cyclanoline.

■ **CS 1.206** *N*-methyl stepholidine.

■ **CS 1.207** Corydalmine.

■ **CS 1.208** Jatrorrhizine.

■ **CS 1.209** Corydaline.

■ **CS 1.210** Groenlandicine.

■ **CS 1.211** Epiberberine.

from *Corydalis cava* Schweigg. & Kort. (family Papaveraceae Juss) inhibited the enzymatic activity of AChE with IC_{50} values equal to $15\,\mu M$,[346] implying that a methoxyl at C10 is beneficial.

Berberine, palmatine, groenlandicine (CS 1.210), jateorrhizine, coptisine, and epiberberine (CS 1.211) isolated from a member of the genus *Coptis* Salisb. (family Ranunculaceae Juss.) inhibited the enzymatic activity of AChE with IC_{50} values equal to $0.4\,\mu M$, $0.5\,\mu M$, $0.5\,\mu M$, $0.5\,\mu M$, $0.8\,\mu M$, and $1\,\mu M$, respectively.[347] Astonishingly, groenlandicine and epiberberine inhibited the enzymatic activity of β-secretase with IC_{50} values at $19.6\,\mu M$ and $8.5\,\mu M$, respectively.[347] Because the amyloid precursor protein (APP) is cleaved by β-secretase and γ-secretase into insoluble β-amyloid peptide, which aggregates to form neurotoxic plaques, it is conceivable that protoberberines and congeners may, by virtue of synergistical β-secretase and AChE activities, produce exquisite molecular templates for the development of leads for the treatment of AD.

1.3.2 *Macleaya cordata* (Willd.) R. Br.

History The plant was first described by Robert Brown in *Narrative of Travels and Discoveries in Northern and Central Africa* published in 1826.

Synonyms *Bocconia cordata* Willd., *Macleaya cordata* var. *yedoensis* (André) Fedde, *Macleaya yedoensis* André

Family Papaveraceae Juss., 1789

Common Name Bo luo hui (Chinese)

Habitat and Description This perennial herb grows in China, Korea, Taiwan, and Japan. The main stem grows to 3 m tall and exudes a yellow latex upon incision. The leaves are simple, alternate, and exstipulate. The petiole is of variable length and reaches 10 cm long. The blade is ovate, glaucous below, 25 cm across, with 8 serrate and irregular lobes; it exposes 1–3 pairs of secondary nerves. The inflorescence is a massive and terminal panicle, which grows to 40 cm long. The perianth comprises 4 tepals, which are whitish and 1 cm long. The androecium includes 30 stamens, which are 0.5 cm long. The gynaocium consists of 2 carpels fused into a unilocular ovary, which is 0.3 cm across. The fruit is a capsule less than 1 cm across, which contains 4–8 seeds (Figure 1.10).

Medicinal Uses In China, the plant is used to treat inflammation and pain.

Phytopharmacology The plant abounds with the benzophenanthridine alkaloids such as angoline,[348] 8-*O*-demethylchelerythrine,[349] sanguinarine[350] (CS 1.212), protopine, allocryptopine,[351] chelerythrine[352] (CS 1.213), 6-methoxy-dihydrosanguinarine, norsanguinarine, 6-acetonyl-dihydrochelerythrine, 6-acetonyl-dihydrosanguinarine, sanguidimerine, chelidimerine, (±)-bocconarborine A, bocconarborine B, cryptopine, dihydrosanguinarine, and dihydrochelerythrine.[353]

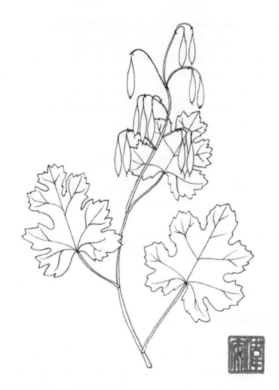

■ **FIGURE 1.10** *Macleaya cordata* (Willd.) R. Br.

■ **CS 1.212** Sanguinarine.

■ **CS 1.213** Chelerythrine.

Proposed Research Pharmacological study of sanguinarine and congeners for the treatment of neurodegenerative diseases.

Rationale Oxidation of C6 of the protoberberine framework followed by cyclization of C6 to C13 produces benzo[c]phenanthridine alkaloids such as sanguinarine, which inhibited the binding of ^{3}H-candesartan to angiotensin (AT1) receptor with an IC_{50} value equal to 4.3 μM.[354] In neurophysiological conditions, renin converts angiotensinogen into angiotensin I, which is the substrate for angiotensin-converting enzyme (ACE), which yields angiotensin II, the canonical function of which is the regulation of blood pressure.[355] However, cerebral ACE and AT1 receptors are involved in cognition,[356] as shown by Georgiev and Yonkov,[357] whereby rodents treated with 1 μg of angiotensin II displayed enhanced learning abilities. Oddly enough, Barnes et al.[358] showed that angiotensin II reduced the release of acetylcholine. In line with these results, angiotensin IV reduced the secretion of hippocampal acetylcholine after 60 min at a dose of 1 nmol via AT1 receptors.[359] Acute activation of AT1 receptors by angiotensin II improves learning and memory, but chronic activation of angiotensin AT1 receptors is deleterious to cognition.[359] Savaskan et al. (2001) demonstrated that Alzheimer patients had the tendency to have a higher ACE activity in their cortex,[356] which implies some yet-unknown mechanism of defense. Rodents demented with scopolamine had improved cognition when given a single dose of angiotensin IV.[360] It is therefore conceivable that benzo[c]phenanthridine alkaloids inhibiting ACE or eliciting angiotensin II or IV antagonist properties may be of value for the treatment of AD. Angiotensin II is converted by aminopeptidase N into angiotensin IV, which binds to the AT4 receptor, which is an insulin-regulated aminopeptidase (IRAP) that inhibits neuropeptide catabolism.[361,362] Hence, chelerythrine and sanguinarine from *Macleaya cordata* (Willd.) R. Br. (family Papaveraceae Juss.), by inhibiting the enzymatic activity of aminopeptidase N by 82% at a dose of 50 μM,[354] may increase angiotensin II in the brain and ameliorate cognition. Furthermore, chelerythrine and sanguinarine inhibited the enzymatic activity of dipeptidyl peptidase IV by 38% and 62%, respectively.[363] Dipeptidyl peptidase IV catabolizes several peptides, of which glucagon-like peptide-1 (GLP-1) and its inhibition have been shown to have beneficial effects against AD.[364]

Indeed, the canonical function of glucagon-like peptide-1 (GLP-1) is to impose the secretion of insulin by the pancreas during prandium, but GLP-1 receptors exist in the brain. For instance, 10 nM of GLP-1 protected hippocampal neurons against glutamate-induced apoptosis insults and maintained the cholinergic activity of rodents poisoned with ibotenic acid,[364] supporting the contention that GLP-1 agonists may potentially be of interest to develop leads for the treatment of AD. This contention was further demonstrated as GLP-1 at a dose of 33 ng/mL saved PC12 cell neurons against β-amyloid (Aβ_{1-42}) peptide and hampered the generation of Aβ_{1-42} peptide by 36% at a dose of 6.6 ng in rodents.[365] Likewise, demented rodents treated with the dipeptidyl peptidase IV inhibitor sitagliptin at a dose of 20 mg/kg experienced some levels of cognitive improvement as a result of hippocampal Aβ_{1-42} peptide depletion to 6.2 pg/mg.[366] The precise neuroprotective mechanism responsible for the aforementioned evidence is yet undelineated, but GLP-1 activates Akt.[367] Note that chelerythrine inhibits the enzymatic activity of protein kinase C (PKC) with an IC$_{50}$ value equal to 0.6 μM.[368] PKC is activated by cytoplasmic Ca^{2+} and diacylglycerol (DAG) and is of immense importance in the pathophysiology of AD because it favors the enzymatic activity of α-secretase and encumbers the phosphorylation of microtubule-associated protein taus.[369] Indeed, the stimulation of PKC by Phorbol 12-Myristate 13-Acetate (PMA) in astrocytes resulted in a boost of a secreted form of APP (sAPPα), which was abolished by forskolin.[370] At a dose of 1 μM, forskolin enhanced the enzymatic activity of adenylate cyclase (AC), hence causing inhibition of α-secretase by 50% in C6 cells,[370] implying that cyclic adenosine monophosphate (cAMP) contravenes the non-amyloidogenic activity of PKC due to c-APM accumulation.[371] Likewise, an elevation of cAMP in rat glioma cells (C-6) stimulated the production of Aβ_{1-42} peptide.[372,373] Note that 8-hydroxydihydrochelerythrine (CS 1.214) and 8-hydroxydihydrosanguinarine (CS 1.215) isolated from *Chelidonium majus* L. (family Papaveraceae Juss.) inhibited AChE.[374] The results reported here indicate that sanguinarine and derivatives might be of value to develop leads to treat AD.

1.3.3 *Nandina domestica* Thunb.

History The plant was first described by Carl Peter Thunberg in *Nova Genera Plantarum* published in 1781.

Synonym *Nandina domestica* var. *linearifolia* C.Y. Wu ex S.Y. Bao

Family Berberidaceae Juss., 1789

Common Name Nan tian zhu (Chinese)

Habitat and Description This ornamental shrub grows to 2 m tall in China, Korea, and Japan. The stems are reddish, and the leaves are bi- or tri-pinnate, red, exstipulate, and alternate. The rachis is 30 cm–50 cm long

■ **CS 1.214** 8-hydroxydihydrochelerythrine.

■ **CS 1.215** 8-hydroxydihydrosanguinarine.

and bears 3–4 pairs of folioles, which are lanceolate 2 cm–10 cm × 0.5 cm–2 cm, with inconspicuous secondary nerves. The inflorescence is a many flowered, terminal panicle, which grows to 30 cm long. The flowers are pure white. The perianth comprises 6 tepals, which are broadly lanceolate, 0.5 cm long, and curved. The androecium consists of 6 showy stamens. The gynoecium is oblong and unilocular and topped with a red stigma. The fruits are globose berries, which are 1 cm across (Figure 1.11).

Medicinal Uses The plant is used to treat asthma and cough in Japan and to check bleeding.

Phytopharmacology The few phytochemical studies devoted to this plant produced the isolation of triterpenes,[375] higenamine,[376] nantenine,[377] dehydronantenine, lignan, and (−)-episyringaresinol.[378]

Proposed Research Pharmacological study of nandenine and congeners for the treatment of neurodegenerative diseases.

Rationale The formation of a covalent bond between C8 and C10 benzylisoquinoline forms a bewildering array of aporphine alkaloids, which have the tendency to bind to dopaminergic receptors and impinge either agonism or antagonism due mainly to the conformation of C6a. Indeed,

■ **FIGURE 1.11** *Nandina domestica* Thunb.

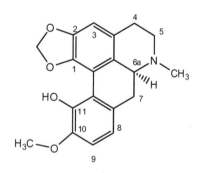

■ **CS 1.216** Bulbocapnine.

aporphines with an alpha C6a hydrogen such as bulbocapnine (CS 1.216), boldine, and glaucine are dopamine receptor antagonists, such as the dopaminergic receptor subtypes 1/2 (D_1/D_2),[379,380] whereas the presence of a beta C6a hydrogen produces agonists, such as notably D_1/D_2.[381,382] Agonists of D_2 are useful for the treatment of PD, and in fact, apomorphine (CS 1.217) is used with levopoda to treat PD despite heavy side effects including vomiting, hypotension, and tolerance.[382] Furthermore, the aporphine alkaloids asimilobine (CS 1.218) and lirinidine (CS 1.219) from *Nelumbo nucifera* Gaertn. (family Nelumbonaceae A. Rich.) displayed serotoninergic antagonistic properties with pA_2 values equal to 5.7 and 7.3, respectively.[383] Later, evidence indicated that asimilobine, nornuciferine (CS 1.220), and annonaine (CS 1.221) from *Annona muricata* L. (family Annonaceae Juss.) bind to and stimulate 5-HT_{1A} with IC_{50} values equal to $3\,\mu M$, $9\,\mu M$, and $5\,\mu M$, respectively.[384,385] In this context, it is important to mention that 5-HT_{1A} agonists impair cognition,[386] improve motor function in PD,[387] increase the secretion of dopamine in the medial prefrontal cortex[388] and *striatum*,[389] and, although controversial, might be of potential value for the treatment of PD. In striatal preparations, boldine from *Peumus boldus* Molina (family Monimiaceae Juss.) and glaucine (CS 1.222) from *Glaucium flavum* Crantz (family Papaveraceae Juss.) elicited binding activity toward D_1 with IC_{50} values equal to $0.4\,\mu M$ and $3.9\,\mu M$, respectively.[390] In the same experiment, boldine (CS 1.223) and glaucine elicited binding activity toward D_2 with IC_{50} values equal to $0.5\,\mu M$ and $3\,\mu M$.[390] With the premise that lipophilic aporphines may provoke a potent dopaminergic effect, series of halogenated boldine derivatives were engineered by Sobarzo-Sanchez et al.,[391] including 3-iodoboldine (CS 1.224), which elicited *in vitro* robust-specific binding to dopaminergic receptors subtype 1 (D_1) and 2 (D_2) with IC_{50} values equal to $0.003\,\mu M$ and $0.09\,\mu M$, respectively.[391] Later, Asencio et al.[392] synthetized series of halogenated aporphines and showed that 3-bromo-predicentrine (CS 1.225) and 3-iodopredicentrine (CS 1.226) had the ability to dock in D_1 with K_i values equal to 15 nM and 6 nM by virtue of the hydroxyl in C2 and a lipophilic group in C3. Other receptors with affinities for aporphines are the adrenergic receptors α subtype 1 ($\alpha 1$), the stimulation of which induces dementia.[393] Of note, dicentrine (CS 1.227) from *Lindera macrophylla* Boerl. (family Lauraceae Juss.) and boldine are both $\alpha 1$ antagonists with a pA2 value equal to 8.3 and 5.6, respectively.[394,395] In this regard, it is noteworthy that the stimulation of cortical $\alpha 1$ by noradrenaline at a dose of $0.2\,\mu g/\mu L$ induced cognitive impairments in rodents which were reversed by $0.01\,\mu g$ of $\alpha 1$ antagonist uripidil.[393] The aporphine boldine at a dose of $100\,\mu M$ protected PC12 cells against dopamine-induced apoptosis by nullifying oxidative mitochondrial insults.[396] Furthermore, PC12 cells exposed to $10\,\mu M$ of the aporphine liriodenine isolated from a member of the genus *Magnolia* L. (family Magnoliaceae Juss.) experienced a collapse in cytoplasmic concentration of dopamine with an IC_{50} value equal to $8.4\,\mu M$ by hindering the enzymatic activity of tyrosine hydroxylase (TH) and aromatic L-amino acid decarboxylase (AADC).[397] The aporphine crebanine (CS 1.228) at a dose of 25 mg/kg protected rodents against scopolamine by inhibiting $\alpha 7$ nicotinic receptors with an IC_{50} value equal to $19.1\,\mu M$,[398] showing the possibility that aporphines with tertiary amine in C6 and a methoxyl moiety in C8 dock into the $\alpha 7$ nicotinic receptor, thus inhibiting β-amyloid peptide-induced neuroapoptosis. Nantenine, from *Nandina domestica* Thunb. (family Lauraceae Juss.), inhibited serotonin-induced aorta contraction *in vitro*,[399] suggesting some possible serotoninergic potencies. Furthermore, nantenine at a dose of $10^{-3}\,M$ inhibited the enzymatic activity of synaptosomal Ca^{2+}-ATPase and Na^+,K^+-ATPase by 34% and 90%, respectively, and induced convulsions with an LD_{50} equal to 120 mg/kg.[400] Nantenine (CS 1.229) at a dose of $30\,\mu M$ relaxed aortic rings challenged with noradrenaline by hampering extracellular influx of Ca^{2+} in the smooth muscle cells,[401] implying the possible inhibition of $\alpha 1$. In fact, nantenine

■ **CS 1.217** (R)-apomorphine.

■ **CS 1.218** Asimilobine.

■ **CS 1.219** Lirinidine.

■ **CS 1.220** Nornuciferine.

■ **CS 1.221** Annonaine.

■ **CS 1.222** S-(+)-glaucine.

binds to serotoninergic receptor subtype 2A (5-HT$_{2A}$), α1, and D$_2$ with K_i values equal to 0.4 μM, 2.1 μM, and 1.7 μM, respectively,[402] and administration of nantenine at a dose of 30 mg/kg protected rodents against 5-hydroxy-L-tryptophan/clorgyline-induced head twitches, implying a blockade of 5-HT$_{2A}$.[402] Because 5-HT$_{2A/2C}$ antagonists enhance cognition,[403] interest has therefore turned toward nantenine derivatives further antagonizing 5-HT$_{2A}$ receptors; Chaudhary et al.[404] showed that the substitution of C1 by lipophylic alkyls increased the 5-HT$_{2A}$ antagonist potency. Furthermore, nantenine inhibited the enzymatic activity of AChE.[405]

1.3.4 *Stephania tetrandra* S. Moore

History The plant was first described by Spencer Moore in *Journal of Botany, British and Foreign* published in 1875.

Family Menispermaceae Juss., 1789

Common Name Fen fang ji (Chinese)

Habitat and Description This slender climber grows wild in China; it develops from tuberous roots to a length of 3 m. The leaves are simple, exstipulate, and spiral. The petiole is slender, 3 cm–7 cm long, and inflated at both apices. The blade is peltate, deltoid, 5 cm–8 cm × 5 cm–10 cm, papery, and it displays 10 whitish nerves. The inflorescence is umbelliform and axillary. The male flowers present 5 sepals, 5 petals, and 4 stamens fused into a synandrium. The female flowers consist of 5 sepals; 5 petals, which are minute; and a gynaocium consisting of a single carpel forming a unilocular ovary. The fruit is a drupe, which is globose and red; the seed is horse-shoe-shaped (Figure 1.12).

Medicinal Uses In China, the plant is used to treat rheumatism.

Phytopharmacology The plant contains the bis-benzylisoquinolines tetrandrine, 2-*N*-methyltetrandrine, fangchilonine, and 2-*N*-methylfanchilonine.[406]

Proposed Research Pharmacological study of tetrandrine and synthetic derivatives for the treatment of neurodegenerative diseases.

Rationale The dimerization of benzylisoquinolines produces bis-benzylisoquinoline alkaloids, which have the ability to paralyze skeletal muscles, hence their uses to make arrow poisons. In fact, bis-benzylisoquinoline alkaloids not only block Ca^{2+} influx in skeletal muscle, thus hindering

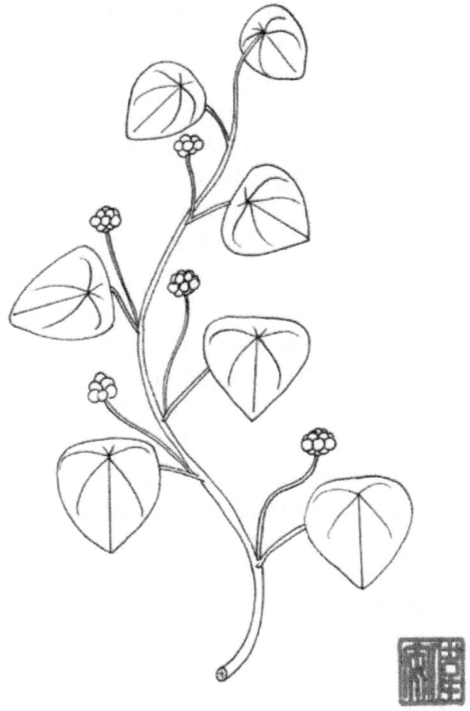

■ **FIGURE 1.12** *Stephania tetrandra* S. Moore.

contraction, but also hinder Ca^{2+} influx in neurons and glial cells, thus combatting neuroapoptosis and neuroinflammation.[407,408] Indeed, cepharanthine (CS 1.230) from a member of the genus *Stephania* Lour. (family Menispermaceae Juss.), at a dose of $1\,\mu g/mL$, inhibited the secretion of tumor necrosis factor-α (TNF-α), interleukin-1β (IL-1β), and interleukin-8 (IL-8) by 48%, 49%, and 41%, respectively, by U937 macrophages challenged with the PKC activator phorbol 12-myristate 13-acetate (PMA).[409] Furthermore $0.2\,\mu g/mL$ of cepharanthine protected SK-N-MC neurons against TNF-α insults.[409] Furthermore, cepharanthine at a dose of $10\,\mu M$ inhibited the enzymatic activity of acid sphingomyelinase (ASM), which converts sphingomyelin into ceramide[410] in the presence of Ca^{2+}.[411] Ceramide induces neuronal apoptosis via mitochondrial insults upon exposure to β-amyloid peptide and/or reactive oxygen species (ROS)-induced stimulation of ASM and serine palmitoyl transferase (SPT).[412] It is therefore plausible to suggest that cepharanthine, by blocking Ca^{2+} channels and consequently ceramide production, may sustain the viability of neurons challenged with pro-apoptic stimuli. In this light, fangchinoline (CS 1.231) and tetrandrine (CS 1.232) from *Stephania tetrandra* S. Moore (family

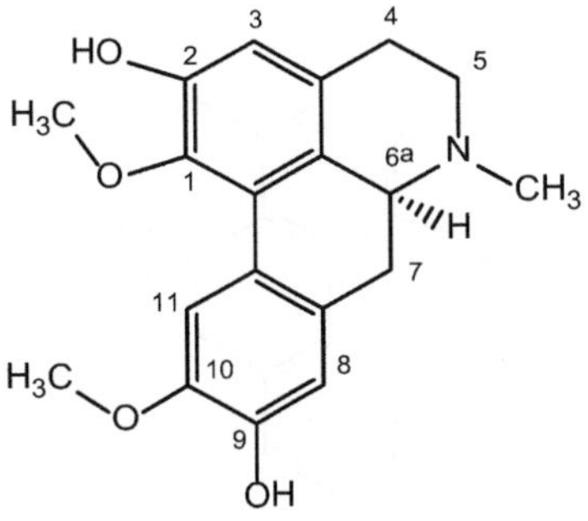

■ **CS 1.223** *S*-(+)-boldine.

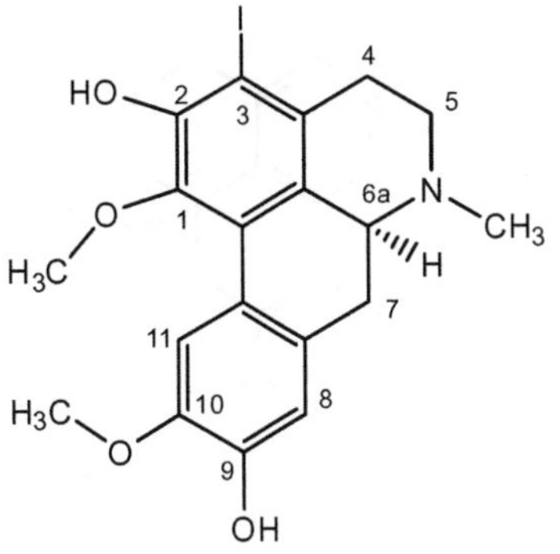

■ **CS 1.224** 3-iodoboldine.

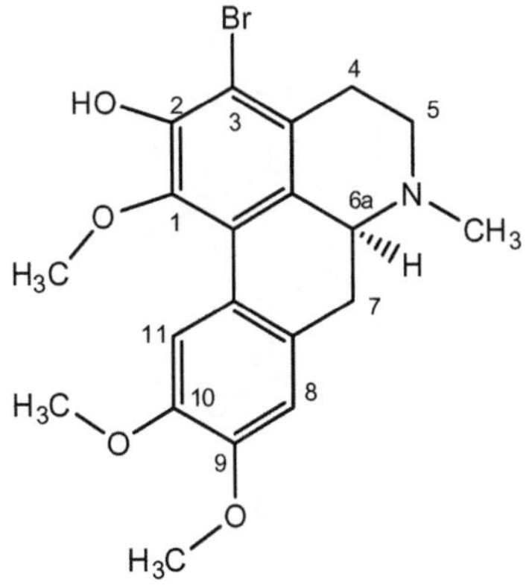

■ **CS 1.225** 3-bromopredicentrine.

■ **CS 1.226** 3-iodopredicentrine.

■ **CS 1.227** (+)-dicentrine.

■ **CS 1.228** (-)-crebanine.

Menispermaceae Juss.) at a dose of $1 \mu M$ safeguarded cerebral granule neurons against H_2O_2 by blocking Ca^{2+} influx.[413] Indeed, tetrandrine at a dose of $3 \mu M$ abated by 70% the secretion of TNF-α spurred by lipopolysaccharide (LPS) by blocking Ca^{2+} influx and therefore NF-κB.[414] Additionally, tetrandrine at a dose of 40 mg/kg protected rodents against β-amyloid (Aβ_{1-42}) peptide–induced dementia in the Morris water maze test with a reduction in cerebral IL-1β and TNF-α.[415] Correspondingly, the Ca^{2+} channels blockade of guattegaumerine (CS 1.233) from *Menispermum dauricum* DC. (family Menispermaceae Juss.) could explain the survival of cortical neurons exposed to H_2O_2 with a marked decrease in malofialdehyse (MDA), increase in the levels of anti-apoptotic protein (Bcl-2), and reduction in pro-apoptotic Bcl-2-associated X protein (Bax), at a dose of $2.5 \mu M$.[416] This paradigm was shown further with dauricine (CS 1.234), which at a dose of 84 mg/kg lessened neuronal apoptosis of rodent models of stroke.[417,418] Note that (R,S)-2-*N*-norberbamunine (CS 1.235) and (S,S)-*O*-4″-methyl,*O*-6′-demethyl-(+)-curine (CS 1.236) from *Abuta grandifolia* (Mart.) Sandwith (family Menispermaceae Juss.) inhibited the enzymatic activity of AChE with IC_{50} values equal to $34.6 \mu M$ and $44.5 \mu M$, respectively,[419] suggesting exciting synergistic Ca^{2+} channel blocking/AChE properties that could be of value to repress neurodegeneration.

1.3.5 *Corydalis stricta* **Stephan ex Fisch.**

History The plant was first described by Christian Friedrich Stephan in *Regni Vegetabilis Systema Naturale* published in 1821.

Synonyms *Corydalis pseudostricta* Popov, *Corydalis schlagintweitii* Fedde, *Corydalis stricta* subsp. *holosepala* Michajlova, *Corydalis stricta* subsp. *spathosepala* Michajlova, *Corydalis stricta* var. *potaninii* Fedde, *Corydalis transalaica* Popov

Family Papaveraceae Juss., 1789

Common Name Zhi jing huang jin (Chinese)

Habitat and Description This perennial herb grows in China, Mongolia, Russia, Pakistan, India, and Nepal. The whole plant is glaucous and grows to 50 cm tall. The leaves are simple, alternate, and exstipulate. The petiole is 5 cm–10 cm long. The blade is bipinnately incised, forming 4–7 pairs of lobes, which are ovate and dentate. The inflorescence is a many-flowered raceme, which is 10 cm long. The sepals are petaloid, dentate, and minute. The petals are yellow and assembled into an elongated corolla, which grows to 2 cm long. The androecium comprises 2 stamens. The fruit is a capsule, which is 1.5 cm–2 cm long and conceals 10 tiny seeds (Figure 1.13).

Medicinal Uses In China, the plant used to regulate menses.

Phytopharmacology The plant engineers an interesting array of isoquinoline alkaloids, including the phthalides hydrastine, adlumidine, and bicuculline; the protoberberines protopine and coreximine; the benzophenanthridine sanguinarine; the aporphines isoboldine and wilsonirine; the benzyl-isoquinoline juziphine; and the simple isoquinoline corypalline.[420]

Proposed Research Pharmacological study of hydrastine derivatives for the treatment of ALS and/or multiple sclerosis (MS).

■ **CS 1.229** (+)-nantenine.

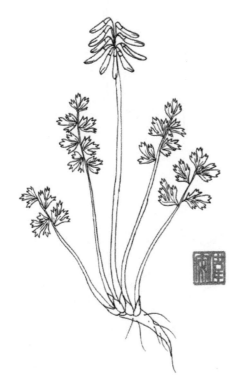

■ **FIGURE 1.13** *Corydalis stricta* Stephan ex Fisch.

■ **CS 1.230** Cepharanthine.

■ **CS 1.231** Fangchinoline.

■ **CS 1.232** Tetrandrine.

■ **CS 1.233** Guattegaumerine.

■ **CS 1.234** Dauricine.

■ **CS 1.235** (R,S)-2-*N*-norberbamunine.

■ **CS 1.236** *(S,S)-O-4″-methyl, O-6′-demethyl-* *(+)-curine.*

■ **CS 1.237** γ-amino butyric acid (GABA).

■ **CS 1.238** (+)-hydrastine.

Rationale The canonical function of γ-amino butyric acid (GABA) (CS 1.237) is to inhibit neurotransmission in the brain, but some evidence is available to demonstrate GABA acts on the immune system. In effect, Bhat et al. (2010) showed the presence of γ-amino butyric acid receptors subtype A ($GABA_A$) on the surface of macrophages and inhibition of cytokine production by splenocytes challenged with myelin oligodendrocyte glycoprotein (MOG) in the presence of topiramate.[421] Paradoxically, the injection of GABA at a dose of 200 mg/kg daily to C57Bl/6J EAE mice immunized with MOG_{35-55} fostered both secretion of TNF-α and IL-6 and paralysis.[422] In this context, it is important to mention that phthalide isoquinoline alkaloids (+)-hydrastine (CS 1.238) from *Corydalis stricta* Stephan ex Fisch. (family Papaveraceae Juss.) and bicuculline (CS 1.239) from *Dicentra cucullaria* (L.) Bernh. (family Papaveraceae Juss.) are $GABA_A$ antagonists with pA2 values equal to 6.5 and 6.1, respectively.[423] Consequently, these alkaloids are powerful poisons that enforce seizures in rodents with CD_{50} values equal to 0.1 mg/kg and 0.3 mg/kg, respectively,[423] and which are not possibly usable in therapeutics. However, one may draw an inference that the aforementioned alkaloids offer frameworks that might be used as starting points for the synthesis of leads for the treatment of ALS, MS, and seizures.

1.3.6 *Sinomenium acutum* (Thunb.) Rehder & E.H. Wilson

History The plant was first described by Alfred Rehder and Ernest Henry Wilson in *Plantae Wilsonianae* published in 1913.

Synonyms *Cocculus diversifolius* Miq., *Cocculus diversifolius* var. *cinereus* Diels., *Cocculus heterophyllus* Hemsl. & E.H. Wilson, *Menispermum acutum* Thunb., *Menispermum diversifolium* (Miq.) Gagnep.,

Menispermum diversifolium var. molle Gagnep., *Sinomenium acutum* var. cinereum (Diels) Rehder & E.H. Wilson, *Sinomenium diversifolium* Diels

Family Menispermaceae Juss., 1789

Common Name Feng long (Chinese)

Habitat and Description This massive woody climber grows in Nepal, Thailand, China, and Korea. The main stem is corky, and the lateral ones are straight and longitudinally fissured. The petiole is 5 cm–15 cm, serpentiform, long, and inflated near the leaf. The blade is broadly cordate to deltoid, 5 cm–15 cm across and displays 2–3 pairs of secondary nerves. The inflorescence is an axillary panicle, which is 25 cm long. The male flowers present 6 sepals and 6 petals, which are minute. The androecium consists of 12 stamens. The female flowers present 6 sepals and 6 petals, which are minute and exhibit a gynoecium of 3 carpels. The fruit is a drupe, which is red, 0.5 cm across, and contains a horse-shoe-shaped seed (Figure 1.14).

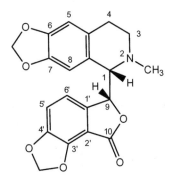

■ **CS 1.239** Bicuculline.

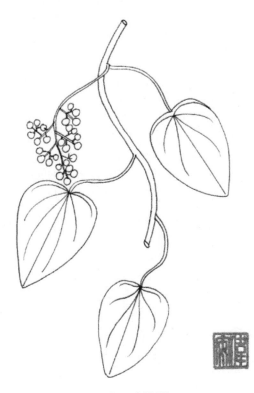

■ **FIGURE 1.14** *Sinomenium acutum* (Thunb.) Rehder & E.H. Wilson.

Medicinal Use In China, the plant is used to treat inflammation and pain.

Phytopharmacology The plant produces series of aporphine alkaloids, including dauriporphine, bianfugecine, dauriporphinoline, menisporphine, and magnoflorine; the morphinan alkaloid sinomenine; as well as the lignan (−)-syringaresinol and N-feruloyltyramine.[424]

■ **CS 1.240** Longanine.

Proposed Research Pharmacological study of sinomenine and synthetic derivative for the treatment of PD.

Rationale Other isoquinoline alkaloids found in the family Menispermaceae Juss. and Papaveraceae Juss. are the morphinan and hasubanan alkaloids, which derive from the benzylisoquinoline reticuline. Hasubanans such as longanine (CS 1.240), N-methylstephuline (CS 1.241), N-methylstephisoferulin (CS 1.242), and 6-cinnamoylhernandine (CS 1.243) elicited opioid receptor subtype δ (δ-opioid) binding activity with IC_{50} values equal to 0.7 μM, 2.1 μM, 2.5 μM, and 5.1 μM, respectively,[425] and are theoretically neuroprotective. Furthermore, the morphinan alkaloid sinomenine from *Sinomenium acutum* (Thunb.) Rehder & E.H. Wilson (family Menispermaceae Juss.) elicited opioid receptor subtype μ (μ-opioid) binding activity in Chinese hamster ovary (CHO) cells with an IC_{50} value equal to 109.5 nM and induced opioid receptor subtype μ phosphorylation activation, hence analgesic effects at a dose of 30 mg/kg in rodents subjected to the tail-flick test.[426]

One should recall that μ-opioid agonists, such as morphine (CS 1.244), protect neurons,[427] and in fact, sinomenine (CS 1.245) protected dopaminergic neurons against LPS insults at 10^{-5} M via inhibition of microglial NADPH oxidase, collapse of ROS, reduction of TNF-α and prostaglandin

■ **CS 1.241** N-methylstephuline.

■ **CS 1.242** N-methylstephisoferulin.

■ **CS 1.243** 6-cinnamoylhernandine.

E2 (PGE2),[428] and probable inhibition of NF-κB. In effect, sinomenine at a dose of 100 μg/mL inhibited the production of vascular cell adhesion molecule (VCAM) and IL-6 by fibroblast-like synoviocytes cells challenged with TNF-α by 29.3% and 36.4%, respectively, via hypophosphorylation of nuclear factor of kappa light polypeptide gene enhancer in B-cell inhibitor α (IκBα).[429] In the same experiment, dopaminergic neurons resisted to 0.2 μM of 1-methyl-4-phenylpyridinium ion (MPP+) via microglial attenuation incurred by sinomenine at a dose of 10^{-14} M.[428] Likewise, sinomenine at a dose of 5 μM prolonged by 84.1% the survival of PC12 cells deprived of oxygen and glucose *in vitro* and rescued rodents against middle cerebral artery occlusion by blocking acid-sensing ion channels 1a (ASIC1a) and voltage-gated calcium channels (VGCC).[430] It therefore seems reasonable to warrant further research on the clinical value of sinomenine and synthetic derivatives for treatment of PD.

■ **CS 1.244** Morphine.

1.3.7 *Crinum latifolium* L.

History The plant was first described by Carl von Linnaeus in *Species Plantarum* published in 1753.

Synonyms *Crinum esquirolii* H. Lév., *Crinum ornatum* var. *latifolium* (L.) Herb.

Family Amaryllidaceae J. St.-Hil., 1805

Common Name Xi nan wen shu lan (Chinese)

Habitat and Description This robust perennial herb grows on the riverbanks of China, Thailand, Laos, Burma, Vietnam, and India. The leaves are basal, ligulate, ensiform, and up to 1 m long. The inflorescence is an umbel

■ **CS 1.245** Sinomenine.

of white, pendulous, magnificent, and ephemeral flowers with involucres, which are 10 cm long. The perianth comprises a 10 cm long tube from which develops 6 elliptic lobes that are 7.5 cm × 1.5 cm. The androecium includes 6 stamens inserted on the inner wall of the corolla. The anthers are linear, 2 cm long, on filiform filaments. The gynoecium consist of 3 carpels fused into a 3-locular ovary. The fruit is a dehiscing capsule (Figure 1.15).

Medicinal Uses In India, the plant is used as counter-irritant for rheumatism.

Phytopharmacology The plant shelters Amaryllidaceae alkaloids, including lycorine, ambelline, hippadine, pratorinine, pratorimine, pratosine,[431] 1-*O*-acetylambelline, 11-*O*-acetyl-1,2-*β*-epoxyambelline,[432] 2-epi-lyocorine, and 2-epipancrassidine.[433]

Proposed Research Pharmacological study of 1-*O*-acetylambelline synthetic derivatives for the treatment of AD.

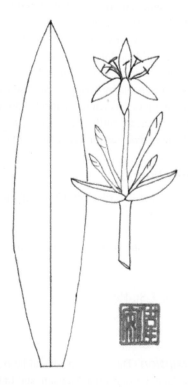

■ **FIGURE 1.15** *Crinum latifolium* L.

Rationale In the annals of AChE inhibition, perhaps no single class of natural product has enjoyed more ingenious speculation than the isoquinoline alkaloids engineered by the family Amaryllidaceae J. St.-Hil. A notable example of such an alkaloid is galanthamine, first isolated from *Galanthus woronowii* Losinsk. (family Amaryllidaceae J. St.-Hil.), which is currently used to delay the progression of AD.[434] The discovery of galanthamine has encouraged the search for more efficient and safer derivatives; López et al.,[435] for instance, tested several alkaloids from the genus *Narcissus* L. (family Amaryllidaceae J. St.-Hil.) and found that sanguinine (CS 1.246), galanthamine (CS 1.247), 12-hydroxygalanthamine (CS 1.248), epinorgalanthamine (CS 1.249), oxoassoanine, assoanine (CS 1.250), and pseudolycorine (CS 1.251) inhibited the enzymatic activity of AChE with IC_{50} values equal to $0.1\,\mu M$, $1.07\,\mu M$, $1.6\,\mu M$, $9.6\,\mu M$, $47.2\,\mu M$, $3.8\,\mu M$, and $152.3\,\mu M$, respectively, whereas lycoramine (CS 1.252) was inactive, implying that Δ_{1-2} and a tertiary amine are necessary for the activity of galanthamine and lycorine derivatives. However, AChE inhibitors delay the progression of AD, which inexorably progresses, hence the need for leads that would not only inhibit the hydrolysis of acetylcholine (CS 1.253) but impart synergistic neuroprotective effect. In this light, it is conceivable that Amaryllidaceae confronting both AChE and neuroinflammation await pharmaceutical development and such leads could be derived from ambelline (CS 1.254).

■ **CS 1.246** Sanguinine.

■ **CS 1.247** Galanthamine.

■ **CS 1.248** 12-hydroxygalanthamine.

CS 1.249 Epinorgalanthamine.

CS 1.250 Assoanine.

CS 1.251 Pseudolycorine.

CS 1.252 Lycoramine.

CS 1.253 Acetylcholine.

1.3.8 *Saururus chinensis* (Lour.) Baill.

History The plant was first described by Henri Ernest Baillon in *Adansonia* published in 1871.

Synonyms *Saururopsis chinensis* (Lour.) Turcz., *Saururopsis cumingii* C. DC., *Saururus cernuus* Thunb., *Saururus cumingii* C. DC., *Saururus loureiri* Decne., *Spathium chinense* Lour.

Family Saururaceae Rich. ex T. Lestib, 1826

Common Names Chine lizard's tail, san bai cao (Chinese), hangesho (Japanese), sam baek cho (Korea)

Habitat and Description This perennial and rhizomatous herb grows wild in China, Korea, and Japan. The leaves are simple and present a stipular sheath, which is 0.2–1 cm long. The petiole is 1–3 cm long. The blade is cordate-lanceolate or aristate, 10 cm × 5 cm–20 cm × 10 cm, papery, and dull light green with a conspicuous white patch; it displays 5–7 pairs of parallel nerves. The inflorescence is a raceme, which is 12–20 cm long, curved, and whitish. The flowers and fruits are minute (Figure 1.16).

Medicinal Uses In China, the plant is used to heal wounds, to assuage inflammation, and to promote diuresis.

Phytopharmacology The plant contains the lignan sauchinone[436]; saucerneols A–C, di-*O*-methyltetrahydrofuriguaiacin B, machilin D and machilin D 4-methyl ether,[437,438] saucerneol D,[439] saucerneol G,[440] saucerneol F,[441] saucerneol I,[442] and manassantin A and B[443]; flavonoids isoquercitrin and quercitrin[444]; aristolactams such as aristolactam AII[444] and aristolactam BII[445]; diterpenes saurufuran A and B[446]; and the sesquiterpene meso-dihydroguaiaretic acid.[447]

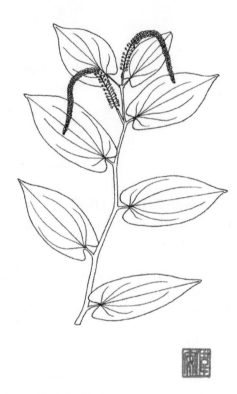

■ **FIGURE 1.16** *Saururus chinensis* (Lour.) Baill.

Proposed Research Pharmacological study of aristolacatam BII and derivatives for the treatment of neurodegenerative diseases.

Rationale A common pathophysiological feature of AD, PD, and ALS is neuroinflammation; therefore, isoquinoline alkaloids that inhibit the enzymatic activity of phospholipase A_2 (PLA$_2$), lipoxygenase (LOX), and/ or cyclo-oxygenase (COX) are of neuropharmacological interest. Within the past decades, a considerable body of evidence has accumulated to indicate that the families Aristolochiaceae Juss., Piperaceae Giseke, and Saururaceae Rich. ex T. Lestib. engineer series of isoquinoline aporphine derivatives, including aristolochic acids and aristolactams, that inhibit the enzymatic activity of phospholipase A_2 (PLA$_2$),[448,449] hence the traditional uses of these poisonous herbs to treat inflammation, pain, and snakebites. Indeed, the venoms of certain snakes abound with phospholipases A2 (PLA$_2$) such as β-bungarotoxin.[450–452] Intriguingly, cerebellar granule neurons exposed to β-bungarotoxin became permeable to Ca^{2+}, hence, nitric oxide synthase (NOS) activation and nitric oxide (NO) cumulation via blockade of K^+ channels[453] or/and store-operated Ca^{2+} entry-like mechanism as described by Putney.[454] NO is critically involved in neuroinflammation, and saurolactam (CS 1.255) and aristolactam BII (CS 1.256) from *Saururus chinensis* (Lour.) Baill. (family Saururaceae Rich. ex T. Lestib.) curbed the production of NO by 30% in macrophages (RAW264.7) challenged with LPS.[455] Likewise, aristolactam AII (CS 1.257), piperolactam A (CS 1.258), and piperolactam B (CS 1.259) from *Piper kadsura* (Choisy) Ohwi (family Piperaceae Giseke) dampened the production of NO by glial cells exposed to LPS with IC_{50} values equal to $9.1 \mu M$, $6.3 \mu M$, and $16.5 \mu M$, respectively.[456]

Note that phospholipase A_2 (PLA_2) hinders the synaptic clearance of glutamate by astrocytes and magnifies the neurotoxicity of β-amyloid peptide.[457,458] Further, β-amyloid ($A\beta_{1-42}$) insults in neurons were attenuated via phospholipase A_2 (PLA_2) inhibition.[459,460] These lines of evidence pave the way for aristolochic acids and aristolactams inhibiting phospholipase A2 (PLA_2) to be viewed as original chemical frameworks from which one may derive leads for the treatment of neurodegenerative diseases and particularly Alzheimer's disease. AD results from the accumulation of β-amyloid peptide obtained from the chimeric cleavage of APP by β- and γ-secretases. β-amyloid peptide binds to C1q protein, which initiates the secretion of TNF-α from microglia, hence microglial mobilization and neuroinflammation.[459] Intriguingly, some evidence suggests that β-amyloid peptide favors the aggregation of platelets,[460] which in turn activate β- and γ-secretase via the secretion of platelet-derived growth factor (PDGF) via Src and Rac.[461] Piperolactam C (CS 1.260), aristolactam BIII (CS 1.261), aristolactam AIIIa (CS 1.262) from *Fissistigma balansae* (Aug. DC.) Merr. (family Annonaceae Juss.), and piperolactam B from *Piper taiwanense* Lin & Lu (family Piperaceae Giseke) inhibited the aggregation of platelets[462,463] and might have a beneficial role for the development of leads for the treatment of AD.

Another noteworthy feature of AD is the accumulation of Polo-like kinases in the cytoplasm of stressed hippocampal and cortical neurons as an attempt to initiate mitosis.[464] In effect, when the viability of hippocampal neurons is challenged, cytoplasmic Ca^{2+} increases and activates calcineurin and Polo-like kinase 2 (Plk2).[465,466] Plk2 requires cyclin-dependent kinase (CDK5), which is of importance for the genesis of neurons.[467] A particularly important finding is that aristolactam AIIIa from *Aristolochia manshuriensis* Kom. (family Aristolochiaceae Juss.) inhibited the enzymatic activity of CDK2 and CDK4 with IC_{50} values of 140 nM and 142 nM, respectively,[468] and dampened the enzymatic activity of Plk1 with an IC_{50} value equal to $47.5 \mu mol/L$.[469] A structure activity relationship

■ **CS 1.255** Saurolactam.

■ **CS 1.256** Aristolactam BII.

■ **CS 1.257** Aristolactam AII.

■ **CS 1.258** Piperolactam A.

■ **CS 1.259** Piperolactam B.

■ **CS 1.260** Piperolactam C.

revealed that the lactam ring is essential for potent dependent kinase 2 (CDK2), hydroxyl groups at C6 or C8 positions enhance CDK2, and a substitution of hydroxyl groups at the C3 and C4 positions and N methylation group reduces activity.[470] In hippocampal neurons, the activation of Polo-like kinase results in reduced levels of spine-associated Rap guanosine triphosphatase-activating protein (SPAR) and PSD-95, which are necessary for the maintenance of dendrites and synapses.[465] Furthermore, Plk2 phosphorylates α-synuclein.[471] The collective evidence suggests that aristolactams inhibiting Polo-like kinase activity may be of critical value for the treatment of neurodegenerative diseases and particularly PD.

Paradoxically, a conflicting body of evidence also exists concerning the role of phospholipase A_2 (PLA$_2$), suggesting it enhances neurotransmission and neuroprotection. For instance, the activation of phospholipase A_2 (PLA$_2$) by melittin at a dose of $100\,\mu M$ favored the regeneration of noradrenergic neurons.[457] Dopamine, serotonin, norepinephrine, N-Methyl-D-aspartate (NMDA), and glutamate stimulated the enzymatic activity of phospholipase A_2 (PLA$_2$)[472] possibly due to Ca^{2+}[473] and/or G proteins.[474] In this context, it is important to mention that cytoplasmic Ca^{2+} activates Ca^{2+}/calmodulin-dependent protein kinase II, which catalyzes the phosphorylation of synapsin I into phosphosynapin I, hence exocytosis and the release of neurotransmitters[475,476] such as acetylcholine. Furthermore, Liu et al.[477] showed that melittin stimulated glycinergic activity at a dose of $1\,\mu M$, which was nullified by the phospholipase A_2 (PLA$_2$) inhibitor 4-BPB and the lipoxygenase (LOX) inhibitor nordihydroguaiaretic acid (NDGA) at doses of $10\,\mu M$ and $100\,\mu M$, respectively.[477]

Some evidence demonstrates that aristolactams inhibit NF-κB, which commands the production of cytokines and pro-inflammatory enzymes such as iNOS and COX-2. In effect, saurolactam from *Saururus chinensis* (Lour.) Baill. (family Saururaceae Rich. ex T. Lestib.) inhibited the enzymatic activity of COX-2 with an IC$_{50}$ value equal to $2\,\mu M$ in macrophages (RAW264.7) challenged with LPS and abrogated the secretion of IL-1β, IL-6, and TNF-α by confronting NF-κB.[455] Likewise, piperolactam A from *Piper longum* L. (family Piperaceae Giseke) inhibited NF-κB.[478] Furthermore, saurolactam inhibited receptor activator of nuclear factor-kappaB ligand (RANKL)-induced tartrate-resistance acid phosphatase activity.[479] RANKL plays a key role in the physiology of bones and is surprisingly found on neurons and astrocytes located in the lateral septal nucleus.[480] Because the binding of osteoclast receptor activator of nuclear factor-kappaB (RANK) to osteoblast RANKL induces the activation of Akt and mTOR to induce osteoclast survival and growth,[480] one may suggest that a RANK/RANKL system may be targeted by aristolactams and congeners for the development of neurodegenerative diseases.

1.3.9 *Dictamnus dasycarpus* Turcz.

History The plant was first described by Nicolai Stepanowitsch Turczaninow in *Bulletin de la Société Impériale des Naturalistes de Moscou* (15:637) published in 1842.

Synonyms *Aquilegia fauriei* H. Lév., *Dictamnus albus* subsp. *dasycarpus* (Turcz.) Kitag., *Dictamnus albus* subsp. *dasycarpus* (Turcz.) L. Winter, *Dictamnus albus* var. *dasycarpus* (Turcz.) T.N. Liou & Y.H. Chang

Family Rutaceae Juss., 1789

Common Name Bai xian (Chinese)

Habitat and Description This perennial herb grows to 1 m tall in Russia, Mongolia, China, and Korea. The stems are hairy and present tiny oil glands. The leaves are imparipinnate, exstipulate, and alternate. The rachis is angled and winged; supports 5–6 pairs of folioles, which are elliptic, serrate, acute at the base and apex, 2–10 cm × 1–5 cm, asymmetrical at the base; and presents 4–5 pairs of secondary nerves and tiny oil glands. The inflorescence is a terminal raceme, which is 30 cm tall. The calyx includes 5 sepals, which are 0.5 cm long. The corolla consists of 5 oblanceolate, 1 cm long, white petals, dotted with oil glands; a single petal is curved. The androecium includes 10 stamens, which are positioned around a disc. The gynoecium comprises 5 carpels, which are partially fused. The fruit consists of 5 membranaceous follicles containing a few seeds that are black and glossy (Figure 1.17).

Medicinal Uses In China, the plant is used to treat rheumatism, jaundice, and cough and is a counter-irritant.

Phytopharmacology The plant contains series of limonoid derivatives fraxinellone[481] and isofraxinellone[482]; 6β-hydroxyfraxinellone[483]; the limonoid obacunone; the quinolines dictamine[481] and haplopine[483]; dasycarine[484]; skimmianine; γ-fagarine[485]; the guaiane sesquiterpene dictamnol[486]; the eudesmane sesquiterpenes dictamnaindiol and dictamnadiol[487]; a series of sesquiterpene glycosides such as dictamnosides A–E[488]; and the sterol pregnenolone.[486]

Proposed Research The pharmacological study of skimmianine and congeners for the treatment of AD.

Rationale A compelling body of experimental evidence indicates that the quinoline scaffold occurs in several synthetic molecules which have

■ **CS 1.261** Aristolactam BIII.

■ **CS 1.262** Aristolactam AIIIa.

■ **FIGURE 1.17** *Dictamnus dasycarpus* Turcz.

the common ability to fight neurodegeneration. They include the transient receptor potential vanilloid subtype 1 (TRPV1) antagonists such as compound 3 (CS 1.263) and compound-46 (CS 1.264) with pK_b values equal to 6.5 and 8.4, respectively.[489,490] Other examples of TRPV1 antagonists are Compound-46, which protected rodents against the pain inflicted by plantar injection of Complete Freund's Adjuvant (CFA) by 32% when given at a dose of 30 mg/kg,[489] and compound-18 (CS 1.265) with a pK_b value equal to 8.5, which reduced CFA-induced pain by 42% when administered at a dose of 5 mg/kg.[490] Although the TRPV1 accounts for the feeling of pain in peripheral sensory neurons, one should recall that TRPV1 antagonists in the brain block the influx of Ca^{2+} into neurons, hence deactivation of protein kinase A (PKA) and arachidonic acid cascade, preservation of mitochondrial integrity, and neuroprotection. In neurophysiological conditions, TRPV1 are stimulated by heat, capsaicin, and positively modulated by cytokines such as TNF-α.[491] TNF-α is released by TNF-α converting enzyme (TACE), which is in fact inhibited by quinoline derivatives. One such compound is BMS-561392 (CS 1.266), which repressed the

■ **CS 1.263** Compound 3.

■ **CS 1.264** Compound-46.

■ **CS 1.265** Compound-18.

■ **CS 1.266** BMS-561392.

production of TNF-α by Chinese hamster ovary (CHO) cells by 50% at a dose of 0.1 μM.[492] TNF-α is of special interest in the pathophysiology of neurodegenerative disease, where it dictates the fate of neurons according to the types of receptors stimulated and the types of cells, namely neurons, astrocytes, or microglial cells. The stimulation of TNF-α receptor type 2 activates TNF receptor-associated factor 2 protein (TRAF2), which imposes a sustained activation of NF-κB and therefore the survival of neurons[493,494] via the synthesis of anti-apoptotic proteins such as Baculoviral IAP repeat-containing protein (cIAP), Bcl-2, and B-cell lymphoma-extra large (Bcl-xL), as well as neurotrophic factors (NTF).[495] Paradoxically, the stimulation of TNF-α receptor type 1, which is associated with death domain (DD) and coupled with the adaptor protein TNF receptor-associated death domain (TRADD), activates FAS-associated death domain (FADD), hence caspase 8 activation, cleavage of caspase 3, and apoptosis.[494] In parallel, TRADD activates TRAF2, which activates kappa light polypeptide gene enhancer in B-cell inhibitor (I-κB) kinase (IKK) and therefore NF-κB, which in glial cells compels the synthesis of pro-inflammatory cytokines such as interleukins and enzymes, including inducible iNOS and COX-2.[494,495]

COX-2 is the target of quinoline derivatives such as compounds 4, 9a, 9b, 9c, 9d, and 9e (CS 1.267–1.272), which inhibited the activity of this enzyme with IC$_{50}$ values equal to 0.09 μM, 0.08 μM, 0.07 μM, 0.07 μM, 0.05 μM, and 0.04 μM, respectively.[496] Furthermore, compounds 4a, 5, 6, 7, and 8 (CS 1.273–1.277) inhibited the enzymatic activity of COX-2 with IC$_{50}$ values equal to 0.1 μM, 0.1 μM, 0.1 μM, 0.08 μM, and 0.07 μM, respectively.[497] COX-2 catalyzes the conversion of arachidonic acid into prostaglandin G2 (PGG$_2$), from which PGE2 originates.[498] It is important here to note that the pathophysiology of MS and ALS, particularly, involves COX-2 and iNOS, which produce superoxide anion and NO, respectively, hence myelin insults.[499–500] In regard to ALS, COX-2 is

■ **CS 1.267** Compound 4.

■ **CS 1.268** Compound 9a.

■ **CS 1.269** Compound 9b.

■ **CS 1.270** Compound 9c.

■ **CS 1.271** Compound 9d.

■ **CS 1.272** Compound 9e.

■ **CS 1.273** Compound 4a.

■ **CS 1.274** Compound 5.

■ **CS 1.275** Compound 6.

■ **CS 1.276** Compound 7.

■ **CS 1.277** Compound 8.

■ **CS 1.278** D4418.

hyperactive, hence the generation of superoxide anions, neuroinflammation, and production of PGE2, which imposes the release of glutamate[501] and the death of neurons via apoptosis.

Note that neuroapoptosis and dementia are confronted by cAMP, which activates PKA and consequently the phosphorylation of cAMP response element binding protein (CREB).[502,503] Phosphodiesterase (PDE) destroys cAMP, and thus PDE inhibitors are beneficial for memory[504]; in fact, the quinoline derivatives D4418 (CS 1.278) and L454560 (CS 1.279) inhibit this enzyme with IC_{50} values equal to $0.1\,\mu M$[505] and $1.4\,nM$,[506] respectively. Likewise, SCH365351 (CS 1.280) inhibited the enzymatic activity of PDE4 with an IC_{50} value equal to $0.05\,\mu M$.[507] It is of interest to mention that cytoplasmic elevation of cAMP in astrocytes propels CREB activity and the synthesis of IL-6.[508] Although it is a chromone derivative, pranlukast (CS 1.281) also is a cysteinyl leukotriene 1 receptor (CysLT1) antagonist commercially used to treat asthma, which at a dose of 0.1 mg/kg

■ **CS 1.279** L454560.

■ **CS 1.280** SCH365351.

■ **CS 1.281** Pranlukast.

protected rodents against the cognitive dysfunction incurred by transient middle cerebral artery occlusion by reducing the cerebral infarct by 37.1% through probable blockade of CysLT1.[509] Hence, one may contemplate the possibility of synthetizing quinoline derivatives with CysLT1 antagonist properties or CysLT4 agonist properties.

Other quinoline derivatives of neuropharmacological interest are JLK1472, JLK1486, JLK1522, and JLK1535 (CS 1.282–1.285), which protected mouse hippocampal (HT-22) cells from glutamate insults by 90%, 60%, 90%, and 70%, respectively, by increasing glutathione (GSH) and stimulating peroxisome proliferator-activated receptor-γ (PPAR-γ).[510] Additional

■ **CS 1.282** JLK1472.

■ **CS 1.283** JLK1486.

■ **CS 1.284** JLK1522.

support to the contention that the quinoline scaffold could be viewed as a beneficial pharmacophore is shown by the fact that the brain itself produces kynurenic acid (CS 1.286) and xanthurenic acid (CS 1.287), which inhibited vesicular glutamate transporter (VGLUT) with Ki values equal to 1.3 μM and 0.1 μM, respectively.[511] These metabolites were further developed into 6-biphenyl-4-yl-quinoline-2,4-dicarboxylic acid (CS 1.288) and 6-(4′-phenylstyryl)-quinoline-2,4-dicarboxylic acid, which exhibited Ki values of 41 μM and 64 μM, respectively.[512]

Oddly enough, there is considerably less information concerning the neuroprotective potential of quinoline alkaloids, which abound especially in members of the family Rutaceae Juss. Dictamnine (CS 1.289). At a dose of 500 μmol/L, they inhibited the contractions of aortic rings evoked by KCl or noradrenaline via a mechanism, implying the Ca^{2+} influx blockade[482]; and pteleprenine (CS 1.290), from *Ptelea trifoliata* L. (family Rutaceae Juss.), at a dose of 10 μM, relaxed preparations of ileum exposed to acetylcholine, nicotine, or 1,1-dimethyl-4-phenylpiperazinium iodide (DMPP) by blocking nicotinic receptors with a pA2 value equal to 6.6.[513] Skimmianine (CS 1.291), kokusaginine (CS 1.292), and confusameline (CS 1.293) from *Euodia merrillii* Kaneh. & Sasaki (family Rutaceae Juss.) displayed serotoninergic receptor subtype 2 (5-HT$_2$) antagonistic properties *in vitro*.[514] Skimmianine and arborinine (CS 1.294) from *Zanthoxylum simulans* Hance (family Rutaceae Juss.). nullified the aggregation of platelets induced by arachidonic acid.[515] Skimmianine from *Adiscanthus fusciflorus* Ducke (family Rutaceae Juss.) inhibited the enzymatic activity of adenine phosphoribosyltransferase (APRT).[516] Because APRT converts adenine into adenosine monophosphate (AMP), it is not difficult to imagine that APRT inhibitors may increase adenine levels and therefore neuroprotection.[517] From *Oricia suaveolens* (Engl.) Verd. (family Rutaceae Juss.), skimmianine and kokusaginine inhibited the generation of ROS by polymorphoneutrophils challenged with zymosan with IC_{50} values equal to 35.5 μM and 61.8 μM, respectively.[518] Skimmianine mitigated the production of superoxide by

■ **CS 1.285** JLK1535.

■ **CS 1.286** Kynurenic acid.

■ **CS 1.287** Xanthurenic acid.

■ **CS 1.288** 6-biphenyl-4-yl-quinoline-2,4-dicarboxylic acid.

■ **CS 1.289** Dictamnine.

■ **CS 1.290** Pteleprenine.

■ **CS 1.291** Skimmianine.

■ **CS 1.292** Kokusaginine.

human neutrophils and the release of elastase by fMLP/CB with IC_{50} values equal to $20.9\,\mu$M and $14.4\,\mu$M, respectively.[519] Likewise, skimmianine from *Dictamnus dasycarpus* Turcz. (family Rutaceae Juss.) nullified the production of NO by microglial cells (BV-2) challenged within LPS.[520] *Melicope triphylla* (Lam.) Merr. (family Rutaceae Juss.) produces the quinoline alkaloids dictamine, evolitrine (CS 1.295), pteleine (CS 1.296), skimmianine, and kokusaginine, which altered the aggregation of platelets induced by arachidonic acid (AA).[521] In the same experiment, dictamine and evolitrine at a dose of $500\,\mu$mol/L inhibited the contractions of aortic rings evoked by K^+/Ca^{2+} by 100% and 91%, respectively (216). Wu et al.[522] isolated from *Ruta graveolens* L. (family Rutaceae Juss.) arborinine, dictamine, and graveolinine, which at a dose of $50\,\mu$g/mL inhibited the aggregation of platelets induced by AA by 51.6%, 69.8%, and 100%, whereas kokusaginine and skimmianine at a dose of $100\,\mu$g/mL inhibited the aforementioned aggregation by 42.2% and 81.5%.[522] Kokusaginine, skimmianine, evolitrine, and confusameline from *Melicope confusa* (Merr.) P.S. Liu (family Rutaceae Juss.) and haplopine (CS 1.297), robustine (CS 1.298), dictamnine, and γ-fagarine (CS 1.299) from *Dictamnus albus* L. (family Rutaceae Juss.) at a dose of $0.1\,\mu$M inhibited the enzymatic activity of PDE by 17.1%, 16.6%, 52.7%, 14.6%, 14.7%, 17.1%, 21.6%, and 67%, respectively, implying that cAMP might be boosted by Rutaceous quinoline alkaloids.[523,524] Evolitrine from *Euodia lunu-ankenda* (Gaertn.) Merr. (family Rutaceae Juss.) at a dose of $60\,$mg/kg inhibited carrageenan paw edema in rodents by 78%,[525] suggesting anti-neuroinflammatory potencies. Some evidence exists to demonstrate that quinoline alkaloids inhibit the catabolism of acetylcholine as ribalinine (CS 1.300) and methyl isoplatydesmine from *Skimmia laureola* (DC.) Siebold & Zucc. ex Walp. (family Rutaceae Juss.), which inhibited the enzymatic activity of AChE with K_i values equal to $30\,\mu$M.[526] Likewise, leiokinine A (CS 1.301) and skimmianine from *Esenbeckia leiocarpa* Engl. (family Rutaceae) inhibited the enzymatic activity of AChE with IC_{50} values of $0.2\,\mu$M and $1.4\,\mu$M, respectively,[527] while leptomerine (CS 1.302) and kokusaginine showed IC_{50} values of $2.5\,\mu$M and $46\,\mu$M, respectively.[527] *Conchocarpus fontanesianus* (A. St.-Hil.) Kallunki & Pirani (family Rutaceae Juss.) produces dictamnine and γ-fagarine, skimmianine, and 2-phenyl-1-methyl-4-quinolone (CS 1.303), which at a dose of $100\,\mu$g/mL inhibited the enzymatic activity of acetylcholinesterase by about 50%.[528] Considering evidence that, first, rutaceous alkaloids: hamper platelet aggregation by blocking serotoninergic receptor subtype 2A (5-HT$_{2A}$)[529] and may act as serotoninergic receptor subtype 2A/2C (5-HT$_{2A/2C}$) receptor antagonists; inhibit the enzymatic activities of AChE and PDE; and abrogate inflammatory oxidative burst, one could reasonably frame the hypothesis that skimmianine and congeners may very well represent a vast and untapped arsenal of leads for the treatment of AD.

■ **CS 1.293** Confusameline.

■ **CS 1.294** Arborinine.

■ **CS 1.295** Evolitrine.

■ **CS 1.296** Pteleine.

■ **CS 1.297** Haplopine.

■ **CS 1.298** Robustine.

■ **CS 1.299** γ-fagarine.

■ **CS 1.300** (-)-ribalinine.

■ **CS 1.301** Leiokinine A.

■ **CS 1.302** Leptomerine.

■ **CS 1.303** 2-phenyl-1-methyl-4-quinolone.

REFERENCES

[306]   Hasegawa M, Shirato T. Two new flavonoid glycosides from the leaves of Phellodendron amurense ruprecht. J Am Chem Soc 1953;75(22):5507–11.

[307]   Leu CH, Li CY, Yao X, Wu TS. Constituents from the leaves of Phellodendron amurense and their antioxidant activity. Chem Pharm Bull 2006;54(9):1308–11.

[308]   Tomita M, Nakano T. Studies on the alkaloids of Rutaceous plants. I. Alkaloids of Phellodendron amurense Rupr. (1). J Pharm Bull 1957;5(1):10–12.

[309]   Yong DM, Hak CK, Min CY, Kyu HL, Sang UC, Kang RL. Isolation of limonoids and alkaloids from Phellodendron amurense and their multidrug resistance (MDR) reversal activity. Arch Pharm Res 2007;30(1):58–63.

[310]   Ikuta A, Nakamura T. Canthin-6-one from the roots of Phellodendron amurense. Planta Med 1995;61(6):581–2.

[311]   Ikuta A, Urabe H. 7-Hydroxyrutaecarpine from fruit of Phellodendron amurense. Nat Med 1999;53(6):333.

[312]   Wada K, Yagi M, Matsumura A, Sasaki K, Sakata M, Haga M. Isolation of limonin and obacunone from Phellodendri cortex shorten the sleeping time induced in mice by alpha-chloralose-urethane. Chem Pharm Bull 1990;38(8):2332–4.

[313]   Kishi K, Yoshikawa K, Arihara S. Limonoids and protolimonoids from the fruits of Phellodendron amurense. Phytochemistry 1992;31(4):1335–8.

[314]   Lis A, Boczek E, Góra J. Chemical composition of the essential oils from fruits, leaves and flowers of the Amur cork tree (Phellodendron amurense Rupr.). Flavour Fragr J 2004;19(6):549–53.

[315] Nimit Y, Schulze I, Cashaw JL, Ruchirawat S, Davis VE. Interaction of catechol-amine-derived alkaloids with central neurotransmitter receptors. J Neurosci Res 1983;10(2):175–89.

[316] Umegaki H, Munoz J, Meyer RC, Spangler EL, Yoshimura J, Ikari H, et al. Involvement of dopamine D(2) receptors in complex maze learning and acetyl-choline release in ventral hippocampus of rats. Neurosci 2001;103(1):27–33.

[317] Liu GQ, Han BY, Wang EH. Blocking actions of l-stephanine, xylopine and 7 other tetrahydroisoquinoline alkaloids on alpha adrenoceptors. Zhongguo Yao Li Xue Bao 1989;10(4):302–6.

[318] Di Monte DA, Lavasani M, Manning-Bog AB. Environmental factors in Parkinson's disease. Neurotoxicol 2002;23(4–5):487–502.

[319] Langston JW, Ballard P, Tetrud JW, Irwin I. Chronic Parkinsonism in humans due to a product of meperidine-analog synthesis. Science 1983;219(4587):979–80.

[320] Langston JW, Forno LS, Tetrud J, Reeves AG, Kaplan JA, Karluk D. Evidence of active nerve cell degeneration in the substantia nigra of humans years after 1-methyl-4-phenyl-1,2,3,6-tetrahydropyridine exposure. Ann Neurol 1999;46(4):598–605.

[321] Suzuki K, Mizuno Y, Yoshida M. Inhibition of mitochondrial respiration by 1,2,3,4-tetrahydroisoquinoline-like endogenous alkaloids in mouse brain. Neurochem Res 1990;15(7):705–10.

[322] Storch A, Ott S, Hwang YI, Ortmann R, Hein A, Frenzel S, et al. Selective dopa-minergic neurotoxicity of isoquinoline derivatives related to Parkinson's disease: studies using heterologous expression systems of the dopamine transporter. Biochem Pharmacol 2002;163(5):909–20.

[323] Miao YS, Zhang AZ, Lin C, Jiang MH, Jin GZ. Effects of l-stepholidine on iso-lated rabbit basilar artery, mesenteric artery, and thoracic aorta. Zhongguo Yao Li Xue Bao 1991;12(3):260–2.

[324] Miao YS, Zhang AZ, Lin C, Jiang MH, Jin GZ. Effects of l-stepholidine on iso-lated rabbit basilar artery, mesenteric artery, and thoracic aorta. Zhongguo Yao Li Xue Bao 1991;12(3):260–2.

[325] Zou LL, Liu J, Jin GZ. Involvement of receptor reserve in D1 agonistic action of (−)-stepholidine in lesioned rats. Biochem Pharmacol 1997;54(2):233–40.

[326] Zhang L, Zhou R, Xiang G. Stepholidine protects against H2O2 neurotoxicity in rat cortical neurons by activation of Akt. Neurosci Lett 2005;383(3):328–32.

[327] Blanchet PJ, Fang J, Gillespie M, Sabounjian L, Locke KW, Gammans R, et al. Effects of the full dopamine D1 receptor agonist dihydrexidine in Parkinson's disease. Clin Neuropharmacol 1998;21(6):339–43.

[328] Gao M, Chu HY, Jin GZ, Zhang ZJ, Wu J, Zhen XC. l-Stepholidine-induced excitation of dopamine neurons in rat ventral tegmental area is associated with its 5-HT(1A) receptor partial agonistic activity. Synapse 2011;65(5):379–87.

[329] Ma ZZ, Xu W, Jensen NH, Roth BL, Liu-Chen LY, Lee DY. Isoquinoline alka-loids isolated from Corydalis yanhusuo and their binding affinities at the dopa-mine D1 receptor. Molecules 2008;13(9):2303–12.

[330] Chang CK, Chueh FY, Hsieh MT, Lin MT. The neuroprotective effect of DL-tetrahydropalmatine in rat heatstroke. Neurosci Lett 1998;267(2):109–12.

[331] Marcenac F, Jin GZ, Gonon F. Effects of tetrahydropalmatine on dopa-mine release and metabolism in the rat striatum. Psychopharmacology 1986;89(1):89–93.

[332] Liu GQ, Algeris A, Garattini S. dl-Tetrahydropalmatine as a monoamine depletor. Arch Int Pharmacodyn Ther 1982;258(1):39–50.

[333] Wu C, Yang K, Liu Q, Wakui M, Jin GZ, Zhen X, et al. Tetrahydroberberine blocks ATP-sensitive potassium channels in dopamine neurons acutely-dissociated from rat substantia nigra pars compacta. Neuropharmacology 2010;59(7–8):567–72.

[334] Schulz JB, Gerhardt E. Apoptosis: its relevance to Parkinson's disease. Clin Neurosci Res 2001;1(6):427–33.

[335] Liss B, Haeckel O, Wildmann J, Miki T, Seino S, Roeper J. K-ATP channels promote the differential degeneration of dopaminergic midbrain neurons. Nat Neurosci 2005;8(12):1742–51.

[336] Shigeta K, Ootaki K, Tatemoto H, Nakanishi T, Inada A, Muto N. Potentiation of nerve growth factor-induced neurite outgrowth in PC12 cells by a Coptidis Rhizoma extract and protoberberine alkaloids. Biosci Biotechnol Biochem 2002;66(11):2491–4.

[337] Cui H,S, Matsumoto K, Murakami Y, Hori H, Zhao Q, Obi R. Berberine exerts neuroprotective actions against in vitro ischemia-induced neuronal cell damage in organotypic hippocampal slice cultures: involvement of B-cell lymphoma 2 phosphorylation suppression. Biol Pharm Bull 2009;32(1):79–85.

[338] Hu J, Chai Y, Wang Y, Kheir MM, Li H, Yuan Z, et al. PI3K p55γ promoter activity enhancement is involved in the anti-apoptotic effect of berberine against cerebral ischemia-reperfusion. Eur J Pharmacol 2012;674(2–3):132–42.

[339] Wang J, Zhang Y, Du S, Zhang M. Protective effects of berberine against amyloid beta-induced toxicity in cultured rat cortical neurons. Neural Regener Res 2011;6(3):183–7.

[340] Hsu YY, Chen CS, Wu SN, Jong YJ, Lo YC. Berberine activates Nrf2 nuclear translocation and protects against oxidative damage via a phosphatidylinositol 3-kinase/Akt-dependent mechanism in NSC34 motor neuron-like cells. Eur J Pharm Sci 2012;46(5):415–25.

[341] Lee B, Sur B, Shim I, Lee H, Hahm DH. Phellodendron amurense and its major alkaloid compound, berberine ameliorates scopolamine-induced neuronal impairment and memory dysfunction in rats. Korean J Physiol Pharmacol 2012;16(2):79–89.

[342] Lee GE, Lee WG, Lee SY, Lee CR, Park CS, Chang S, et al. Characterization of protoberberine analogs employed as novel human P2X7 receptor antagonists. Toxicol Appl Pharmacol 2011;252(2):192–200.

[343] Skaper SD, Debetto P, Giusti P. The P2X7 purinergic receptor: from physiology to neurological disorders. FASEB J 2010;24(2):337–45.

[344] Kim DK, Lee KT, Baek NI, Kim SH, Park HW, Lim JP, et al. Acetylcholinesterase inhibitors from the aerial parts of Corydalis speciosa. Arch Pharm Res 2004;27(11):1127–31.

[345] Ingkaninan K, Phengpa P, Yuenyongsawad S, Khorana N. Acetylcholinesterase inhibitors from Stephania venosa tuber. J Pharm Pharmacol 2006;58(5):695–700.

[346] Chlebek J, Macáková K, Cahlíková L, Kurfürst M, Kuneš J, Opletal L. Acetylcholinesterase and butyrylcholinesterase inhibitory compounds from Corydalis cava (Fumariaceae). Nat Prod Commun 2011;6(5):607–10.

[347] Jung HA, Min BS, Yokozawa T, Lee JH, Kim YS, Choi JS. Anti-Alzheimer and antioxidant activities of Coptidis Rhizoma alkaloids. Biol Pharm Bull 2009;32(8):1433–8.

[348] Lee SK, Qing WG, Mar W, Luyengi L, Mehta RG, Kawanishi K, et al. Angoline and chelerythrine, benzophenanthridine alkaloids that do not inhibit protein kinase C. J Biol Chem 1998;273(31):19829–33.

[349] Tolkachev ON, Savina AA, Sheichenko VI, Proskudina VV. 8-O-demethylchelerythrine from Macleaya cordata. Pharm Chem J 1999;33(2):86–7.

[350] Fonin VS, Tsybulko NS, Tolkachev ON. Spectrophotometric method of sanguinarine detection in the biomass of Macleaya cordata cells. Appl Biochem Microbiol 1999;35(4):422–5.

[351] Luo XB, Chen B, Yao SZ. Rapid determination of protopine, allocryptopine, sanguinarine and chelerythrine in fruits of Macleaya cordata by microwave-assisted solvent and HPLC-ESI/MS. Phytochem Anal 2006;17(6):431–8.

[352] Pi G, Ren P, Yu J, Shi R, Yuan Z, Wang C. Separation of sanguinarine and chelerythrine in Macleaya cordata (Willd) R. Br. based on methyl acrylate-co-divinylbenzene macroporous adsorbents. J Chromatogr A 2008;1192(1):17–24.

[353] Ye F, Feng F, Liu W. Alkaloids from Macleaya cordata. Zhongguo Zhongyao Zazhi 2009;34(13):1683–6.

[354] Caballero-George C, Vanderheyden PML, Solis PN, Gupta MP, Pieters L, Vauquelin G, et al. In vitro effect of sanguinarine alkaloid on binding of [3H]candesartan to the human angiotensin AT1 receptor. Eur J Pharmacol 2003;458(3):257–62.

[355] Turner AJ, Hooper NM. The angiotensin-converting enzyme gene family: genomics and pharmacology. Trends Pharmacol Sci 2002;23(4):177–83.

[356] Savaskan E, Hock C, Olivieri G, Bruttel S, Rosenberg C, Hulette C, et al. Cortical alterations of angiotensin converting enzyme, angiotensin II and AT1 receptor in Alzheimer's dementia. Neurobiol Aging 2001;22(4):541–6.

[357] Georgiev V, Yonkov D. Participation of angiotensin II in learning and memory: I. Interaction of angiotensin II with saralasin. Methods Find Exp Clin Pharmacol 1985;7(8):415–18.

[358] Barnes JM, Barnes NM, Costall B, Horovitz ZP, Naylor RJ. Angiotensin II inhibits the release of [3H]acetylcholine from rat entorhinal cortex in vitro. Brain Res 1989;491:136–43.

[359] De Bundel D, Demaegdt H, Lahoutte T, Caveliers V, Kersemans K, Ceulemans AG, et al. Involvement of the AT1 receptor subtype in the effects of angiotensin IV and LVV-haemorphin 7 on hippocampal neurotransmitter levels and spatial working memory. J Neurochem 2010;112(5):1223–34.

[360] Pederson ES, Krishnan R, Harding JW, Wright JW. A role for the angiotensin AT4 receptor subtype in overcoming scopolamine-induced spatial memory deficits. Regul Pept 2001;102(2–3):147–56.

[361] Albiston AL, McDowall SG, Matsacos D, Sim P, Clune E, Mustafa T, et al. Evidence that the angiotensin IV (AT(4)) receptor is the enzyme insulin-regulated aminopeptidase. J Biol Chem 2001;276(52):48623–6.

[362] Mentlein R. Dipeptidyl-peptidase IV (CD26)-role in the inactivation of regulatory peptides. Regul Pept 1999;85(1):9–24.

[363] Sedo A, Vlasicová K, Barták P, Vespalec R, Vicar J, Simánek V, et al. Quaternary benzo[c]phenanthridine alkaloids as inhibitors of aminopeptidase N and dipeptidyl peptidase IV. Phytother Res 2002;16(1):84–7.

[364] Perry T, Haughey NJ, Mattson MP, Egan JM, Greig NH. Protection and reversal of excitotoxic neuronal damage by glucagon-like peptide-1 and exendin-4. J Pharmacol Exp Ther 2002;302(3):881–8.

[365] Perry T, Lahiri DK, Sambamurti K, Chen D, Mattson MP, Egan JM, et al. Glucagon-like peptide-1 decreases endogenous amyloid-beta peptide (Abeta) levels and protects hippocampal neurons from death induced by Abeta and iron. J Neurosci Res 2003;72:603–12.

[366] D'Amico M, Di Filippo C, Marfella R, Abbatecola AM, Ferraraccio F, Rossi F, et al. Long-term inhibition of dipeptidyl peptidase-4 in Alzheimer's prone mice. Exp Gerontol 2010;45(3):202–7.

[367] Wang Q, Li L, Xu E, Wong V, Rhodes C, Brubaker PL. Glucagon-like peptide-1 regulates proliferation and apoptosis via activation of protein kinase B in pancreatic INS-1 beta cells. Diabetologia 2004;47(3):478–87.

[368] Herbert JM, Augereau JM, Gleye J, Maffrand JP. Chelerythrine is a potent and specific inhibitor of protein kinase C. Biochem Biophys Res Commun 1990;172(3):993–9.

[369] Alkon DL, Sun MK, Nelson TJ. PKC signaling deficits: a mechanistic hypothesis for the origins of Alzheimer's disease. Trends Pharmacol Sci 2007;28(2):51–60.

[370] Efthimiopoulos S, Punj S, Manolopoulos V, Pangalos M, Wang GP, Refolo LM, et al. Intracellular cyclic AMP inhibits constitutive and phorbol ester-stimulated secretory cleavage of amyloid precursor protein. J Neurochem 1996;67(2):872–5.

[371] Ochi H, Kume N, Nishi E, Kita T. Elevated levels of cAMP inhibit protein kinase C-independent mechanisms of endothelial platelet-derived growth factor-B chain and intercellular adhesion molecule-1 gene induction by lysophosphatidylcholine. Circ Res 1995;77(3):530–5.

[372] Kumar A, La Rosa FG, Hovland AR, Cole WC, Edwards-Prasad J, Prasad KN. Adenosine 3′,5′-cyclic monophosphate increases processing of amyloid precursor protein (APP) to beta-amyloid in neuroblastoma cells without changing APP levels or expression of APP mRNA. Neurochem Res 1999;24(10):1209–15.

[373] Lee RK, Wurtman RJ. Metabotropic glutamate receptors increase amyloid precursor protein processing in astrocytes: inhibition by cyclic AMP. J Neurochem 1997;68(5):1830–5.

[374] Cho KM, Yoo ID, Kim WG. 8-Hydroxydihydrochelerythrine and 8-hydroxydihydrosanguinarine with a potent acetylcholinesterase inhibitory activity from Chelidonium majus L. Biol Pharm Bull 2006;29(11):2317–20.

[375] Kodai T, Horiuchi Y, Nishioka Y, Noda N. Novel cycloartane-type triterpenoid from the fruits of Nandina domestica. J Nat Med 2010;64(2):216–18.

[376] Tsukiyama M, Ueki T, Yasuda Y, Kikuchi H, Akaishi T, Okumura H, et al. Beta2-adrenoceptor-mediated tracheal relaxation induced by higenamine from Nandina domestica Thunberg. Planta Med 2009;75(13):1393–9.

[377] Tsuchida H, Ohizumi Y. (+)-Nantenine isolated from Nandina domestica Thunb. inhibits adrenergic pressor responses in pithed rats. Eur J Pharmacol 2003;477(1):53–8.

[378] Kunitomo JI, Ju-ichi M, Ando Y, Yoshikawa Y, Nakamura S. Isolation of new base, dehydronantenine and lignan, (−)-episyringaresinol from Nandina domestica Thunb. Yakugaku Zasshi 1975;95(4):445–7.

[379] Zhang A, Zhang Y, Branfman AR, Baldessarini RJ, Neumeyer JL. Advances in development of dopaminergic aporphinoids. J Med Chem 2007;50(2):171–81.

[380] Miller RJ, Kelly PH, Neumeyer JL. Aporphines. 15. Action of aporphine alkaloids on dopaminergic mechanisms in rat brain. Eur J Pharmacol 1976;35(1):77–83.

[381] Goldman ME, Kebabian JW. Aporphine enantiomers. Interactions with D-1 and D-2 dopamine receptors. Mol Pharmacol 1984;25(1):18–23.

[382] Montastruc JL, Llau ME, Senard JM, Tran MA, Rascol O, Montastruc P. A study of tolerance to apomorphine. Br J Pharmacol 1996;117(5):781–6.

[383] Shoji N, Umeyama A, Saito N, Iuchi A, Takemoto T, Kajiwara A, et al. Asimilobine and lirinidine, serotonergic receptor antagonists, from Nelumbo nucifera. J Nat Prod 1987;50(4):773–4.

[384] Hasrat JA, Peters L, De Backer JP, Vauquelin G, Vlietinck AJ. Screening of medicinal plants from Suriname for 5-HT(1A) ligands: bioactive isoquinoline alkaloids from the fruit of Annona muricata. Phytomedicine 1997;4(2):133–40.

[385] Hasrat JA, De Bruyne T, De Backer JP, Vauquelin G, Vlietinck AJ. Isoquinoline derivatives isolated from the fruit of Annona muricata as 5-HTergic 5-HT1A receptor agonists in rats: unexploited antidepressive (lead) products. J Pharm Pharmacol 1997;49(11):1145–9.

[386] King MV, Marsden CA, Fone KC. A role for the 5-HT1A, 5-HT4 and 5-HT6 receptors in learning and memory. Trends Pharmacol Sci 2008;29(9):482–92.

[387] Bibbiani F, Oh JD, Chase TN. Serotonin 5-HT1A agonist improves motor complications in rodent and primate parkinsonian models. Neurology 2001;57(10):1829–34.

[388] Liu T, Fujita T, Nakatsuka T, Kumamoto E. Phospholipase A2 activation enhances inhibitory synaptic transmission in rat substantia gelatinosa neurons. J Neurophysiol 2008;99(3):1274–84.

[389] Ichikawa J, Meltzer HY. The effect of serotonin(1A) receptor agonism on antipsychotic drug-induced dopamine release in rat striatum and nucleus accumbens. Brain Res 2008;858(2):252–63.

[390] Asencio M, Delaquerrière B, Cassels BK, Speisky H, Comoy E, Protais P. Biochemical and behavioral effects of boldine and glaucine on dopamine systems. Pharmacol Biochem Behav 1999;62(1):7–13.

[391] Sobarzo-Sanchez EM, Arbaoui J, Protais P, Cassels BKJ. Halogenated boldine derivatives with enhanced monoamine receptor selectivity. Nat Prod 2000;63(4):480–4.

[392] Asencio M, Hurtado-Guzmán C, López JJ, Cassels BK, Protais P, Chagraoui A. Structure-affinity relationships of halogenated predicentrine and glaucine derivatives at D1 and D2 dopaminergic receptors: halogenation and D1 receptor selectivity. Bioorg Med Chem 2005;13(11):3699–704.

[393] Arnsten AF, Mathew R, Ubriani R, Taylor JR, Li BM. Alpha-1 noradrenergic receptor stimulation impairs prefrontal cortical cognitive function. Biol Psychiatry 1999;45(1):26–31.

[394] Chuliá S, Moreau J, Naline E, Noguera MA, Ivorra MD, D'Ocón MP, et al. The effect of S-(+)-boldine on the alpha 1-adrenoceptor of the guinea-pig aorta. Br J Pharmacol 1996;119(7):1305–12.

[395] Teng CM, Yu SM, Ko FN, Chen CC, Huang YL, Huang TF. Dicentrine, a natural vascular a1-adrenoceptor antagonist, isolated from Lindera megaphylla. Br J Pharmacol 1991;104(3):651–6.

[396] Youn YC, Kwon OS, Han ES, Song JH, Shin YK, Lee CS. Protective effect of boldine on dopamine-induced membrane permeability transition in brain mitochondria and viability loss in PC12 cells. Biochem Pharmacol 2002;63(3):495–505.

[397] Jin CM, Lee JJ, Yang YJ, Kim YM, Kim YK, Ryu SY, et al. Liriodenine inhibits dopamine biosynthesis and L-DOPA-induced dopamine content in PC12 cells. Arch Pharm Res 2007;30(8):984–90.

[398] Rojsanga P, Boonyarat C, Utsintong M, Nemecz Á, Yamauchi JG, Talley TT, et al. The effect of crebanine on memory and cognition impairment via the alpha-7 nicotinic acetylcholine receptor. Life Sci 2012;91(3–4):107–14.

[399] Shoji N, Umeyama A, Takemoto T, Ohizumi Y. Serotonergic receptor antagonist from Nandina domestica Thunberg. J Pharm Sci 1984;73(4):568–70.

[400] De A, Ribeiro R, Rodríguez De Lores Arnaiz G. Nantenine and papaverine differentially modify synaptosomal membrane enzymes. Phytomedicine 2000;7(4):313–23.

[401] Orallo F, Alzueta AF. Preliminary study of the vasorelaxant effects of (+)-nantenine, an alkaloid isolated from Platycapnos spicata, in rat aorta. Planta Med 2001;67(9):800–6.

[402] Indra B, Tadano T, Nakagawasai O, Arai Y, Yasuhara H, Ohizumi Y, et al. Suppressive effect of nantenine, isolated from Nandina domestica Thunberg, on the 5-hydroxy-L-tryptophan plus clorgyline-induced head-twitch response in mice. Life Sci 2002;70(22):2647–56.

[403] Meneses A. 5-HT system and cognition. Neurosci Biobehav Rev 1999;23(8):1111–25.

[404] Chaudhary S, Pecic S, LeGendre O, Navarro HA, Harding WW. (+/−)-Nantenine analogs as antagonists at human 5-HT 2A receptors: C1 and flexible congeners. Bioorg Med Chem Lett 2009;19(9):2530–2.

[405] Pecic S, McAnuff MA, Harding WW. Nantenine as an acetylcholinesterase inhibitor: SAR, enzyme kinetics and molecular modeling investigations. J Enzyme Inhib Med Chem 2011;26(1):46–55.

[406] Ogino T, Katsuhara T, Sato T, Sasaki H, Okada M, Maruno M. New alkaloids from the root of Stephania tetrandra (Fen-Fang-Ji). Heterocycles 1998;48(2):311–17.

[407] Kaltschmidt B, Widera D, Kaltschmidt C. Signaling via NF-kappaB in the nervous system. Biochim Biophys Acta 2005;1745(3):287–99.

[408] Yang Z, Li C, Wang X, Zhai C, Yi Z, Wang L, et al. Dauricine induces apoptosis, inhibits proliferation and invasion through inhibiting NF-kappaB signaling pathway in colon cancer cells. J Cell Physiol 2010;225(1):266–75.

[409] Okamoto M, Ono M, Baba M. Suppression of cytokine production and neural cell death by the anti-inflammatory alkaloid cepharanthine: a potential agent against HIV-1 encephalopathy. Biochem Pharmacol 2001;62(6):747–53.

[410]   Kornhuber J, Muehlbacher M, Trapp S, Pechmann S, Friedl A, Reichel M, et al. Identification of novel functional inhibitors of acid sphingomyelinase. PLOS ONE 2011;6(8):e23852.

[411]   Li X, Gulbins E, Zhang Y. Oxidative stress triggers Ca-dependent lysosome trafficking and activation of acid sphingomyelinase. Cell Physiol Biochem 2012;30(4):815–26.

[412]   Cutler RG, Kelly J, Storie K, Pedersen WA, Tammara A, Hatanpaa K, et al. Involvement of oxidative stress-induced abnormalities in ceramide and cholesterol metabolism in brain aging and Alzheimer's disease. Proc Natl Acad Sci U S A 2004;101(7):2070–5.

[413]   Koh SB, Ban JY, Lee BY, Seong YH. Protective effects of fangchinoline and tetrandrine on hydrogen peroxide-induced oxidative neuronal cell damage in cultured rat cerebellar granule cells. Planta Med 2003;69(6):506–12.

[414]   Wang B, Yang L, Yan HL, Wang M, Xiao JG. Effect of tetrandrine on calcium-dependent tumour necrosis factor-alpha production in glia-neurone mixed cultures. Basic Clin Pharmacol Toxicol 2005;97(4):244–8.

[415]   He FQ, Qiu BY, Zhang XH, Li TK, Xie Q, Cui DJ, et al. Tetrandrine attenuates spatial memory impairment and hippocampal neuroinflammation via inhibiting NF-κB activation in a rat model of Alzheimer's disease induced by amyloid-β (1–42). Brain Res 2010;1384:89–96.

[416]   Lü Q, Xu XL, He Z, Huang XJ, Guo LJ, Wang HX. Guattegaumerine protects primary cultured cortical neurons against oxidative stress injury induced by hydrogen peroxide concomitant with serum deprivation. Cell Mol Neurobiol 2009;29(3):355–64.

[417]   Li YH, Gong PL. Neuroprotective effects of dauricine against apoptosis induced by transient focal cerebral ischaemia in rats via a mitochondrial pathway. Clin Exp Pharmacol Physiol 2007;34(3):177–84.

[418]   Guo DL, Zhou ZN, Zeng FD, Hu CJ. Dauricine inhibited L-type calcium current in single cardiomyocyte of guinea pig. Zhongguo Yao Li Xue Bao 1997;18(5):419–21.

[419]   Cometa MF, Fortuna S, Palazzino G, Volpe MT, Rengifo Salgado E, Nicoletti M, et al. New cholinesterase inhibiting bisbenzylisoquinoline alkaloids from Abuta grandifolia. Fitoterapia 2012;83(3):476–80.

[420]   Irgashev T, Israilov IA, Batsurén D, Yunusov MS. Alkaloids of Corydalis stricta. Chem Nat Comp 1983;19(4):461–3.

[421]   Bhat R, Axtell R, Mitra A, Miranda M, Lock C, Tsien RW, et al. Inhibitory role for GABA in autoimmune inflammation. Proc Natl Acad Sci U S A 2010;107(6):2580–5.

[422]   Carmans S, Hendriks JJ, Slaets H, Thewissen K, Stinissen P, Rigo JM, et al. Systemic treatment with the inhibitory neurotransmitter γ-aminobutyric acid aggravates experimental autoimmune encephalomyelitis by affecting proinflammatory immune responses. J Neuroimmunol 2012;255(1–2):45–53.

[423]   Huang JH, Johnston GA. (+)-Hydrastine, a potent competitive antagonist at mammalian GABAA receptors. Br J Pharmacol 1990;99(4):727–30.

[424]   Min YD, Sang UC, Kang RL. Aporphine alkaloids and their reversal activity of multidrug resistance (MDR) from the stems and rhizomes of Sinomenium acutum. Arch Pharm Res 2006;29(8):627–32.

[425] Carroll AR, Arumugan T, Redburn J, Ngo A, Guymer GP, Forster PI, et al. Hasubanan alkaloids with delta-opioid binding affinity from the aerial parts of Stephania japonica. J Nat Prod 2010;73(5):988–91.

[426] Wang MH, Chang CK, Cheng JH, Wu HT, Li YX, Cheng JT. Activation of opioid mu-receptor by sinomenine in cell and mice. Neurosci Lett 2008;443(3):209–12.

[427] Iglesias M, Segura MF, Comella JX, Olmos G. Mu-opioid receptor activation prevents apoptosis following serum withdrawal in differentiated SH-SY5Y cells and cortical neurons via phosphatidylinositol 3-kinase. Neuropharmacology 2003;44:482–92.

[428] Qian L, Xu Z, Zhang W, Wilson B, Hong JS, Flood PM. Sinomenine, a natural dextrorotatory morphinan analog, is anti-inflammatory and neuroprotective through inhibition of microglial NADPH oxidase. J Neuroinflammation 2007;4:23.

[429] Chen DP, Wong CK, Leung PC, Fung KP, Lau CB, Lau CP, et al. Anti-inflammatory activities of Chinese herbal medicine sinomenine and Liang Miao San on tumor necrosis factor-α-activated human fibroblast-like synoviocytes in rheumatoid arthritis. J Ethnopharmacol 2011;137(1):457–68.

[430] Wu WN, Wu PF, Chen XL, Zhang Z, Gu J, Yang YJ, et al. Sinomenine protects against ischaemic brain injury: involvement of co-inhibition of acid-sensing ion channel 1a and L-type calcium channels. Br J Pharmacol 2011;164(5):1445–59.

[431] Ghosal S, Saini KS, Frahm AW. Alkaloids of Crinum latifolium. Phytochemistry 1983;22(10):2305–9.

[432] Ghosal S, Rao PH, Saini KS. Natural occurrence of 11-O-acetylambelline and 11-O-acetyl-1,2-β-epoxyambelline in Crinum latifolium: immuno-regulant alkaloids. Pharm Res 1985;5:251–2.

[433] Ghosal S, Unnikrishnan S, Singh SK. Occurrence of two epimeric alkaloids and metabolism compared with lycorine in Crinum latifolium. Phytochemistry 1989;28(9):2535–7.

[434] Heinrich M, Lee Teoh H. Galanthamine from snowdrop—the development of a modern drug against Alzheimer's disease from local Caucasian knowledge. J Ethnopharmacol 2004;92(2–3):147–62.

[435] López S, Bastida J, Viladomat F, Codina C. Acetylcholinesterase inhibitory activity of some Amaryllidaceae alkaloids and Narcissus extracts. Life Sci 2002;71(21):2521–9.

[436] Wang EC, Shih MH, Liu MC, Chen MT, Lee GH. Studies on constituents of saururus chinensis. Heterocycles 1996;43(5):969–75.

[437] Sung SH, Kim YC. Hepatoprotective diastereomeric lignans from Saururus chinensis herbs. J Nat Prod 2000;63(7):1019–21.

[438] Sang HS, Huh MS, Kim YC. New tetrahydrofuran-type sesquilignans of Saururus chinensis root. Chem Pharm Bull 2001;49(9):1192–4.

[439] Moon TC, Kim JC, Song SE, Suh SJ, Seo CS, Kim YK, et al. Saucerneol D, a naturally occurring sesquilignan, inhibits LPS-induced iNOS expression in RAW264.7 cells by blocking NF-kappaB and MAPK activation. Int Immunopharmacol 2008;8(10):1395–400.

[440] Lu Y, Hong TG, Jin M, Yang JH, Suh SJ, Piao DG, et al. Saucerneol G, a new lignan, from Saururus chinensis inhibits matrix metalloproteinase-9

induction via a nuclear factor κB and mitogen activated protein kinases in lipopolysaccharide-stimulated RAW264.7 cells. Biol Pharm Bull 2010;33(12):1944–8.

[441]   Lu Y, Suh SJ, Kwak CH, Kwon KM, Seo CS, Li Y, et al. Saucerneol F, a new lignan, inhibits iNOS expression via MAPKs, NF-κB and AP-1 inactivation in LPS-induced RAW264.7 cells. Int Immunopharmacol 2012;12(1):175–81.

[442]   Lee Y-K, Seo C-S, Lee C-S, Lee K-S, Kang S-J, Jahng Y, et al. Inhibition of DNA topoisomerases I and II and cytotoxicity by lignans from Saururus chinensis. Arch Pharm Res 2009;32(10):1409–15.

[443]   Rho MC, Kwon OE, Kim K, Lee SW, Chung MY, Kim YH, et al. Inhibitory effects of manassantin A and B isolated from the roots of Saururus chinensis on PMA-induced ICAM-1 Expression. Planta Med 2003;69(12):1147–9.

[444]   Li R, Ren L, Chen Y. Chemical constituents of Saururus chinensis (Lour.) Bail. Zhongguo Zhong Yao Za Zhi 1999;24(8):479–81.

[445]   Luo Y, Zhang Z, Zuo Y. Studies on chemical constituents from the aerial parts of Saururus chinensis. Adv Mat Res 2012;393–395:1427–30.

[446]   Hwang BY, Lee JH, Nam JB, Kim HS, Hong YS, Lee JJ. Two new furanoditerpenes from Saururus chinenesis and their effects on the activation of peroxisome proliferator-activated receptor gamma. J Nat Prod 2002;65(4):616–17.

[447]   Moon TC, Seo CS, Haa K, Kim JC, Hwang NK, Hong TG, et al. Meso-dihydroguaiaretic acid isolated from Saururus chinensis inhibits cyclooxygenase-2 and 5-lipoxygenase in mouse bone marrow-derived mast cells. Arch Pharm Res 2008;31(5):606–10.

[448]   Rosenthal MD, Sannanaik Vishwanath B, Franson RC. Effects of aristolochic acid on phospholipase A2 activity and arachidonate metabolism of human neutrophils. Biochim Biophys Acta 1989;1001(1):1–8.

[449]   Chandra V, Jasti J, Kaur P, Srinivasan A, Betzel C, Singh TP. Structural basis of phospholipase A2 inhibition for the synthesis of prostaglandins by the plant alkaloid aristolochic acid from a 1.7 A crystal structure. Biochemistry 2002;41(36):10914–19.

[450]   Chen YJ. Phospholipase A(2) activity of beta-bungarotoxin is essential for induction of cytotoxicity on cerebellar granule neurons. J Neurobiol 2005;64(2):213–23.

[451]   Strong PN, Goerke J, Oberg SG, Kelly RB. beta-Bungarotoxin, a pre-synaptic toxin with enzymatic activity. Proc Natl Acad Sci U S A 1976;73(1):178–82.

[452]   Chen YJ. Phospholipase A(2) activity of beta-bungarotoxin is essential for induction of cytotoxicity on cerebellar granule neurons. J Neurobiol 2005;64(2):213–23.

[453]   Benishin CG. Potassium channel blockade by the B subunit of beta-bungarotoxin. Mol Pharmacol 1990;38(2):164–9.

[454]   Putney JW. Pharmacology of store-operated calcium channels. Mol Interv 2010;10(4):209–18.

[455]   Lee SU, Choi YH, Kim YS, Min YK, Rhee M, Kim SH. Anti-resorptive saurolactam exhibits in vitro anti-inflammatory activity via ERK1/2-NF-kappaB signaling pathway. Int Immunopharmacol 2009;10(3):298–303.

[456]   Kim KH, Choi JW, Choi SU, Ha SK, Kim SY, Park HJ, et al. The chemical constituents of Piper kadsura and their cytotoxic and anti-neuroinflammatory activities. J Enzyme Inhib Med Chem 2011;26(2):254–60.

[457] Nakamura S. Involvement of phospholipase A2 in axonal regeneration of brain noradrenergic neurones. Neuroreport 1993;4:371–4.

[458] Kanfer JN, Singh IN, Pettegrew JW, McCartney DG, Sorrentino G. Phospholipid metabolism in Alzheimer's disease and in a human cholinergic cell. J Lipid Mediat Cell Signal 1996;14(1–3):361–3.

[459] Singh IN, Sorrentino G, Sitar DS, Kanfer JN. (−)Nicotine inhibits the activations of phospholipases A2 and D by amyloid beta peptide. Brain Res 1998;800(2):275–81.

[460] Halliday G, Robinson SR, Shepherd C, Kril J. Alzheimer's disease and inflammation: a review of cellular and therapeutic mechanisms. Clin Exp Pharmacol Physiol 2000;27(1–2):1–8.

[461] Gianni D, Zambrano N, Bimonte M, Minopoli G, Mercken L, Talamo F, et al. Platelet-derived growth factor induces the beta-gamma-secretase-mediated cleavage of Alzheimer's amyloid precursor protein through a Src-Rac-dependent pathway. J Biol Chem 2003;278(11):9290–2.

[462] Chia YC, Chang FR, Teng CM, Wu YC. Aristolactams and dioxoaporphines from Fissistigma balansae and Fissistigma oldhamii. J Nat Prod 2000;63(8):1160–3.

[463] Chen YC, Chen JJ, Chang YL, Teng CM, Lin WY, Wu CC, et al. A new aristolactam alkaloid and anti-platelet aggregation constituents from Piper taiwanense. Planta Med 2004;70(2):174–7.

[464] Harris PL, Zhu X, Pamies C, Rottkamp CA, Ghanbari HA, McShea A, et al. Neuronal polo-like kinase in Alzheimer disease indicates cell cycle changes. Neurobiol Aging 2000;21(6):837–41.

[465] Seeburg DP, Pak D, Sheng M. Polo-like kinases in the nervous system. Oncogene 2005;24(2):292–8.

[466] Lee KJ, Lee Y, Rozeboom A, Lee JY, Udagawa N, Hoe HS, et al. Requirement for Plk2 in orchestrated ras and rap signaling, homeostatic structural plasticity, and memory. Neuron 2011;69(5):957–73.

[467] Jessberger S, Gage FH, Eisch AJ, Lagace DC. Making a neuron: Cdk5 in embryonic and adult neurogenesis. Trends Neurosci 2009;32(11):575–82.

[468] Hegde VR, Borges S, Patel M, Das PR, Wu B, Gullo VP, et al. New potential antitumor compounds from the plant Aristolochia manshuriens is as inhibitors of the CDK2 enzyme. Bioorg Med Chem Lett 2010;20(4):1344–6.

[469] Li L, Wang X, Chen J, Ding H, Zhang Y, Hu TC, et al. The natural product aristolactam AIIIa as a new ligand targeting the polo-box domain of polo-like kinase 1 potently inhibits cancer cell proliferation. Acta Pharmacol Sin 2009;30(10):1443–53.

[470] Hegde VR, Borges S, Pu H, Patel M, Gullo VP, Wu B, et al. Semi-synthetic aristolactams—inhibitors of CDK2 enzyme. Bioorg Med Chem Lett 2010;20(4):1384–7.

[471] Inglis KJ, Chereau D, Brigham EF, Chiou SS, Schöbel S, Frigon NL, et al. Polo-like kinase 2 (PLK2) phosphorylates alpha-synuclein at serine 129 in central nervous system. Biol Chem 2009;284(5):2598–602.

[472] Farooqui AA, Yang HC, Rosenberger TA, Horrocks LA. Phospholipase A2 and its role in brain tissue. J Neurochem 1997;69(3):889–901.

[473] Ross BM, Kim DK, Bonventre JV, Kish SJ. Characterization of a novel phospholipase A2 activity in human brain. J Neurochem 1995;64:2213–21.

[474] Nishio E, Nakata H, Arimura S, Watanabe Y. α1-adrenergic receptor stimulation causes arachidonic acid release through pertussis toxin-sensitive GTP-binding

protein and JNK activation in rabbit aortic smooth muscle cells. Biochem Biophys Res Commun 1996;219(2):277–82.

[475] Moskowitz N, Schook W, Puszkin S. Interaction of brain synaptic vesicles induced by endogenous Ca^{2+}-dependent phospholipase A2-dependent phospholipase A2. Science 1982;216:305–7.

[476] Piomelli D, Wang JK, Sihra TS, Nairn AC, Czernik AJ, Greengard P. Inhibition of Ca^{2+}/calmodulin-dependent protein kinase II by arachidonic acid and its metabolites. Proc Natl Acad Sci U S A 1989;86:8550–4.

[477] Liu T, Fujita T, Nakatsuka T, Kumamoto E. Phospholipase A2 activation enhances inhibitory synaptic transmission in rat substantia gelatinosa neurons. J Neurophysiol 2008;99(3):1274–84.

[478] Sala E, Guasch L, Iwaszkiewicz J, Mulero M, Salvadó MJ, Bladé C, et al. Identification of human IKK-2 inhibitors of natural origin (Part II): In Silico prediction of IKK-2 inhibitors in natural extracts with known anti-inflammatory activity. Eur J Med Chem 2011;46(12):6098–103.

[479] Kim MH, Ryu SY, Choi JS, Min YK, Kim SH. Saurolactam inhibits osteoclast differentiation and stimulates apoptosis of mature osteoclasts. J Cell Physiol 2009;221(3):618–28.

[480] Raju R, Balakrishnan L, Nanjappa V, Bhattacharjee M, Getnet D, Muthusamy B, et al. A comprehensive manually curated reaction map of RANKL/RANK-signaling pathway. Database (Oxford) 2011 doi:10.1093/database/bar021.

[481] Yu SM, Ko FK, Su MJ, Wu TS, Wang ML, Huang TF, et al. Vasorelaxing effect in rat thoracic aorta caused by fraxinellone and dictamine isolated from the Chinese herb Dictamnus dasycarpus Turcz: comparison with cromakalim and Ca^{2+} channel blockers. Naunyn Schmiedebergs Arch Pharmacol 1992;345(3):349–55.

[482] Miyazawa M, Shimamura H, Nakamura SI, Kameoka H. Antimutagenic activity of isofraxinellone from Dictamnus dasycarpus. J Agri Food Chem 1995;43(6):1428–31.

[483] Zhao W, Wolfender JL, Hostettmann K, Li HY, Stoeckli-Evans H, Xu R, et al. Sesquiterpene glycosides from Dictamnus dasycarpus. Phytochem 1998;47(1):63–8.

[484] Chen J, Tang JS, Tian J, Wang YP, Wu FE. Dasycarine, a new quinoline alkaloid from Dictamnus dasycarpus. Chin Chem Lett 2000;11(8):707–8.

[485] Du CF, Yang XX, Tu PF. Studies on chemical constituents in bark of dictamnus dasycarpus. Zhongguo Zhong Yao Za Zhi 2005;30(21):1663–6.

[486] Takeuchi N, Fujita T, Goto K, Morisaki N, Osone N, Tobinaga S. Dictamnol, a new trinor-guaiane type sesquiterpene, from the roots of Dictamnus dasycarpus Turcz. Chem Pharm Bull 1993;41(5):923–5.

[487] Guo LN, Pei YH, Chen G, Cong H, Liu JC. Two new compounds from Dictamnus dasycarpus. J Asian Nat Prod Res 2012;14(2):105–10.

[488] Zhao W, Wolfender JL, Hostettmann K, Xu R, Qin G. Antifungal alkaloids and limonoid derivatives from Dictamnus dasycarpus. Phytochemistry 1998;47(1):7–11.

[489] Westaway SM, Chung YK, Davis JB, Holland V, Jerman JC, Medhurst SJ, et al. N-Tetrahydroquinolinyl, N-quinolinyl and N-isoquinolinyl biaryl carboxamides as antagonists of TRPV1. Bioorg Med Chem Lett 2006;16(17):4533–6.

[490] Westaway SM, Thompson M, Rami HK, Stemp G, Trouw LS, Mitchell DJ, et al. Design and synthesis of 6-phenylnicotinamide derivatives as antagonists of TRPV1. Bioorg Med Chem Lett 2008;18(20):5609–13.

[491] Hensellek S, Brell P, Schaible HG, Brauer R, Segond von Banchet G. The cytokine TNFalpha increases the proportion of DRG neurones expressing the TRPV1 receptor via the TNFR1 receptor and ERK activation. Mol Cell Neurosci 2007;36(3):381–91.

[492] Kim ML, Zhang B, Mills IP, Milla ME, Brunden KR, Lee VMY. Effects of TNFalpha-converting enzyme inhibition on amyloid β production and APP processing in vitro and in vivo. J Neurosci 2008;28(46):12052–61.

[493] Marchetti L, Klein M, Schlett K, Pfizenmaier K, Eisel UL. Tumor necrosis factor (TNF)-mediated neuroprotection against glutamate-induced excitotoxicity is enhanced by N-methyl-D-aspartate receptor activation. Essential role of a TNF receptor 2-mediated phosphatidylinositol 3-kinase-dependent NF-kappa B pathway. J Biol Chem 2004;279(31):32869–81.

[494] Feng X. Regulatory roles and molecular signaling of TNF family members in osteoclasts. Gene 2005;350(1):1–13.

[495] Tamatani M, Che YH, Matsuzaki H, Ogawa S, Okado H, Miyake S, et al. Tumor necrosis factor induces Bcl-2 and Bcl-x expression through NF-kappaB activation in primary hippocampal neurons. J Biol Chem 1999;274(13):8531–8.

[496] Zarghi A, Ghodsi R, Azizi E, Daraie B, Hedayati M, Dadrass OG. Synthesis and biological evaluation of new 4-carboxyl quinoline derivatives as cyclooxygenase-2 inhibitors. Bioorg Med Chem 2009;17(14):5312–17.

[497] Ghodsi R, Zarghi A, Daraei B, Hedayati M. Design, synthesis and biological evaluation of new 2,3-diarylquinoline derivatives as selective cyclooxygenase-2 inhibitors. Bioorg Med Chem 2010;18(3):1029–33.

[498] Rose JW, Hill KE, Watt HE, Carlson NG. Inflammatory cell expression of cyclooxygenase-2 in the multiple sclerosis lesion. J Neuroimmunol 2004;149(1–2):40–9.

[499] Bezzi P, Carmignoto G, Pasti L, Vesce S, Rossi D, Rizzini BL, et al. Prostaglandins stimulate calcium-dependent glutamate release in astrocytes. Nature 1998;391(6664):281–5.

[500] Trotti D, Rossi D, Gjesdal O, Levy LM, Racagni G, Danbolt NC, et al. Peroxynitrite inhibits glutamate transporter subtypes. J Biol Chem 1996;271(11):5976–9.

[501] Minghetti L. Cyclooxygenase-2 (COX-2) in inflammatory and degenerative brain diseases. J Neuropathol Exp Neurol 2004;63(9):901–10.

[502] Lonze BE, Riccio A, Cohen S, Ginty DD. Apoptosis, axonal growth defects, and degeneration of peripheral neurons in mice lacking CREB. Neuron 2002;34(3):371–85.

[503] Arnsten AF, Ramos BP, Birnbaum SG, Taylor JR. Protein kinase A as a therapeutic target for memory disorders: rationale and challenges. Trends Mol Med 2005;11(3):121–8.

[504] Barad M, Bourtchouladze R, Winder DG, Golan H, Kandel E. Rolipram, a type IV-specific phosphodiesterase inhibitor, facilitates the establishment of long-lasting long-term potentiation and improves memory. Proc Natl Acad Sci U S A 1998;95(25):15020–5.

[505] Buckley GM, Cooper N, Dyke HJ, Galleway FP, Gowers L, Haughan AF, et al. 8-Methoxyquinoline-5-carboxamides as PDE4 inhibitors: a potential treatment for asthma. Bioorg Med Chem Lett 2002;12(12):1613–15.

[506] Huang Z, Dias R, Jones T, Liu S, Styhler A, Claveau D, et al. L-454,560, a potent and selective PDE4 inhibitor with in vivo efficacy in animal models of asthma and cognition. Biochem Pharmacol 2007;73(12):1971–81.

[507] Billah M, Cooper N, Cuss F, Davenport RJ, Dyke HJ, Egan R, et al. Synthesis and profile of SCH351591, a novel PDE4 inhibitor. Bioorg Med Chem Lett 2002;12(12):1621–3.

[508] Spooren A, Kooijman R, Lintermans B, Van Craenenbroeck K, Vermeulen L, Haegeman G, et al. Cooperation of NFkappaB and CREB to induce synergistic IL-6 expression in astrocytes. Cell Signal 2010;22(5):871–81.

[509] Yu GL, Wei EQ, Wang ML, Zhang WP, Zhang SH, Weng JQ, et al. Pranlukast, a cysteinyl leukotriene receptor-1 antagonist, protects against chronic ischemic brain injury and inhibits the glial scar formation in mice. Brain Res 2005;1053(1–2):116–25.

[510] Madonna S, Maher P, Kraus JL. N,N-Bis-(8-hydroxyquinoline-5-yl methyl)-benzyl substituted amines (HQNBA): peroxisome proliferator-activated receptor (PPAR-γ) agonists with neuroprotective properties. Bioorg Med Chem Lett 2010;20(23):6966–8.

[511] Bartlett RD, Esslinger CS, Thompson CM, Bridges RJ. Substituted quinolines as inhibitors of L-glutamate transport into synaptic vesicles. Neuropharmacology 1998;37(7):839–46.

[512] Carrigan CN, Bartlett RD, Esslinger CS, Cybulski KA, Tongcharoensirikul P, Bridges RJ, et al. Synthesis and in vitro pharmacology of substituted quinoline-2,4-dicarboxylic acids as inhibitors of vesicular glutamate transport. J Med Chem 2002;45(11):2260–76.

[513] Seya K, Miki I, Murata K, Junke H, Motomura S, Araki T, et al. Pharmacological properties of pteleprenine, a quinoline alkaloid extracted from Orixa japonica, on guinea-pig ileum and canine left atrium. J Pharm Pharmacol 1998;50(7):803–7.

[514] Cheng JT, Chang TK, Chen IS. Skimmianine and related furoquinolines function as antagonist of 5-hydroxytryptamine receptors in animals. J Autonomic Pharmacol 1994;14(5):365–74.

[515] Chen IS, Tsai IW, Teng CM, Chen JJ, Chang YL, Ko FN, et al. Pyranoquinoline alkaloids from Zanthoxylum simulans. Phytochemistry 1997;46(3):525–9.

[516] Napolitano HB, Silva M, Ellena J, Rocha WC, Vieira PC, Thiemann OH, et al. Redetermination of skimmianine: a new inhibitor against the Leishmania APRT enzyme. Acta Crystallogr Sect E: Struct Rep Online 2003;59(10):1503–5.

[517] Watanabe S, Yoshimi Y, Ikekita M. Neuroprotective effect of adenine on purkinje cell survival in rat cerebellar primary cultures. J Neurosci Res 2003;74(5):754–9.

[518] Wansi JD, Mesaik MA, Chiozem DD, Devkota KP, Gaboriaud-Kolar N, Lallemand MC, et al. Oxidative burst inhibitory and cytotoxic indoloquinazoline and furoquinoline alkaloids from Oricia suaveolens. J Nat Prod 2008;71(11):1942–5.

[519] Wang TY, Wu JB, Hwang TL, Kuo YH, Chen JJ. A new quinolone and other constituents from the fruits of Tetradium ruticarpum: effects on neutrophil pro-inflammatory responses. Chem Biodivers 2010;7(7):1828–34.

[520] Yoon JS, Jeong EJ, Yang H, Kim SH, Sung SH, Kim YC. Inhibitory alkaloids from Dictamnus dasycarpus root barks on lipopolysaccharide-induced nitric oxideproduction in BV2 cells. J Enzyme Inhib Med Chem 2012;27(4):490–4.

[521] Su TL, Lin FW, Teng CM, Chen KT, Wu TS. Antiplatelet aggregation principles from the stem and root bark of Melicope triphylla. Phytother Res 1998;12(Suppl. 1):S74–6.

[522] Wu TS, Shi LS, Wang JJ, Iou SC, Chang HC, Chen YP, et al. Cytotoxic and antiplatelet aggregation principles of Ruta graveolens. J Chin Chem Soc 2003;50(1):171–8.

[523] Rybalkin SD, Rybalkina IG, Shimizu-Albergine M, Tang XB, Beavo JA. PDE5 is converted to an activated state upon cGMP binding to the GAF A domain. EMBO J 2003;22(3):469–78.

[524] Nam KW, Je KH, Shin YJ, Kang SS, Mar W. Inhibitory effects of furoquinoline alkaloids from Melicope confusa and Dictamnus albus against human phospho-diesterase 5 (hPDE5A) in vitro. Arch Pharm Res 2005;28(6):675–9.

[525] Lal B, Bhise NB, Gidwani RM, Lakdawala AD, Joshi K, Patvardhan S. Isolation, synthesis and biological activity of Evolitrine and analogs. Arkivoc 2005;2:77–97.

[526] Rahman AU, Khalid A, Sultana N, Nabeel Ghayur M, Ahmed Mesaik M, RiazKhan M, et al. New natural cholinesterase inhibiting and calcium channel blocking quinoline alkaloids. J Enzyme Inhib Med Chem 2006;21(6):703–10.

[527] Cardoso-Lopes EM, Maier JA, Da Silva MR, Regasini LO, Simote SY, Lopes NP, et al. Alkaloids from stems of Esenbeckia leiocarpa Engl. (Rutaceae) as potential treatment for Alzheimer disease. Molecules 2010;15(12):9205–13.

[528] Cabral RS, Sartori MC, Cordeiro I, Queiroga CL, Eberlin MN, Lago JHG, et al. Anticholinesterase activity evaluation of alkaloids and coumarin from stems of Conchocarpus fontanesianus. Braz J Pharmacogn 2012;22(2):374–80.

[529] Broos K, De Meyer SF, Feys HB, Vanhoorelbeke K, Deckmyn H. Blood platelet biochemistry. Thromb Res 2012;129(3):245–9.

Topic **1.4**

Terpene Alkaloids

1.4.1 *Dendrobium catenatum* Lindl.

History The plant was first described by John Lindley in *The Genera and Species of Orchidaceous Plants* published in 1830.

Synonyms *Callista stricklandiana* (Rchb. f.) Kuntze, *Dendrobium funiushanense* T.B. Chao, Zhi X. Chen & Z.K. Chen, *Dendrobium* *huoshanense* C.Z. Tang & S.J. Cheng, *Dendrobium officinale* Kimura & Migo, *Dendrobium perefauriei* Hayata, *Dendrobium stricklandianum* Rchb. f., *Dendrobium tosaense* Makino, *Dendrobium tosaense* var. perefauriei (Hayata) Masam.

Family Orchidaceae Juss., 1789

Common Name Huang shi hu (Chinese)

Habitat and Description This little orchid grows on rocks in the mountains of China, Japan, and Korea. The stem is fleshy, up to 30 cm long, and articulated with internodes 4 cm long. The leaves are simple, thin, and alternate. The blade is elliptic, 3–20 cm × 0.5–2 cm, and acute at the apex. The flowers are showy, light green, and terminal. The labellum is white with red marks. The dorsal sepal is ovate, 12–15 cm × 0.5 cm, and tri-nerved. The leaves are lanceolate and as long as the sepals. The petals are oblong, as long as the sepals, and penta-nerved. The lip is broadly ovate, 13–20 cm × 1 cm, somewhat 3-lobed and wavy, 1.5–3 mm long, and nearly glabrous; the apex is subacute and bilobed, and the lobes are sharply toothed (Figure 1.18).

Medicinal Uses In China, the plant is used to treat rheumatism.

Phytopharmacology The plant accumulates a series of sesquiterpene dendrobane alkaloids[530] such as dendrobine and flavonoids including naringenin.[531]

Proposed Research Pharmacological study of dendrobane sesquiterpene alkaloids for the treatment of amyotrophic lateral sclerosis (ALS).

Rationale Accumulating data shows that sesquiterpene alkaloids are immunosuppressive and therefore anti-neuroinflammatory. In effect, *Nuphar pumila* (Timm) DC. (family Nymphaeaceae Salisb.) produces the sesquiterpene alkaloids 6-hydroxythiobinupharidine (CS 1.304), 6,6′-dihydroxythiobinupharidine (CS 1.305), 6-hydroxythionuphlutine B (CS 1.306), 6′-hydroxythionuphlutine B, and 6,6′-dihydroxythionuphlutine B (CS 1.307), which at a dose of 10^{-6} M nullified anti-sheep erythrocyte plaque forming cells in mouse splenocytes and therefore antibody formation.[532,533] Furthermore, the hydroxywilfordate sesquiterpene alkaloids wilfortrine (CS 1.308) and the wilfordate sesquiterpene pyridine alkaloid euonine (CS 1.309) from *Tripterygium wilfordii* Hook. f. (family Celastraceae R.Br.) exhibited immunosuppressive activities in the

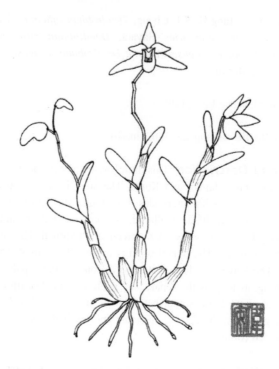

■ **FIGURE 1.18** *Dendrobium catenatum* Lindl.

■ **CS 1.304** 6-hydroxythiobinupharidine.

hemolysin reaction in rodents at doses equal to 40 mg/kg and 80 mg/kg, respectively,[534] and delayed the clearance of particles of charcoal and the mass of spleen and thymus at a dose of 80 mg/kg.[534] In line with these results, the hydroxywilfordate sesquiterpene alkaloid wilfortrine and the

■ **CS 1.305** 6,6′-dihydroxythiobinupharidine.

■ **CS 1.306** 6-hydroxythionuphlutine B.

■ **CS 1.307** 6,6′-dihydroxythionuphlutine B.

wilfordate sesquiterpene pyridine alkaloid euonine from *Tripterygium wilfordii* Hook. f. (family Celastraceae R. Br.) suppressed the immunity of rodents.[535] In addition, *Tripterygium wilfordii* Hook. f. (family Celastraceae R. Br.) shelters the evoninate sesquiterpene alkaloids ebenifoline E-II (CS 1.310) and cangorinine E-I (CS 1.311), which nullified

■ **CS 1.308** Wilfortrine.

■ **CS 1.309** Euonine.

■ **CS 1.310** Ebenifoline E-II.

■ **CS 1.311** Cangorinine E-I.

■ **CS 1.312** (−)-dendrobine.

■ **CS 1.313** Nobilonine.

■ **CS 1.314** 3-hydroxy-2-oxodendrobine.

the secretion of interleukin-2β (IL-2β) and interleukin-4β (IL-4β) at a dose of $10\,\mu g/mL$.[536] Wilfordate sesquiterpene pyridine alkaloids euonine and wilforine and the evoninate sesquiterpene alkaloids cangorinine E-I from the same plant reduced the transcription of inducible nitric oxide synthetase (iNOS) with IC_{50} values equal to $19\,\mu M$, $36\,\mu M$, and $12\,\mu M$, respectively, in macrophages exposed to lipopolysaccharides (LPS).[537] In the same experiment, the wilfordate sesquiterpene pyridine alkaloid wilforine reduced the levels of cyclo-oxygenase-2 (COX-2) and interleukin-1β (IL-1β) with IC_{50} values equal to $34\,\mu M$ and $46\,\mu M$[537]. This evidence suggests the possibility that the inhibition of nuclear factor kappa-light-chain-enhancer of activated B cells (NF-κB) may very well be incurred by sesquiterpene alkaloids. One should recall that the development of neurodegenerative diseases and especially Alzheimer's disease (AD) results from the excessive expression of iNOS, COX-2, and cytokines such as tumor necrosis factor-α (TNF-α) by activated microglia, which magnify and mediate neuroinflammation.[538] In the brain, the stimulation of TNF-α receptor type 1, which is associated with death domain (DD) and coupled with the adaptor protein TNF receptor-associated death domain (TRADD), activates TRAF2 and RIP, which activates kappa light polypeptide gene enhancer in B-cell inhibitor (I-κB) kinase (IKK) and therefore NF-κB, which in glial cells commands the synthesis of cytokines via iNOS and COX-2. Intriguingly, a mixture of dendrobane sesquiterpene alkaloids including dendrobine (CS 1.312), nobilonine (CS 1.313), and 3-hydroxy-2-oxodendrobine (CS 1.314) from *Dendrobium nobile* Lindl. (family Orchidaceae Juss.) prevented LPS-induced dementia in rodents in the Morris water maze at a dose of 80 mg/kg with concomitant reduction of hippocampal TNF-α receptor type 1.[539] Additionally, a mixture of dendrobine, dendramine (CS 1.315), and noliboline isolated from a member of the genus *Dendrobium* Sw. (family Orchidaceae Juss.) at a dose of 0.2 mg/L protected neurons by 38.6% against apoptosis imposed by deprivation of glucose and oxygen by reducing the cytoplasmic concentration of Ca^{2+}, maintaining mitochondrial integrity, and reducing levels of activated caspase 3.[540] Note that the chemical framework of dendrobine is closely related with picrotoxinin (CS 1.316), hence its ability to block the glycinergic receptor at the picrotoxin binding site.[539] In the spinal cord, the Renshaw cells secrete glycine (CS 1.317), which stimulates of glycinergic receptors, resulting in inhibition of motoneurons; in fact, glycinergic activity decreases during ALS. Given the ability of dendrobane sesquiterpene alkaloids to interact with glycinergic receptors, the bewildering array of dendrobane sesquiterpene alkaloids, and vast number of species within the genus *Dendrobium* Sw., it is tempting to speculate that dendrobane sesquiterpene alkaloids may produce leads for the treatment of ALS.

1.4.2 *Aconitum kongboense* Lauener

History The plant was first described by Lucien André Lauener in *Notes from the Royal Botanic Garden, Edinburgh* published in 1963.

Family Ranunculaceae Juss., 1789

Common Name Gong bu wu tou (Chinese)

Habitat and Description This perennial herb grows in the grasslands of China. The stem is erect and 1.5 m tall. The leaves are simple, exstipulate, and alternate. The petiole is of variable length and reaches 10 cm long. The blade is palmate, 15 cm across, cuneate at the base, and it develops 5 lobes, which are deeply incised. The inflorescence is a terminal raceme, which is many flowered and 50 cm long. The calyx comprises 5 whitish and 1.5 cm long sepals, including 2 lower, 2 lateral, and an upper one. The corolla consists of a pair of petals that are 1 cm long and spurred. The androecium includes numerous stamens, which are dark purple. The gynoecium comprises 10 carpels, and the fruits are follicles (Figure 1.19).

Medicinal Uses In China, the plant is used to treat rheumatism.

Phytopharmacology The plant engineers a series of aconitane diterpene alkaloids, including kongboenine, chasmaconitine, talatisamine, aconitine (CS 1.318),[541] pyrochasmaconitine, pyrocrassicauline A, and 14-benzoyltalatisamine[542]; as well as the napelline-type diterpene alkaloids songorine, 12-*epi*-15-*O*-acetylnapelline, songoramine, and 12-*epi*-19-dehydronapelline.

Proposed Research Pharmacological study of songorine and derivatives for the treatment of AD.

Rationale Members of the genus *Aconitum* L. (family Ranunculaceae Juss.) produce a series of dreadfully toxic diterpene alkaloids which impose death by ventricular tachycardia.[546] Indeed, *Aconitum* alkaloids

■ **CS 1.315** Dendramine.

■ **CS 1.316** Picrotoxinin.

■ **CS 1.317** Glycine.

■ **FIGURE 1.19** *Aconitum kongboense* Lauener.

■ **CS 1.318** Aconitine.

have a binding affinity to voltage-dependent Na$^+$ channels (VDSC), which are either activated or blocked according to the diterpene framework they are exposed to.[543] The aconitane alkaloid aconitine which bears a benzoyl ester in the C14 position binds to voltage-dependent Na$^+$ channels and delays its inactivation,[544] whereas lappaconitine (CS 1.319) and

N-deacetyllappaconitine (CS 1.320) with a benzoyl ester at the C4 position block the voltage-dependent Na⁺ channel.[545] Sustained activation of VDSC disrupts the generation of action potentials and axonal conductance in hippocampal neurons, as shown with aconitine at a dose of $1\,\mu M$,[546] as well as *N*-desacetyllappaconitine and lappaconitine with IC_{50} values equal to $4.7\,\mu M$ and $12.6\,\mu M$.[547] Here, it is worthwhile to emphasize that the chronic activation of voltage-dependent Na⁺ channel favors glutamate release, activates *N*-methyl-D-aspartate (NMDA) receptors, induces an influx of extracellular Ca^{2+},[548] and accounts for apoptosis of motoneurons;

■ **CS 1.319** Lappaconitine.

■ **CS 1.320** *N*-deacetyllappaconitine.

derivatives of lappaconitine and *N*-deacetyllappaconitine may produce new treatments for ALS.[549] One might argue that because of their toxicity, aconitane alkaloids are of no use in therapeutics, but synthetic modification may very well offer safer chemical entities from which to develop neuroprotective agents. The aconitane alkaloid mesaconitine (CS 1.321) from a member of the genus *Aconitum* L. (family Ranunculaceae Juss.) assuaged, via intracerebral injection at a dose of 80 ng, plantar inflammation and pain induced by carrageenan,[550] suggesting central analgesic activity. Additional evidence supporting the hypothesis that *Aconitum* alkaloids are central analgesics was obtained using the tail-flick test, whereby lappaconitine at a dose of 6 mg/kg exhibited a dose-dependent analgesic activity which was increased by co-administrating serotonin, raising the exciting contention that *Aconitum* alkaloids may interfere with the serotoninergic system.[551] In fact, the napellin diterpene songorine (CS 1.322) from *Aconitum baicalense* Turcz. ex Rapaics (family Ranunculaceae Juss.) at a dose of 0.02 mg/kg displayed antidepressant effects in the tail suspension test and assuaged the paw edema induced by plantar injection of serotonin.[552] A speculative mechanism by which *Aconitum* alkaloids may reduce pain is through stimulation of serotoninergic receptor subtype 3 (5-HT$_3$).[553] It should be recalled that the stimulation of ionotropic 5-HT$_3$

■ **CS 1.321** Mesaconitine.

■ **CS 1.322** Songorine.

inhibits the release of acetylcholine in the cortex,[554] so 5-HT$_3$ antagonist may be of value to treat AD. Such antagonists may very well be synthesized from *Aconitum* alkaloids. Note that songorine from *Aconitum leucostomum* Vorosch. (family Ranunculaceae Juss.) blocked γ-amino butyric acid subtype A (GABA$_A$) receptors of hippocampal CA1 pyramidal neurons of rodents with an IC$_{50}$ value equal to 19.6 μM, resulting in increased discharge,[555] and may improve cognition because hippocampal pyramidal neurons are involved in memory acquisition.[556] There is evidence indicating that the aconitane diterpene methyllycaconitine (CS 1.323) from members of the genera *Aconitum* L. (family Ranunculaceae Juss.) and *Delphinium* L. (family Ranunculaceae) has affinity and blocks the $\alpha3\beta4$ acetylcholine receptors expressed in human embryonic kidney (HEK293) cells with a K_i value equal to 0.4 μM.[557] This fact preliminarily suggests the possibility that methyllycaconitine might antagonize $\alpha4\beta2$ nicotinic receptors[558] and is used therefore as a starting point for the synthesis of $\alpha4\beta2$ nicotinic receptor agonists for the treatment of AD. Other compelling evidence is provided by the finding that the aconitane diterpene alkaloid talatisamine (CS 1.324) from *Aconitum kongboense* Lauener blocked voltage-gated K$^+$ channels with an IC$_{50}$ value of 146 μM[559] and consequently protected CA1 hippocampal pyramidal neurons against amyloid-β_{1-40}-induced apoptosis via reduction of pro-apoptotic Bcl-2-associated X protein (Bax), hence preservation of mitochondrial integrity and therefore inactivation of caspase 3 and 9.[560] In effect, the opening of voltage-gated K$^+$ channels provokes a massive efflux of K$^+$ out neurons, hence hyperpolarization and apoptosis by reduced activity of pro-apoptotic protein (p53), Forkhead, and cAMP response element binding protein (CREB).[561]

■ **CS 1.323** Methyllycaconitine.

Delphinium linearilobum N. Busch (family Ranunculaceae Juss.) shows the atidane diterpene alkaloid cochlearenine which scavenged free radicals *in vitro* by 32%.[562]

Members of the genus *Spiraea* L. (family Rosaceae Juss.) produce unusual series of diterpene alkaloids which are neuroprotective. One such alkaloid is the atisine-type diterpene spiramine T (CS 1.325), which protected rodents against experimental ischemia and reperfusion at a dose of 2 mg/kg with reduction of cerebral enzymatic activity of nitric oxide synthetase (NOS), hence collapse in nitric oxide (NO), and interestingly, increase of cerebral enzymatic activity of glutathione peroxidase (GPx), hence lower cytoplasmic membrane insults as shown by a decrease in malonaldehyde (MDA) by 50%.[563] Furthermore, the atisine-type diterpene alkaloids spiramine A, spiramine C, and spiradine F (CS 1.326–1.328) from *Spiraea japonica* L.f. (family Rosaceae) inhibited the aggregation of platelets induced by platelet activating factor (PAF) with IC_{50} values equal to 6.7 μM, 32.6 μM, and 138.9 μM, respectively,[564] implying that the presence of an oxygen atom at C15 and the occurrence of the oxazolidine moiety favors the antiplatelet activity and therefore blood perfusion. Accumulating data shows that low cerebral blood perfusion is an early event in the pathophysiology of AD. Indeed, early cerebrovascular insults may result from the production of amyloid β peptide by platelets that accumulate amyloid precursor protein (APP) and convert it into β-amyloid peptide, which deposits into the cerebral vasculature, resulting in cerebral amyloid angiopathy.[565,566]

■ **CS 1.324** Talatisamine.

1.4.3 *Cynanchum wilfordii* (Maxim.) Maxim. ex Hook. f.

History The plant was first described by Carl Johann Maximowicz in *The Flora of British India* published in 1883.

Family Asclepiadaceae Borkh., 1797

■ **CS 1.325** Spiramine T.

■ **CS 1.326** Spiramine A.

Synonyms *Cynoctonum wilfordii* Maxim., *Seutera wilfordii* (Maxim.) Pobed., *Vincetoxicum wilfordii* (Maxim.) Franch. & Sav.

Common Names Koikema (Japanese), ge shan xiao (Chinese)

Habitat and Description This climber grows in Russia, China, Korea, and Japan. The roots are tuberous, and the stem is terete, flexuous, and up to 2 m long. The leaves are simple, opposite, and exstipulate. The plant exudes a white latex upon incision. The petiole grows to 4 cm long. The blade is cordate, 4–6 cm × 2–4 cm, membranaceous, acuminate at the apex, and has 4 pairs of secondary nerves. The inflorescence is an axillary raceme. The calyx comprises 5 sepals, which are triangular and minute. The corolla is yellow, tubular, and produces 5 lobes that are 0.5 cm long. The receptacle produces a column embedded into a corona. The fruit is a follicle, which is lanceolate, 10 cm, and stuffed with comose seeds that are 2 cm long (Figure 1.20).

■ **CS 1.327** Spiramine C.

■ **CS 1.328** Spiradine F.

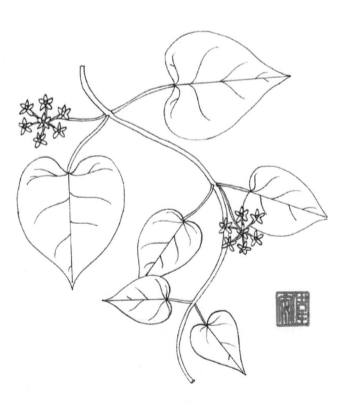

■ **FIGURE 1.20** *Cynanchum wilfordii* (Maxim.) Maxim. ex Hook. f.

Medicinal Uses In Korea, the plant is used to treat diabetes and to invigorate. In China, the plant is used to invigorate.

Phytopharmacology The plant shelters a broad array of pregnane glycosides with aglycones such as sarcostin, deacylcynanchogenin, deacyl-metaplexigenin, and the steroidal alkaloid gagaminine.[567,568]

Proposed Research Pharmacological study of gagaminine and derivatives for the treatment of AD.

Rationale A large body of experimental evidence demonstrates that steroidal alkaloids from members of the family Buxaceae Dumort inhibit the enzymatic activity of acetylcholinesterase (AChE). In effect, *Sarcococca saligna* (D. Don) Müll. Arg. (family Buxaceae Dumort.) produces a series of Buxus type steroidal alkaloids such as salignenamide E, salignenamide F, axillaridine A, sarsalignenone,[569] salonine C, dictyophlebine, iso-*N*-for-mylchonemorphine,[570] 2,3-dehydrosarsalignone,[571] isosarcodine, sarcorine, sarcodine, sarcocine, and alkaloid-C[572] (CS 1.329–1.341) which inhibited the enzymatic activity of AChE with IC_{50} values equal to 6.2 μM, 6.3 μM, 5.2 μM, 5.8 μM, 7.8 μM, 6.2 μM, 6.3 μM, 7 μM, 10.3 μM, 69.9 μM, 49.7 μM, 20 μM, and 42.2 μM, respectively, implying that amine moieties in C-3 and C-20 are essential for activity.[573] Note that sarcorine, sarcodine, sarcocine, and alkaloid-C relaxed rabbit jejunum preparations exposed to K^+ with IC_{50} values equal to 63.5 μg/mL, 19.6 μg/mL, 24.3 μg/mL, and 200 μg/mL, respectively, implying voltage-dependent Ca^{2+} channel-blocking potencies.[572]

■ **CS 1.329** Salignenamide E.

■ **CS 1.330** Salignenamide F.

■ **CS 1.331** Axillaridine A.

■ **CS 1.332** Sarsalignenone.

■ **CS 1.333** Salonine C.

■ **CS 1.334** Dictyophlebine.

■ **CS 1.335** iso-*N*-formylchonemorphine.

■ **CS 1.336** 2,3-dehydrosarsalignone.

■ **CS 1.337** Isosarcodine.

■ **CS 1.338** Sarcorine.

■ **CS 1.339** Sarcodine.

■ **CS 1.340** Sarcocine.

■ **CS 1.341** Alkaloid-C.

Likewise, *Sarcococca hookeriana* Baill. (family Buxaceae Dumort.) produces phulchowkiamide A, hookerianamide F, sarcovagenine C, sarcovagine D[573] epipachysamine-E-5-en-4-one, sarcovagine C,[574] hookerianamide H, hookerianamide I, *N*-methylepipachysamine D (CS 1.342–1.350), and dictyophlebine,[575] which inhibited the enzymatic

■ **CS 1.342** Phulchowkiamide A.

■ **CS 1.343** Hookerianamide F.

■ **CS 1.344** Sarcovagenine C.

■ **CS 1.345** Sarcovagine D.

■ **CS 1.346** Epipachysamine-E-5-en-4-one.

■ **CS 1.347** Sarcovagine C.

■ **CS 1.348** Hookerianamide H.

■ **CS 1.349** Hookerianamide I.

■ **CS 1.350** *N*-methylepipachysamine D.

activity of AChE with IC_{50} values equal to $0.5\,\mu M$, $1.6\,\mu M$, $1.5\,\mu M$, $2.2\,\mu M$, $9.9\,\mu M$, $8.1\,\mu M$, $2.9\,\mu M$, $34.1\,\mu M$, $10.1\,\mu M$, and $6.2\,\mu M$, respectively. Funtumafrine C (CS 1.351) from *Sarcococca coriacea* Müll. Arg. (family Buxaceae Dumort.) inhibited acetylcholinesterase with an IC_{50} value

■ **CS 1.351** Funtumafrine C.

■ **CS 1.352** Buxamine B.

■ **CS 1.353** Buxamine C.

of 45.7 μM.[576] Buxamine B and buxamine C (CS 1.352 and 1.353) inhibited AChE with IC$_{50}$ values equal to 74 μM and 7.5 μM, respectively.[577] (+)-*O*6-buxafurandiene (CS 1.354), (+)-7-deoxy-*O*6-buxafurandiene (CS 1.355), (+)-benzoylbuxidienine (CS 1.356), (+)-buxapapillinine (CS 1.357), (+)-buxaquamarine (CS 1.358) and (+)-irehine (CS 1.359) from *Buxus hyrcana* Pojark. (family Buxaceae Dumort.) inhibited AChE activity with IC$_{50}$ values equal to 17 μM, 13 μM, 35 μM, 80 μM, 76 μM, and 100 μM, respectively.[578] (+)-Buxabenzamidienine (CS 1.360) and (+)-buxamidine (CS 1.361) from *Buxus sempervirens* L. (family Buxaceae Dumort.) inhibited the enzymatic activity of AChE with IC$_{50}$ values equal 0.7 μM and 1.7 μM, respectively.[579] Other steroidal alkaloids which inhibit the enzymatic activity of acetylcholinesterase are impericine (CS 1.362) from *Fritillaria imperialis* L. (family Liliaceae Juss.),[580] as well as conessine (CS 1.363), isoconessimine (CS 1.364), conessimine (CS 1.365),

■ **CS 1.354** (+)-*O6*-buxafurandiene.

■ **CS 1.355** (+)-7-deoxy-*O6*-buxafurandiene.

■ **CS 1.356** (-)-benzoylbuxidienine.

■ **CS 1.357** (+)-buxapapillinine.

■ **CS 1.358** (+)-buxaquamarine.

■ CS 1.359 (+)-irehine.

■ CS 1.360 (+)-buxabenzamidienine.

■ CS 1.361 (+)-buxamidine.

conarrhimine, and conimine from *Holarrhena antidysenterica* (L.) Wall. ex A. DC. (family Apocynaceae Juss.) with IC_{50} values equal to $67.9\,\mu M$, $21\,\mu M$, $>300\,\mu M$, $4\,\mu M$, $28\,\mu M$, and $23\,\mu M$, respectively,[581] implying that the precise type of steroidal framework is not of dramatic importance for the activity.

One might argue that steroidal alkaloids being AChE inhibitors *sensu stricto* may have limited efficacy, but some evidence suggests that in fact

■ **CS 1.362** Impericine.

■ **CS 1.363** Conessine.

■ **CS 1.364** Isoconessimine.

■ **CS 1.365** Conessimine.

these alkaloids have multiple targets in the brain. This is exemplified with (+)-(20S)-3-(benzoylamino)-20-(dimethylamino)-5α-pregn-2-en-4β-ol (CS 1.366), (+)-(20S)-20-(dimethylamino)-3α-(methylsenecioylamino)-5α-pregn-12β-ol (CS 1.367), and (+)-(20S)-20-(dimethylamino)-3α-(methylbenzoylamino)-5α-pregn-12β-yl acetate (CS 1.368) from *Pachysandra procumbens* Michx. (family Buxaceae Dumort.), which inhibited the enzymatic activity of estrone sulfatase with IC_{50} values equal to 0.1 μM, 0.4 μM, and 5.7 μM, respectively.[582] Steroid sulfatase is of particular interest in neuropharmacology because its inhibition enhances cognitive function. One such inhibitor is-3-*O*-sulfamate, which at a dose of 10 mg/kg increased the neuroprotective properties of neurosteroids.[583] Gagaminine (CS 1.369), from *Cynanchum wilfordii* (Maxim.) Maxim. ex Hook. f. (family Asclepiadaceae Borkh.), inhibited the enzymatic activity of aldehyde oxidase with an IC_{50} equal to 0.8 μM[584] by virtue of its cinnamoyl moiety[585] and may protect neurons against oxidative insults. In addition, gagaminine has the interesting ability to hinder the enzymatic activity of 5-lipoxygenase (5-LOX) with an IC_{50} value equal to 26 μM.[586] Are steroidal alkaloids vast and unexploited sources of 5-lipoxyhebase (5-LOX), 12-lipoxygenase (12-LOX), and 15-lipoxygenase (15-LOX) inhibitors awaiting therapeutic development? Likewise, buxidin from *Buxus hyrcana* Pojark. (family Buxaceae Dumort.) at a dose of 3.1 μg/mL inhibited by half the production of reactive oxygen species (ROS) by monocytes challenged with zymosan, at a dose of 7.5 μg/mL inhibited phytohemagglutinin-induced T-cell proliferation by 86%, and nullified the secretion of IL-4 and IL-2 at a dose of 16 μg/mL.[587] Additional support to the contention that steroidal alkaloids have multiple neuroprotective effects is provided by the finding that peimisine (CS 1.370) and puqiedinone (CS 1.371) from *Fritillaria unibracteata* P.G. Xiao & K.C. Hsia (family Liliaceae Juss.) at a dose of 10 μM protected rat pheochromocytoma (PC12) cells poisoned with rotenone by 40.8% and 39.9%,[588]

■ **CS 1.366** (+)-(20 S)-3-(benzoylamino)-20-(dimethylamino)-5α-pregn-2-en-4β-ol.

■ **CS 1.367** (+)-(20 S)-20-(dimethylamino)-3α-(methylsenecioylamino)-5α-pregn-12β-ol.

■ **CS 1.368** (+)-(20 S)-20-(Dimethylamino)-3α-(methylbenzoylamino)-5α-pregn-12β-yl acetate.

■ **CS 1.369** Gagaminine.

■ **CS 1.370** Peimisine.

■ **CS 1.371** Puqiedinone.

but the mechanism underlying their neuroprotective potential remains unclear. Another exciting fact about steroidal alkaloids is their ability to block angiotensin-converting enzyme (ACE) and therefore improve cognition by confronting the chronic activation of angiotensin (AT1) receptor.[588] Such alkaloids as puqienine E, puqienine B, and puqienine A (CS 1.372–1.374) from *Fritillaria puqiensis* G.D. Yu & C.Y. Chen (family Liliaceae Juss.) inhibited ACE activity by 70.2%, 24.7%, and 20.4%, respectively, at a dose of 200 μM.[589] Conessine is able to dock into and block histaminergic receptor subtype 3 (H$_3$) with a K_i equal to 8.2[590] and therefore inhibit the histamine H$_3$ receptors in the area of cognition, namely *cortex* and *hippocampus*, by favoring cholinergic transmission through an increase in the secretion of acetylcholine.[591] Taking into account their multiple neuroprotective effects, steroidal alkaloids can be viewed as a conceptually new leads that might contribute to the development of neuroprotective drugs.

In fact, Bordet et al.[592] showed that the synthetic steroidal alkaloid cholest-4-en-3-one,oxime (TRO19622) (CS 1.375), at a dose of 10 μM, allowed

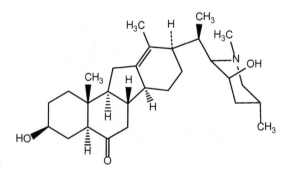

■ **CS 1.372** Puqienine E.

■ **CS 1.373** Puqienine B.

■ **CS 1.374** Puqienine A.

■ **CS 1.375** Cholest-4-en-3-one,oxime.

the survival of 70% of motoneurons deprived of neurotrophic factors *in vitro*[592] via the protection of mitochondria by targeting the voltage-dependent anion channel (VDAC).[593] In ALS models of rodents, cholest-4-en-3-one,oxime maintained 60% of neuromuscular junctions in gastrocnemius muscles, and inhibited muscular atrophy via inhibition of microglial activation.[594] Members of the genus *Veratrum* L. (family Melanthiaceae Batsch ex Borkh.) are extremely toxic and particularly teratogenic due to the presence of steroidal alkaloids such as veratridine (CS 1.376) and cyclopamine (CS 1.377). In effect, *Veratrum californicum* (Durand) produces cyclopamine, which imposes massive fetal malformation by binding to transmembrane protein smoothened (SMO) and blocking the Hedgehog (Hh) signaling pathway.[595] In neurophysiological conditions, activation of the receptor Patched (PTC-1) by sonic hedgehog (Shh) abates the inhibition of the SMO by patched1 (PTCH1), and consequently the SMO induces Gli, which commands the transcription of genes coding for cell survival and differentiation.[596] Intriguingly, Reilly et al.[597] showed that sonic hedgehog (Shh) and nerve growth factor (NGF) cooperate to generate basal forebrain cholinergic neurons, implying a possible use of the hedgehog pathway to fight neurodegeneration.

■ **CS 1.376** Veratridine.

■ **CS 1.377** Cyclopamine.

REFERENCES

[530] Zhu Y, Zhang A, He B, Zhang X, Yu Q, Si J. Quantitative variation of total alkaloids contents in *Dendrobium officinale*. Zhongguo Zhongyao Zazhi 2010;35(18):2388–91.

[531] Zhou G, Lv G. Comparative studies on scavenging DPPH free radicals activity of flavone C-glycosides from different parts of *Dendrobium officinale*. Zhongguo Zhongyao Zazhi 2012;37(11):1536–40.

[532] Yamahara JZ, Shimoda H, Matsuda H, Yoshikawa M. Potent immunosuppressive principles, dimeric sesquiterpene thioalkaloids, isolated from nupharis rhizoma, the rhizoma of *Nuphar pumilum* (nymphaeaceae), structure-requirement of nuphar-alkaloid for immunosuppressive activity. Biol Pharm Bull 1996;1(9):1241–3.

[533] Matsuda H, Shimoda H, Yoshikawa M. Dimeric sesquiterpene thioalkaloids with potent immunosuppressive activity from the rhizome of *Nuphar pumilum*:

structural requirements of *Nuphar* alkaloids for immunosuppressive activity. Bioorg Med Chem 2001;9(4):1031–2.

[534] Ma J, Dey M, Yang H, et al. Anti-inflammatory and immunosuppressive compounds from *Tripterygium wilfordii*. Phytochemistry 2007;68:1172–8.

[535] Zheng YL, Xu Y, Lin JF. Immunosuppressive effects of wilfortrine and euonine. Yao Xue Xue Bao 1989;24(8):568–72.

[536] Duan H, Takaishi Y, Momota H, Ohmoto Y, Taki T, Jia Y, et al. Immunosuppressive sesquiterpene alkaloids from *Tripterygium wilfordii*. J Nat Prod 2001;64:582–7.

[537] Ma J, Dey M, Yang H, Poulev A, Pouleva R, Dorn R, et al. Anti-inflammatory and immunosuppressive compounds from *Tripterygium wilfordii*. Phytochemistry 2007;68(8):1172–8.

[538] Akiyama H, Barger S, Barnum S, Bradt B, Bauer J, Cole GM, et al. Inflammation and Alzheimer's disease. Neurobiol Aging 2000;21(3):383–421.

[539] Rajendra S, Lynch J,W, Schofield PR. The glycine receptor. Pharmacol Ther 1997;73(2):121–46.

[540] Wang Q, Gong Q, Wu Q, Shi J. Neuroprotective effects of *Dendrobium* alkaloids on rat cortical neurons injured by oxygen-glucose deprivation and reperfusion. Phytomedicine 2010;17(2):108–15.

[541] Fu JJ, Qin JJ, Zeng Q, Jin HZ, Zhang WD. Chemical constituents of the aerial parts of *Aconitum kongboense*. Chem Nat Comp 2011;47(5):854–5.

[542] Ameri A. The effects of *Aconitum* alkaloids on the central nervous system. Prog Neurobiol 1998;56(2):211–35.

[543] Catterall WA. Neurotoxins that act on voltage-sensitive sodium channels in excitable membranes. Ann Rev Pharm Toxicol 1980;20:15–43.

[544] Valeev A,E, Verkhratskii AN, Dzhakhangirov FN. Effects of allapinine on sodium currents in neurons isolated from the rat trigeminal ganglion and cardiomyocytes. Neirofiziologiya 1990;22:201–6.

[545] Ameri A, Shi Q, Aschoff J, Peters T. Electrophysiological effects of aconitine in rat hippocampal slices. Neuropharmacology 1996;35:13–22.

[546] Ameri A. Structure-dependent differences in the effects of the *Aconitum* alkaloids lappaconitine, *N*-desacetyllappaconitine and lappaconidine in rat. Brain Res 1997;769(1):36–43.

[547] Benoit E, Escande D. Riluzole specifically blocks inactivated Na channels in myelinated nerve fiber. Pflugers Arch 1991;419(6):603–9.

[548] Taylor CP, Meldrum BS. Na^+ channels as targets for neuroprotective drugs. Trends Pharmacol Sci 1995;16(9):309–16.

[549] Hikino H, Takata H, Fujiwara M, Konno C, Ohuchi K. Mechanism of inhibitory action of mesaconitine in acute inflammations. Eur J Pharmacol 1982;82(1–2):65–71.

[550] Guo X, Tang XC. Effects of reserpine and 5-HT on analgesia induced by lappaconitine and N-deacetyllappaconitine. Zhongguo Yao Li Xue Bao 1990;11(1):14–18.

[551] Nesterova YV, Povetieva TN, Suslov NI, Semenov AA, Pushkarskiy SV. Antidepressant activity of diterpene alkaloids of *Aconitum baicalense* Turcz. Bull Exp Biol Med 2011;151(4):425–8.

[552] Liu Y, Zhang S, Zhou L, Liu L. The toxicity of aconitum alkaloids on cardiocytes and the progress of its research using the methods of molecular toxicology. Chin J Forensic Med 2009;24(6):399–401.

[553] Bardin L, Lavarenne J, Eschalier A. Serotonin receptor subtypes involved in the spinal antinociceptive effect of 5-HT in rats. Pain 2000;86(1–2):11–18.

[554] Barnes JM, Barnes NM, Costall B, Naylor RJ, Tyers MB. 5-HT3 receptors mediate inhibition of acetylcholine release in cortical tissue. Nature 1989;338:762–3.

[555] Zhao XY, Wang Y, Li Y, Chen XQ, Yang HH, Yue JM, et al. Songorine, a diterpenoid alkaloid of the genus *Aconitum*, is a novel GABA(A) receptor antagonist in rat brain. Neurosci Lett 2003;337(1):33–6.

[556] Oh MM, Power JM, Thompson LT, Moriearty PL, Disterhoft JF. Metrifonate increases neuronal excitability in CA1 pyramidal neurons from both young and aging rabbit hippocampus. J Neurosci 1999;19(5):1814–23.

[557] Free RB, von Fischer ND, Boyd RT, McKay DB. Pharmacological characterization of recombinant bovine alpha3beta4 neuronal nicotinic receptors stably expressed in HEK 293 cells. Neurosci Lett 2003;343(3):180–4.

[558] Paterson P, Nordberg A. Neuronal nicotinic receptors in the human brain. Prog Neurobiol 2000;61:75–111.

[559] Song MK, Liu H, Jiang HL, Yue JM, Hu GY, Chen HZ. Discovery of talatisamine as a novel specific blocker for the delayed rectifier K$^+$ channels in rat hippocampal neurons. Neuroscience 2008;155(2):469–75.

[560] Wang Y, Song M, Hou L, Yu Z, Chen H. The newly identified K$^+$ channel blocker talatisamine attenuates beta-amyloid oligomers induced neurotoxicity in cultured cortical neurons. Neurosci Lett 2012;518(2):122–7.

[561] Yang Q, Yan D, Wang Y. K$^+$ regulates DNA binding of transcription factors to control gene expression related to neuronal apoptosis. Neuroreport 2006;17(11):1199–204.

[562] Kolak U, Öztürk M, Özgökçe F, Ulubelen A. Norditerpene alkaloids from *Delphinium linearilobum* and antioxidant activity. Phytochemistry 2006;67(19):2170–5.

[563] Li L, Shen YM, Yang XS, Wu WL, Wang BG, Chen ZH, et al. Effects of spiramine T on antioxidant enzymatic activities and nitric oxide production in cerebral ischemia-reperfusion gerbils. Brain Res 2002;944(1–2):205–9.

[564] Li L, Shen YM, Yang XS, Zuo GY, Shen ZQ, Chen ZH, et al. Antiplatelet aggregation activity of diterpene alkaloids from *Spiraea japonica*. Eur J Pharmacol 2002;449(1–2):23–8.

[565] Bell RD, Zlokovic BV. Neurovascular mechanisms and blood-brain barrier disorder in Alzheimer's disease. Acta Neuropathol 2009;118(1):103–13.

[566] Borroni B, Akkawi N, Martini G, Colciaghi F, Prometti P, Rozzini L, et al. Microvascular damage and platelet abnormalities in early Alzheimer's disease. J Neurol Sci 2002;203–204:189–93.

[567] Xiang WJ, Ma L, Hu LH. C21 steroidal glycosides from *Cynanchum wilfordii*. Helvetica Chimica Acta 2009;92(12):2659–74.

[568] Jiang Y, Choi HG, Li Y, Park YM, Lee JH, Kim H, et al. Chemical constituents of *Cynanchum wilfordii* and the chemotaxonomy of two species of the family Asclepiadaceae, *C. wilfordii* and *C. auriculatum*. Arch Pharm Res 2011;34(12):2021–7.

[569] Atta-ur-Rahman ZH, Khalid A, Anjum S, Khan MR, Choudhary MI. Pregnane-type steroidal alkaloids of *Sarcococca saligna*: a new class of cholinesterase inhibitors. Helv Chim Acta 2002;85(2):678–88.

[570] Atta-Ur-Rahman ZH, Feroz F, Khalid A, Nawaz SA, Khan MR, et al. New cholinesterase-inhibiting steroidal alkaloids from *Sarcococca saligna*. Helv Chim Acta 2004;87(2):439–48.

[571] Atta-Ur-Rahman ZH, Feroz F, Naeem I, Nawaz SA, Khan N, et al. New pregnane-type steroidal alkaloids from *Sarcococca saligna* and their cholinesterase inhibitory activity. Steroids 2004;69(11–12):735–41.

[572] Khalid A, Ghayur MN, Feroz F, Atta-Ur-Rahman ZH, Gilani AH, et al. Cholinesterase inhibitory and spasmolytic potential of steroidal alkaloids. J Steroid Biochem Mol Biol 2004;92(5):477–84.

[573] Atta-ur-Raman ZH, Wellenzohn B, Tonmunphean S, Khalid A, Choudhary MI, Rode BM. 3D-QSAR Studies on natural acetylcholinesterase inhibitors of *Sarcococca saligna* by comparative molecular field analysis (CoMFA). Bioorg Med Chem Lett 2003;13(24):4375–80.

[574] Devkota KP, Lenta BN, Wansi JD, Choudhary MI, Kisangau DP, Naz Q, et al. Bioactive 5alpha-pregnane-type steroidal alkaloids from *Sarcococca hookeriana*. J Nat Prod 2008;71(8):1481–4.

[575] Devkota KP, Lenta BN, Choudhary MI, Naz Q, Fekam FB, Rosenthal PJ, et al. Cholinesterase inhibiting and antiplasmodial steroidal alkaloids from *Sarcococca hookeriana*. Chem Pharm Bull 2007;55(9):1397–401.

[576] Kalauni SK, Choudhary MI, Khalid A, Manandhar MD, Shaheen F, Atta-ur-Rahman ZH. New cholinesterase inhibiting steroidal alkaloids from the leaves of *Sarcococca coriacea* of Nepalese origin. Chem Pharm Bull 2002;50(11):1423–6.

[577] Khalid A, Azim MK, Parveen S, Atta-ur-Rahman ZH, Choudhary MI. Structural basis of acetylcholinesterase inhibition by triterpenoidal alkaloids. Biochem Biophys Res Commun 2005;331(4):1528–32.

[578] Babar ZU, Ata A, Meshkatalsadat MH. New bioactive steroidal alkaloids from *Buxus hyrcana*. Steroids 2006;71(13–14):1045–51.

[579] Orhan IE, Khan MTH, Erdem SA, Kartal M, Şener B. Selective cholinesterase inhibitors from Buxus sempervirens L. and their molecular docking studies. Curr Comput Aided Drug Des 2011;7(4):276–86.

[580] Atta-ur-Rahman ZH, Akhtar MN, Choudhary MI, Tsuda Y, Sener B, Khalid A, et al. New steroidal alkaloids from *Fritillaria imperialis* and their cholinesterase inhibiting activities. Chem Pharm Bull 2002;50(8):1013–16.

[581] Yang Z-D, Duan D-Z, Xue W-W, Yao X-J, Li S. Steroidal alkaloids from *Holarrhena antidysenterica* as acetylcholinesterase inhibitors and the investigation for structure-activity relationships. Life Sci 2012;90(23–24):929–33.

[582] Chang LC, Bhat KPL, Fong HHS, Pezzuto JM, Kinghorn AD. Novel bioactive steroidal alkaloids from *Pachysandra procumbens*. Tetrahedron 2000;56(20):3133–8.

[583] Li PK, Rhodes ME, Jagannathan S, Johnson DA. Reversal of scopolamine induced amnesia in rats by the steroid sulfatase inhibitor estrone-3-O-sulfamate. Brain Res Cogn Brain Res 1995;2(4):251–4.

[584] Lee D-U, Shin U-S, Huh K. Inhibitory effects of gagaminine, a steroidal alkaloid from *Cynanchum wilfordi*, on lipid peroxidation and aldehyde oxidase activity. Planta Med 1996;62(6):485–7.

[585] Lee DU, Shin US, Huh K. Structure-activity relationships of gagaminine and its derivatives on the inhibition of hepatic aldehyde oxidase activity and lipid peroxidation. Arch Pharm Res 1998;21(3):273–7.

[586] Lee D-U, Lee W-C. A 5-lipoxygenase inhibitor isolated from the roots of *Cynanchum wilfordi* Hemsley. Kor J Pharmacogn 1997;28(4):247–51.

[587] Mesaik MA, Halim SA, Ul-Haq Z, Choudhary MI, Shahnaz S, Ayatollahi SAM, et al. Immunosuppressive activity of buxidin and E-buxenone from *Buxus hyrcana*. Chem Biol Drug Design 2010;75(3):310–17.

[588] Zhang QJ, Zheng ZF, Yu DQ. Steroidal alkaloids from the bulbs of *Fritillaria unibracteata*. J Asian Nat Prod Res 2011;13(12):1098–103.

[589] An J-J, Zhou J-L, Li H-J, Jiang Y, Li P. Puqienine E: an angiotensin converting enzyme inhibitory steroidal alkaloid from *Fritillaria puqiensis*. Fitoterapia 2010;81(3):149–52.

[590] Zhao C, Sun M, Bennani YL, Gopalakrishnan SM, Witte DG, Miller TR, et al. The alkaloid conessine and analogues as potent histamine H3 receptor antagonists. J Med Chem 2008;51(17):5423–30.

[591] Esbenshade TA, Browman KE, Miller TR, Krueger KM, Komater-Roderwald V, Zhang M, et al. Pharmacological properties and procognitive effects of ABT-288, a potent and selective histamine H3 receptor antagonist. J Pharmacol Exp Ther 2012;343(1):233–45.

[592] Bordet T, Buisson B, Michaud M, Drouot C, Galéa P, Delaage P, et al. Identification and characterization of cholest-4-en-3-one, oxime (TRO19622), a novel drug candidate for amyotrophic lateral sclerosis. J Pharmacol Exp Ther 2007;322(2):709–20.

[593] Bordet T, Berna P, Abitbol JL, Pruss RM. Olesoxime (TRO19622), a novel mitochondrial-targeted neuroprotective compound. Pharmaceuticals 2010;3:345–68.

[594] Sunyach C, Michaud M, Arnoux T, Bernard-Marissal N, Aebischer J, Latyszenok V, et al. Olesoxime delays muscle denervation, astrogliosis, microglial activation and motoneuron death in an ALS mouse model. Neuropharmacology 2012;62(7):2346–52.

[595] Hovhannisyan A, Matz M, Gebhardt R. From teratogens to potential therapeutics: natural inhibitors of the hedgehog signaling network come of age. Planta Med 2009;75(13):1371–80.

[596] Winkler JD, Isaacs AK, Xiang C, Baubet V, Dahmane N. Design, synthesis, and biological evaluation of estrone-derived hedgehog signaling inhibitors. Tetrahedron 2011;67(52):10261–6.

[597] Reilly JO, Karavanova ID, Williams KP, Mahanthappa NK, Allendoerfer KL. Cooperative effects of Sonic Hedgehog and NGF on basal forebrain cholinergic neurons. Mol Cell Neurosci 2002;19(1):88–96.

Chapter 2

Terpenes

■ INTRODUCTION

Microglial cells form the praetorium of the brain, which account for the clearance of necrotic neurons and debris in the parenchyma via phagocytosis. However, when microglial cells are chronically activated by β-amyloid peptides during Alzheimer's disease (AD) or α-synuclein during Parkinson's disease (PD), these cells release cytokines which are neurotoxic. Therefore, any agents able to assuage neuro-inflammation may be first line candidates for the treatment of AD, PD and Amyotrophic Lateral Sclerosis (ALS) and such agents may very well be found among the terpenes engineered by medicinal plants such citronellol from *Pelargonium graveolens* L'Hér. ex Aiton (2.1), eremanthine from *Laurus nobilis* L. (2.2), 15-methoxypinusolidic acid from *Biota orientalis* (L.) Endl. (family Cupressaceae Gray) (2.3), ajugalide D from *Ajuga ciliata* Bunge (2.3), carnosol from by *Rosmarinus officinalis* L. (2.3), triptolide from *Tripterygium wilfordii* Hook. f. (2.3), and asiatic acid from *Centella asiatica* (L.) Urb. (2.4). The ability of terpenes to go from medicinal plant to being able to counter neuro-inflammation often rests on the inhibition of nuclear factor kappa-light-chain-enhancer of activated B cells (NF–κB) which command the transcription of cyclo-oxygenase-2 (COX-2) and inducible nitric oxide synthetase (iNOS); the enzymatic activity of which results in the production of prostaglandian E2 (PGE2) and nitric oxide (NO). The blockade of nuclear factor NF–κB results from the hypophosphorylation of nuclear factor of kappa light polypeptide gene enhancer in B-cells inhibitor α (IκBα) as upstream effectors such as the kappa light polypeptide gene enhancer in B-cells inhibitor (IκB) kinase (IKK) which are targeted by terpenes.

Additionally, terpenes from medicinal plants have the competency, within what could be termed a "protective window dose", to protect neurons against β-amyloid peptide, glutamate, NO, oxygen and glucose deprivation and other toxic stimuli by hindering apoptosis by targeting numerous kinases, boosting the clearance of reactive oxygen species (ROS) and protecting the mitochondrial intergrity as described in this chapter. Examples of neuroprotective terpenes from medicinal plants are thymoquinone from *Monarda didyma* L. (2.1),

C. Wiart: Lead Compounds from Medicinal Plants for the Treatment of Neurodegenerative Diseases.
DOI: http://dx.doi.org/10.1016/B978-0-12-398373-2.00002-9

bilobalide from *Ginkgo biloba* L. (2.2), isoatriplicolide tiglate from *Paulownia tomentosa* (Thunb.) Steud. (2.2), spirafolide from *Laurus nobilis* L. (2.2), tanshinone IIA from *Salvia miltiorrhiza* Bunge (2.3), asiatic acid from *Centella asiatica* (L.) Urb. (2.4)., deoxygedunin from *Azadirachta indica* A. Juss. (2.4) and obacunone from *Dictamnus dasycarpus* Turcz. (2.4). Another interesting feature of terpenes described in this chapter is their compelling ability, within a "trophic window zone", to induce the sprouting of neurites and even the growth of neurons, thus suggesting a possibility to put an end to ALS, to treat spinal injuries and to replenish neuronal population in loci of neurodegeneration. In light of this, Chapter 3 proposes several plant terpenes for thymoquinone, (2R)-hydroxyneomajucin, ajugaciliatin B, and 11-keto-β-boswellic acid are sources of synthetic derivatives alkaloids for the treatment of these neurodegenerative diseases.

Topic **2.1**

Monoterpenes

2.1.1 *Monarda didyma* **L.**

History The plant was first described by Carl von Linnaeus in *Species Plantarum* published in 1753.

Family Lamiaceae Martinov, 1820

Common Name Mei guo bo he (Chinese)

Habitat and Description It is an aromatic annual herb which grows in China. The stem is quadrangular and sparsely hairy. The leaves are simple, decussate, and exstipulate. The petiole is 2.5 cm long. The blade is broadly lanceolate, serrate, dark green above, thin, 5–10 cm × 2–4 cm, acuminate at the apex, and it presents 6–10 pairs of secondary nerves arching at the margin. The inflorescence is a terminal and showy verticillaster, which is 12 cm across, resembling a primitive capitulum. The calyx is tubular, 1 cm long, and presents 5 tiny lobes. The corolla is purple and bilobed; the upper lobe is straight, and the lower is 3-lobed. The androecium consists of 2 stamens attached to the inner wall of the corolla. The gynoecium is ovoid, minute, and comprises 4 lodges and develops a slender style that is protruding and bifid. The fruit consists of seeds packed into the remnant calyx (Figure 2.1).

■ **FIGURE 2.1** *Monarda didyma* L.

Medicinal Uses In China, the plant is used to facilitate digestion.

Phytopharmacology The plant produces an essential oil comprising numerous monoterpenes, including thymol, γ-terpinene, p-cymene, δ-3-carene, myrcene,[1] thymohydroquinone, and thymoquinone.[2]

Proposed Research Pharmacological study of thymoquinone and derivatives for the treatment of Parkinson's disease (PD).

Rationale It has been commonly accepted that monoterpenes are mere fragrant constituents of essential oils from medicinal plants with applications restricted to perfumery and the food industry. One could argue that their chemical framework is too simple for any robust neuropharmacological activity, but there is compelling data evidence to the contrary: monoterpenes freely cross the blood brain barrier by virtue of their small size and lipophicity and in the brain command neuroprotection by notably stimulating the γ-amino butyric acid (GABA) system and confronting neuroinflammation.

■ **CS 2.1** (−)-citronellol.

■ **CS 2.2** (+)-citronellal.

In fact, the simplest scaffold *id est* acyclic monoterpenoids produce small lipophilic agents with surprising neuropharmacological activities. This is, for instance, exemplified with citronellol (CS 2.1) and citronellal (CS 2.2) from *Melissa officinalis* L. (family Lamiaceae Martinov), which at a dose of 0.5 μM potentiated the effect of GABA on γ-amino butyric acid receptors subtype A (GABA$_A$) expressed by oocyte by 224% and 301%, respectively.[3]

It is worthwhile to emphasize that GABA receptor agonists are able to protect neurons by activating protein kinase B (Akt), hence phosphorylation and inactivation of ASK1 and consequently a decrease of mitogen-activated protein kinase (MAPK) kinase 1 (MKK4) and mitogen-activated protein kinase (MAPK) kinase 7 (MKK7) and therefore neuron survival via c-Jun N-terminal kinase (JNK) and c-Jun inhibition.[4] Hence, monoterpenes with GABA properties may very well be regarded as neuroprotective agents. The GABA-ergic property of citronellol was further shown whereby at a dose of 400 mg/kg this acyclic monoterpene delayed but did not block experimental seizures in rodents induced by paralyzing neuronal conductivity with an IC$_{50}$ of 2.2 μM, suggesting the possible inhibition of voltage-dependent Na$^+$ channels and the involvement of GABAA positive modulation.[5] *Pelargonium graveolens* L'Hér. ex Aiton (family Geraniaceae Juss.) produces citronellol, which assuaged the pain induced by prostaglandin E2 (PGE2) in rodents by 8%[6] by probably limiting the release of cytokines as shown at a dose of 40 μg/mL against microglial cells challenged with lipopolysaccharide (LPS) *in vitro* by 88%.[7] The anti-inflammatory property of citronellol was further shown because 750 μM of this acyclic monoterpene nullified the generation of nitric oxide (NO) by macrophages challenged with LPS by direct interaction with inducible nitric oxide synthetase (iNOS).[8] At a dose of 50 μM, citronellol reduced the expression of cyclo-oxygenase-2 (COX-2) via inhibition of nuclear factor kappa-light-chain-enhancer of activated B cells (NF-κB).[8] Likewise, citronellol at a dose of 400 μM activated peroxisome proliferator-activated receptor-γ (PPAR-γ) and therefore repressed the expression of COX-2 by 51% in BAEC cells challenged with LPS.[9] Citronellal at a dose of 200 mg/kg elicited central nociception via opioid receptors in rodents in the hot-plate test.[10] Opioid receptor stimulation may therefore account for the analgesic activity of citronellol at a dose of 100 mg/kg against formalin, capsaicin, and glutamate in rodents.[11] Note that transient receptor potential vanilloid subtype 1 (TRPV1) was activated by citronellol with EC$_{50}$ values equal to 43 μM,[12] leading to the suggestion that citronellol synthetic derivatives may be antagonists and therefore of value for the treatment of PD. One potential explanation for the neuroapoptotic effect imparted by TRPV1 agonists is an increase of free cytoplasmic Ca^{2+} in dopaminergic neurons, which causes mitochondrial insults, release of cytochrome c, and activation of caspase 9 and 3.[13] Note that citronellol at a dose of 0.1 M relaxed intact and endothelium-denuded rings of mesenteric arteries contracted by phenylephrine or KCl, suggesting a mechanism involving the blockade of voltage-dependent Ca^{2+} channels.[14]

Removal of the hydroxyl group of citronellol in C1, unsaturation between C1 and C2, and insertion of a hydroxyl moiety in C3 produce the acyclic monoterpene linalool, which abounds in the essential oil of the spice *Coriandrum sativum* L. (+)-Linalool (CS 2.3) and (−)-linalool at a dose of 5 μM nullified the binding of L-[3H]glutamate to cortical membranes *in vitro*, at a dose of 350 mg/kg delayed the episodes of seizures induced by NMDA, and at dose of 45 μM abated the seizures induced by the NMDA agonist quinolinic acid, suggesting antagonistic properties against ionotropic glutamate receptor.[15] In parallel, 1 μM of linalool magnified the potencies of GABA against GABA$_A$ by 350%.[16] Note that the activation of NMDA results in a massive influx of Ca^{2+} in neurons, resulting in calpain and caspase activation, a burst of reactive oxygen species (ROS), activation of pro-apoptotic Bcl-2-associated X protein (Bax), mitochondrial insults, and apoptosis.[17] By repressing glutamatergic neurotransmission and facilitating GABA-ergic neurotransmission, linalool is likely to fight neurodegeneration. The Asian culinary lemon grass *Cymbopogon citratus* (DC.) Stapf (family Poaceae Barnhart) produces linalool, which reduced prostaglandine E2 (PGE2)-induced hyperalgesia by 10%[6] via the probable reduction of NO and PGE2 and inhibition of COX-2 as shown at a dose of 10^{-3} M in macrophages challenged with LPS.[18] (−)-Linalool at a dose of 200 mg/kg attenuated glutamate-induced nociception in mice hind paw with an ID$_{50}$ value of 139.1 mg/kg by inhibiting NMDA and transient receptor potential A1 (TRPA1).[19,20] Note that TRPA1 decreases the levels of Ca^{2+} in hippocampal astrocytes and therefore reduces GABA transport by GABA transporters (GAT-3), resulting in elevated levels of synaptic GABA.[21] *Aniba rosaeodora* Ducke (family Lauraceae Juss.) produces (−)-linalool, which diminished the levels of cyclic adenosine monophosphate (cAMP) in isolated chick retina induced by forskolin with an IC$_{50}$ value equal to 310 μM.[22] (−)-Linalool at a dose of 100 mg/kg elicited antinociceptive effects in the hot-plate test via stimulation of the adenosine receptors subtype 1 (A$_1$) and adenosine receptors subtype 2 (A$_2$) because analgesia was abrogated 1,3-dipropyl-8-cyclopentylxanthine and 3,7-dimethyl-1-propargilxanthine.[23] Note that the binding of adenosine to presynaptic adenosine receptors subtype 1 (A$_1$) compels a collapse in cAMP and consequently the inactivation of voltage dependent Ca^{2+} channels, hence lowered secretion of acetylcholine.[24] Paradoxically, the activation of adenosine receptors subtype 2 (A$_2$)

■ CS 2.3 (−)-linalool.

■ **CS 2.4** Myrcene.

■ **CS 2.5** Geranial.

■ **CS 2.6** Neral.

increases the release of acetylcholine in the prefrontal cortex.[25] At a dose of 10 μM, linalyl acetate *Salvia libanotica* Boiss. & Gaill. (family Lamiaceae Martinov) abated the activity of NF-κB induced by tumor necrosis factor-α (TNF-α) in malignant cells.[26]

Removal of the hydroxyl group of linalool in C3 and unsaturation of C3–C10 of linalool produce β-myrcene (CS 2.4) from *Cymbopogon citratus* (DC.) Stapf (family Poaceae Barnhart), which elicited analgesic activity in the hot-plate test at a dose of 20 mg/kg by release of endogenous opioids.[27] Furthermore β-myrcene reduced at a dose of 90 mg/kg PGE2-induced rodent paw hyperalgesia by 88%[6] and protected rodents against the pain inflicted by plantar injection of PGE2.[28] This analgesia was negated by N^G-monomethyl-L-arginine, implying the direct involvement of NO synthesis and subsequent activation of guanylate cyclase (GC).[28] In neurophysiological conditions, the conversion of L-arginine into L-citrulline by nitric oxide synthase (NOS) yields NO,[29] hence activation of guanylyl cyclase (GC) and release of cyclic guanosine 3,5'-monophosphate (cGMP).[30] In neurons, cyclic guanosine 3,5'-monophosphate (cGMP) induces protein kinase G (PKG), which stimulates K$^+$ channels, hence the efflux of K$^+$ out of the cell.[31] Note that low concentrations of NO surprisingly produce neuroprotection via the stimulation of PKG' which stimulates mitochondrial K$^+_{ATP}$ channels, hence mitochondrial depolarization, a brief elevation of free cytoplasmic Ca^{2+}, and phosphorylation of protein kinase C (PKC).[32]

The unsaturation of C2-C3 of citronellal produces citral, which consists of two isomers, *id sunt* geranial (CS 2.5) (trans-citral) and neral (CS 2.6) (cis-citral). α-Citral and β-citral reduced PGE2-induced hyperalgesia by 41% and 40%, respectively.[6] In fact, citral from *Cymbopogon citratus* (DC.) Stapf (family Poaceae Barnhart) at dose of 12 μg/mL blocked the generation of NO by macrophages (RAW 264.7) challenged with LPS by decreasing the expression of iNOS as a result of NF-κB inhibition.[33] Likewise, citral decreased the expression of COX-2 by 70% at a dose of 200 μM via peroxisome proliferator-activated receptor-α (PPAR-α) and PPAR-γ activation in BAEC cells challenged with LPS.[34] Furthermore, citral is a partial agonist of TRPV1[35] and by structural modification may produce antagonists with neuroprotective properties. α−Citral displayed estrogenic properties in the estrogen-inducible yeast screen and displaced ^{3}H-estradiol from neuronal estrogen receptors α and β at doses of 10^7 nM.[36] Note that 17β-estradiol at a dose of 10 nM activated (PKC in cortical neurons, providing protection against β-amyloid (Aβ_{25-35}) peptide via increased levels of Bcl-2 and activation of extracellular signal-regulated kinase (ERK1/2),[37] implying that estrogens are of tremendous importance in the pathophysiology of neurodegenerative diseases.

The reduction of C1 aldehyde of geranial produces geraniol (CS 2.7), which notably occurs in the essential oil of *Ocimum basilicum* L. (family Lamiaceae Martinov). Geraniol protected bovine liposomes against Fe^{3-} ascorbate-induced peroxidation by 29% at dose of $100 \mu M$.[38] Geraniol at dose of $750 \mu M$ nullified the generation of NO by macrophages challenged with LPS by lowering the expression of iNOS, and at a dose of $50 \mu M$ inhibited the levels of COX-2 via inhibition of NF-κB.[39] The anti-inflammatory potency of geraniol was further demonstrated as at a dose of $400 \mu M$ it induced activated (PPAR-γ) and therefore negated the expression of COX-2 by 44% in bovine aortic endothelial cells challenged with LPS.[9] TRPV1 was activated by geraniol with an EC_{50} value equal to $102 \mu M$.[12] At a dose of $1 \mu M$, geraniol increased the potencies of GABA on $GABA_A$ by 350%.[16] Like the aforementioned α-citral, geraniol at a dose of $2 \mu M$ exhibited estrogenic potencies against yeasts expressing human estrogen receptors[33] and displayed estrogenic in the estrogen-inducible yeast screen and displaced ^{3}H-estradiol from estrogen receptors α and β at doses of 10^7 nM.[36]

The intramolecular cyclization of monoterpenes by the formation of a covalent bond between C1 and C6 within the acyclic monoterpene scaffold produce the monocyclic menthane monoterpenoids, including (+)-limonene (CS 2.8) from *Citrus sinensis* (L.) Osbeck (family Rutaceae Juss.), which at a dose of $25 \mu M$ stimulated adenosine receptor subtype 2 (A_2) expressed by Chinese hamster ovary (CHO) cells. Hence, there was an increase in cAMP, protein kinase A (PKA), and subsequent stimulation of cAMP response element binding protein (CREB) and a transient increase in cytoplasmic Ca^{2+} via an opening of the voltage-sensitive calcium channel (VSCC),[40] implying neuroprotective and neurotrophic potencies. *Cymbopogon citratus* (DC.) Stapf (family Poaceae Barnhart) produces (+)-limonene and (−)-limonene (CS 2.9), which reduced at a dose of 90 mg/kg PGE2-induced rodent paw hyperalgesia by 17%.[6]

Hydroxylation and oxidation of C4 on (−)-limonene produce (−)-carvone (CS 2.10) from *Mentha spicata* L. (family Lamiaceae Martinov), which at a dose of 200 mg/kg protected rodent against the pain inflicted by acetic acid and formalin via a non-opioid central mechanism probably involving the blockade of voltage-dependent Na^+ channels.[41] α,β-Epoxy-carvone (CS 2.11) from *Carum carvi* L. (family Apiaceae Lindl.) at a dose of 400 mg/kg protected rodents against the seizures induced by pentylenetetrazol (PTZ) by 87.5% via the inhibition of the voltage-dependent Na^+ channel.[42] The reduction of Δ_{7-8} and the formation of Δ_{4-5} on (+)-limonene forms α-phellandrene (CS 2.12) from *Zingiber officinale* Roscoe (family Zingiberaceae Martinov), which displayed analgesic in rodents against acetic acid, formalin, capsaicin, glutamate, and carrageenan at doses equal

■ **CS 2.7** Geraniol.

■ **CS 2.8** (+)-limonene.

■ **CS 2.9** (−)-limonene.

■ **CS 2.10** (−)-carvone.

■ **CS 2.11** Epoxy-carvone.

■ **CS 2.12** R-(-)-α-phellandrene.

■ **CS 2.13** R-(+)-α-terpineol.

■ **CS 2.14** γ-terpinene.

■ **CS 2.15** (+)-terpinen-4-ol.

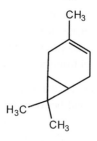

■ **CS 2.16** (+)-3-carene.

to 12.5 mg/kg, 50 mg/kg, 12.5 mg/kg, 25 mg/kg, and 50 mg/kg via opioid receptors.[43] Hydroxylation of (+)-limonene in C7 produces α-terpineol (CS 2.13) from *Salvia libanotica* Boiss. & Gaill. (family Lamiaceae Martinov), which inhibited the enzymatic activity of cyclo-oxygenase (COX) with an IC_{50} value equal to $0.6\,\mu M$[44] and nullified the activity of NF-κB induced by TNF-α at a dose of $10\,\mu M$ in malignant cells.[45] Furthermore, α-terpineol at a dose of $1\,\mu M$ enhanced the potencies of GABA against $GABA_A$ expressed by oocytes by 199%.[46] From *Nigella sativa* L. (family Ranunculaceae Juss.), γ-terpinene (CS 2.14) and terpinen-4-ol (CS 2.15) at a dose of $25\,\mu M$ inhibited the secretion of NO from macrophage (RAW 264.7) challenged with LPS by 28.5% and 20.8%, respectively.[47]

Removal of the hydroxyl group in C7 followed by the aromatization of α-terpinene produces p-cymene which occur in the fixed oil expressed from the aromatic seeds of the spice *Nigella sativa* L. (family Ranunculaceae Juss.), which at a dose of $25\,\mu M$ inhibited the secretion of NO from macrophage (RAW 264.7) with LPS by 32.3%.[47] Formation of a covalent bond in C5 forms the bicyclic monoterpene 3-carene (CS 2.16) from members of the genus *Salvia* L. (family Lamiaceae Martinov), which inhibited the enzymatic activity of acetylcholinesterase (AChE).[48]

p-Cymene (CS 2.17) hydroxylated in C1 gives thymol (CS 2.18), and in C2 yields carvacrol (CS 2.19), which occurs in the essential oil of *Thymus vulgaris* L. (family Lamiaceae Martinov). This aromatic monoterpene activated transient receptor potential vanilloid subtype 3 (TRPV3) expressed by oocytes with EC_{50} equal 0.4 μM.[49] This finding supports the notion that this monoterpene may mitigate the toxic effect of β-amyloid peptide by blocking NO-induced entry of Ca $^{2+}$ into cholinergic neurons.[50] Carvacrol inhibited production of PGE2 catalyzed by COX-2 with an IC_{50} value of 0.8 μM.[51] In fact, carvacrol at a dose of 1 μM from thyme oil inhibited COX-2 promoter activity in bovine aortic endothelial cells challenged with LPS by 80%.[52] From *Nigella sativa* L. (family Ranunculaceae Juss.), carvacrol at a dose of 25 μM inhibited the secretion of NO from RAW 264.7 challenged with LPS by 35.1%.[47] At the same dose, carvacrol activated PPAR-α and PPAR-γ.[52] Carvacrol at a dose of 100 mg/kg protected rodents against the pain incurred by subplantar injection of carrageenan by reducing the secretion of TNF-α, and *in vitro* 100 μg/mL of carvacrol suppressed the production of NO.[53] Thymol inhibited L-type Ca $^{2+}$ channels and K^+ channels expressed by myocytes with EC_{50} values equal to 158 μM and 60.6 μM.[54] Additionally, thymol exhibited positive $GABA_A$ receptor modulator activities as it favored the binding of [^{3}H]flunitrazepam with an EC_{50} equal to 130.9 μM in cortical neurons.[55] Thymol TRPV3 receptor expressed by *Xenopus* oocytes with EC_{50} equal to 0.8 mM.[49]

Hydroxylation of thymol in C4 followed by oxidation produces thymoquinone (CS 2.20), which occurs in the fixed oil of *Nigella sativa* L (family Ranunculaceae Juss.), for which a conflicting body of evidence exists concerning its neuroprotective roles. Thymoquinone at a dose of 10 mg/kg protected rodents against the pain inflicted by formalin via stimulation of opioid receptor subtypes μ and κ.[56] Thymoquinone at a dose of 1 μM nullified the contractions of pulmonary arterial rings challenged with phenylephrine or serotonine.[57] Thymoquinone at a dose of 5 mg/kg increased the survival of hippocampal CA1 neurons following experimental ischemia in rodents with a decrease in lipid peroxidation, an increase of glutathione (GSH), and an increase in the enzymatic activities of superoxide dismutase (SOD) and catalase (CAT).[58] However, thymoquinone inhibited the activation of ERK1/2 and Akt by vascular endothelial growth factor (VEGF) at a dose of 40 nmol/L.[59] Thymoquinone at a dose of 25 μmol/L suppressed NF-κB activation induced by TNF-α, LPS, okadaic acid, or phorbol myristate acetate by negating kappa light polypeptide gene enhancer in B-cells inhibitor (IκB) kinase (IKK) and consequently the phosphorylation of nuclear factor of kappa light polypeptide gene enhancer in B-cells

■ **CS 2.17** p-cymene.

■ **CS 2.18** Thymol.

■ **CS 2.19** Carvacrol.

■ **CS 2.20** Thymoquinone.

■ **CS 2.21** (−)-menthol.

■ **CS 2.22** (-)-isopulegol.

■ **CS 2.23** α-pinene.

■ **CS 2.24** β-pinene.

inhibitor α (IκBα) in malignant cells.[60] In addition, thymoquinone at a dose of 25 μM inhibited the secretion of NO from RAW 264.7 challenged with LPS by 95%.[47] Of note, thymoquinone at a dose of 20 mg/kg displayed anxiolytic activity in rodent by increasing GABA and lowering NO production.[61] Thymoquinone at dose of 1 μM protected dopaminergic neurons cultured *in vitro* against 1-methyl-4-phenylpyridinium ion (MPP$^+$) and rotenone-induced apoptosis by 25% and 79%, respectively.[62] Thymoquinone at a dose of 25 μM protected cortical neurons again ethanol insults by preventing the increase of cytoplasmic Ca^{2+}, preserving mitochondrial integrity, increasing the levels of anti-apoptotic Bcl-2, while decreasing pro-apoptotic Bcl-2-associated X protein (Bax), and inactivating caspase 3.[63]

Reduction of Δ_{1-2} and Δ_{8-10} and hydroxylation of C3 in beta of (+)-limonene yield (−)-menthol (CS 2.21) found in *Menthas piperita* L. (family Lamiaceae Martinov). (−)Menthol at a dose of 0.5 μM blocked Ca $^{2+}$ channel current *in vitro*.[64] Menthol repressed the secretion of leukotriene B4 (LTB4), PGE2, and Interleukin-1β (IL-1β) by monocytes challenged with LPS.[65] Menthol at 100 μM elicited robust positive modulation of GABA$_A$.[56,66] In addition, menthol at a dose of 100 μM enhanced glycine receptor.[56] Menthol at a dose of 30 mg/kg incurred analgesia to rodents subjected to the hot-plate test and acetic acid abdominal constriction test by stimulating opioid receptors subtype κ.[67] The formation of Δ_{8-10} on (−)-menthol forms isopulegol (CS 2.22), which occurs in the essential oil of *Eucalyptus citriodora* Hook. (family Myrtaceae Juss.), and at a dose of 200 mg/kg delayed the seizures induced by pentylenetetrazol (PTZ) in rodents with a decrease in lipid peroxidation, retention of catalase activity, and an increase in GSH via GABA$_A$ activation.[68] Furthermore, isopulegol at a dose 300 μM, enhanced the potencies of GABA against GABA$_A$ expressed by oocytes by 280%.[46]

The removal of C8 followed by the formation of a covalent bond with C2 on α-terpineol yields the bicyclic monoterpenes α-pinene (CS 2.23) and β-pinene (CS 2.24). α-Pinene from *Salvia lavandulaefolia* inhibited the enzymatic activity of AChE with IC$_{50}$ values equal to 0.6 μM.[69] Likewise, α-pinene and β-pinene from members of the genus *Salvia* L. (family Lamiaceae Martinov) inhibited the enzymatic activity of AChE.[48] α-Pinene and β-pinene from *Salvia lavandulifolia* Vahl (family Lamiaceae Martinov) protected bovine liposomes against Fe $^{3+}$-ascorbate–induced peroxidation by 67% and 38%, respectively, at a dose of 100 μM.[38] β-Pinene elicited agonistic properties toward opioid receptor subtype μ-opioid in rodents,[70] suggesting neuroprotective potencies. From *Nigella sativa* L. (family Ranunculaceae Juss.), β-pinene at a dose of 25 μM inhibited the secretion of NO from RAW 264.7 challenged with LPS by 33.8%.[47] Being opioid agonist and anti-inflammatory, β-pinene might deserve further attention in light of its neurological benefits.

Removal of the hydroxyl moiety in C7 followed by the formation of a covalent bond C3–C7 and addition of a hydroxyl group in C4 and reduction of Δ_{2-3} on α-terpineol produces the bicyclic monoterpene borneol, which occurs in the resin of the Southeast Asian timber *Dryobalanops aromatica* C.F. Gaertn. (family Dipterocarpaceae Blume). Of note, (+)-borneol (CS 2.25), (−)-borneol (CS 2.26), and (−)-bornyl acetate (CS 2.27) exhibited positive modulator activities against $GABA_A$ expressed by oocytes.[71] In particular, (+)-borneol and (−)-borneol with EC_{50} equal to $248\,\mu M$ and $237\,\mu M$ boosted the activity of $GABA$[40] and displayed $GABA_A$ agonist activities.[40] (+)-Borneol activated TRPV3 expressed by oocytes with EC_{50} equal to $3.4\,\mu M$.[49] Borneol from *Thymus vulgaris* L. (family Lamiaceae Martinov) inhibited the enzymatic activity of COX-2 promoter activity in bovine aortic endothelial cells challenged with LPS by 55%[52] and inhibited COX-2 promoter activity in bovine aortic endothelial cells challenged with LPS by 55%.[52] Borneol at a dose of $0.3\,\mu M$ protected cortical neurons *in vitro* against deprivation of oxygen and glucose by reduction of ROS, iNOS, NO, inactivation of caspase 3 and 9, hindrance of nuclear factor of kappa light polypeptide gene enhancer in B-cells inhibitor α ($I\kappa B\alpha$) degradation, and therefore the inhibition of NF-κB.[72]

Oxidation of the hydroxyl group in C4 of borneol provides camphor, which at a dose of $300\,\mu M$ nullified the 1,1-dimethyl-4-phenylpiperaziniumiodide (DMPP)-induced secretion of [³H]norepinephrine by chromaffin cells via inhibition of intracellular raise of Ca^{2+} with an IC_{50} value equal to $88\,\mu M$.[73] In addition, camphor inhibited Na$^+$ increase induced by 1,1-dimethyl-4-phenylpiperazinium iodide (DMPP) with an IC_{50} of $19\,\mu M$ by antagonizing nicotinic receptors.[73] Camphor from *Salvia lavandulifolia* Vahl (family Lamiaceae Martinov) mitigated the oxidation of bovine liposomes against Fe^{3+}-ascorbate peroxidation at a dose of $100\,\mu M$.[38] Camphor exhibited positive modulator activity against $GABA_A$ expressed by oocytes.[71] (+)-Camphor (CS 2.28) and (−)-camphor and, to a lesser extent, borneol at $10\,\mu M$ activated the TRPV1 expressed by human embryonic kidney (HEK) cells.[74] Camphor activated TRPV3 expressed by oocytes with EC_{50} equal to $6\,\mu M$.[49]

The formation of an ether C3–C7 on α-terpineol followed by a reduction of Δ_{2-3} yields 1,8-cineole (CS 2.29), which is a major constituent of members of the genus *Eucalyptus* L'Hér. (family Myrtaceae Juss.) and *Salvia* L. (family Lamiaceae Martinov). This epoxide monoterpene from the essential oil of *Salvia lavandulifolia* Vahl (family Lamiaceae Martinov) inhibited the enzymatic activity of AChE.[69] 1,8-Cineole from the same plant protected bovine liposomes against Fe^{3+}-ascorbate-induced peroxidation by 42% at a dose of $100\,\mu M$.[38] 1,8-Cineole at a dose of 10^{-6}M inhibited the secretion of TNF-α and IL-1β by human lymphocytes challenged with ionomycin/PMA and human monocytes exposed to LPS.[75]

■ **CS 2.25** (+)-borneol.

■ **CS 2.26** (−)-borneol.

■ **CS 2.27** (−)-bornyl acetate.

■ **CS 2.28** (1R)-(+)-camphor.

■ **CS 2.29** 1,8-cineole.

■ **CS 2.30** α-thujone.

■ **CS 2.31** β-thujone.

The reduction of Δ_{2-3}, followed by the formation of a covalent bond between C2 and C6, produces thujone, which exists as isomers α-thujone (CS 2.30) and β-thujone (CS 2.31), according to the conformation of the methyl group in C1 in β or α position. α-Thujone and β-thujone found in the essential oil of *Artemisia absinthium* L. (family Asteraceae Bercht. & J. Presl) confer to the plant neurotoxic effects or absinthism, whereas α-thujone and β-thujone block $GABA_A$ receptors.[56] α-Thujone itself blocks $GABA_A$ with an IC_{50} equal to $21\,\mu M$.[61] Thujone from *Salvia lavandulifolia* Vahl (family Lamiaceae Martinov) protected bovine liposomes against Fe^{3+}-ascorbate-induced peroxidation by 22% at dose of $100\,\mu M$.[38]

REFERENCES

[1] Fraternale D, Giamperi L, Bucchini A, Ricci D, Epifano F, Burini G, et al. Chemical composition, antifungal and *in vitro* antioxidant properties of *Monarda didyma* L. essential oil. J Essential Oil Res 2006;18(5):581–5.

[2] Taborsky J, Kunt M, Kloucek P, Lachman J, Zeleny V, Kokoska L. Identification of potential sources of thymoquinone and related compounds in Asteraceae, Cupressaceae, Lamiaceae, and Ranunculaceae families. Central Eur J Chem 2012;10(6):1899–906.

[3] Aoshima H, Hamamoto K. Potentiation of GABAA receptors expressed in Xenopus oocytes by perfume and phytoncid. Biosci, Biotech Biochem 1999;63(4):743–8.

[4] Xu J, Li C, Yin XH, Zhang GY. Additive neuroprotection of GABA A and GABA B receptor agonists in cerebral ischemic injury via PI-3K/Akt pathway inhibiting the ASK1-JNK cascade. Neuropharmacol 2008;54(7):1029–40.

[5] de Sousa DP, Gonçalves JCR, Quintans-Júnior L, Cruz JS, Araújo DAM, de Almeida RN. Study of anticonvulsant effect of citronellol, a monoterpene alcohol, in rodents. Neurosci Lett 2006;401:231–5.

[6] Lorenzetti BB, Souza GEP, Sartic SJ, Filho DS, Ferreira SHJ. Myrcene mimiCS the peripheral analgesic activity of lemongrass tea. J Ethnopharmacol 1991;34:43–8.

[7] Elmann A, Mordechay S, Rindner M, Ravid U. Anti-neuroinflammatory effects of geranium oil in microglial cells. J Functional Foods 2010;2(1):17–22.

[8] Su YW, Chao SH, Lee MH, Ou TY, Tsai YC. Inhibitory effects of citronellol and geraniol on nitric oxide and prostaglandin E_2 production in macrophages. Planta Med 2010;76:1666–71.

[9] Katsukawa M, Nakata R, Koeji S, Hori K, Takahashi S, Inoue H. Citronellol and geraniol, components of rose oil, activate peroxisome proliferator-activated receptor α and γ and suppress cyclooxygenase-2 expression. Biosci Biotech Biochem 2011;75(5):1010–2.

[10] Melo MS, Sena LCS, Barreto FJN, et al. Antinociceptive effect of citronellal in mice. Pharm Biol 2010;48:411–6.

[11] Brito RG, Santos PL, Prado DS, Santana MT, Araújo AAS, Bonjardim LR, et al. Citronellol reduces orofacial nociceptive behaviour in mice—evidence of involvement of retrosplenial cortex and periaqueductal grey areas. Basic Clin Pharmacol Toxicol 2012;112(4):215–21.

[12] Ohkawara S, Tanaka-Kagawa T, Furukawa Y, Nishimura T, Jinno H. Activation of the human transient receptor potential vanilloid subtype 1 by essential oils. Biol Pharm Bull 2010;33(8):1434–7.

[13] Kim SR, Lee DY, Chung ES, Oh UT, Kim SU, Jin BK. Transient receptor potential vanilloid subtype 1 mediates cell death of mesencephalic dopaminergic neurons *in vivo* and *in vitro*. J Neurosci 2005;25(3):662–71.

[14] Bastos JF, Moreira IJ, Ribeiro TP, Medeiros IA, Antondiolli AR, De Sousa DP, et al. Hypotensive and vasorelaxant effects of citronellol, a monoterpene alcohol, in rats. Basic Clin Pharmacol Toxicol 2010;106(4):331–7.

[15] Elisabetsky E, Brum LF, Souza DO. Anticonvulsant properties of linalool in glutamate-related seizure models. Phytomed 1999;6(2):107–13.

[16] Hossain SJ, Hamamoto K, Aoshima H, Hara YJ. Effects of tea components on the response of GABAA receptors expressed in Xenopus oocytes. Agric Food Chem 2002;50:3954–60.

[17] Mattson MP. Calcium and neurodegeneration. Aging Cell 2007;6(3):337–50.

[18] Peana AT, Marzocco S, Popolo A, Pinto A. (−)-Linalool inhibits *in vitro* NO formation: probable involvement in the antinociceptive activity of this monoterpene compound. Life Sci 2006;78(7):719–23.

[19] Batista PA, Werner MF, Oliveira EC, Burgos L, Pereira P, Brum LF, et al. Evidence for the involvement of ionotropic glutamatergic receptors on the antinociceptive effect of (−)-linalool in mice. Neurosci Lett 2008;440:299.

[20] Batista P, Harris E, Werner M, Santos A, Story G. Inhibition of TRPA1 and NMDA channels contributes to anti-nociception induced by (−)-linalool. J Pain 2011;12:30.

[21] Shigetomi E, Tong X, Kwan K,Y, Corey D,P, Khakh B,S. TRPA1 channels regulate astrocyte resting calcium and inhibitory synapse efficacy through GAT-3. Nat Neurosci 2011;15(1):70–80.

[22] Sampaio LDFS, Maia JGS, De Parijós AM, De Souza RZ, Barata LES. Linalool from rosewood (*Aniba rosaeodora* Ducke) oil inhibits adenylate cyclase in the retina, contributing to understanding its biological activity. Phytother Res 2012;26(1):73–7.

[23] Peana A,T, Rubattu P, Piga G,G, Fumagalli S, Boatto G, Pippia P, et al. Involvement of adenosine A1 and A2A receptors in (−)-linalool-induced antinociception. Life Sci 2006;78(21):2471–4.

[24] Sawynok J, Liu XJ. Adenosine in the spinal cord and periphery: release and regulation of pain. Progress Neurobiol 2003;69:313–40.

[25]   Van Dort CJ, Baghdoyan HA, Lydic R. Adenosine A(1) and A(2A) receptors in mouse prefrontal cortex modulate acetylcholine release and behavioral arousal. J Neurosci 2009;29(3):871–81.

[26]   Hassan S, Lindhagen E, Goransson H, Fryknas M, Isaksson A, Gali-Muhtasib H, et al. Gene expression signature based chemical genomiCS and activity pattern in a panel of tumour cell lines propose linalyl acetate as a protein kinase/NF-κB inhibitor. Gene Ther Mol Biol 2008;12:359–70.

[27]   Rao VSN, Menezes AMS, Viana GSB. Effect of myrcene on nociception in mice. J Pharm Pharmacol 1990;42:877–8.

[28]   Duarte ID, dos Santos IR, Lorenzetti BB, Ferreira SH. Analgesia by direct antagonism of nociceptor sensitization involves the arginine-nitric oxide-cGMP pathway. Eur J Pharmacol 1992;217(2–3):225–7.

[29]   Moncada S, Palmer RMJ, Higgs EA. Nitric oxide, physiology, pathophysiology, and pharmacology. Pharm Rev 1991;43:109–42.

[30]   Busse R, Lueckhoff A, Bassenge E. Endothelium-derived relaxant factor inhibits platelet activation. Naunyn-Schmiedeberg's Arch Pharmacol 1987;336:566–71.

[31]   Chai Y, Lin YF. Stimulation of neuronal KATP channels by cGMP-dependent protein kinase: involvement of ROS and 5-hydroxydecanoate-sensitive factors in signal transduction. Am J Physiol-Cell Physiol 2010;298(4):C875–92.

[32]   Busija DW, Gaspar T, Domoki F, Katakam PV, Bari F. Mitochondrial-mediated suppression of ROS production upon exposure of neurons to lethal stress: mitochondrial targeted preconditioning. Adv Drug Delivery Rev 2008;60(13–14):1471–7.

[33]   Lee HJ, Jeong HS, Kim DJ, Noh YH, Yuk DY, Hong JT. Inhibitory effect of citral on NO production by suppression of iNOS expression and NF-κB activation in RAW264.7 cells. Arch Pharm Res 2008;31:342–9.

[34]   Katsukawa M, Nakata R, Takizawa Y, Hori K, Takahashi S, Inoue H. Citral, a component of lemongrass oil, activates PPARα and γ and suppresses COX-2 expression. Biochim Biophys Acta 2010;1801(11):1214–20.

[35]   Stotz SC, Vriens J, Martyn D, Clardy J, Clapham DE. Citral sensing by transient receptor potential channels in dorsal root ganglion neurons. PLoS One 2008;3:1–14.

[36]   Howes MJR, Houghton PJ, Barlow DJ, Pocock VJ, Milligan SR. Assessment of estrogenic activity in some common essential oil constituents. J Pharm Pharmacol 2002;54(11):1521–8.

[37]   Cordey M, Gundimeda U, Gopalakrishna R, Pike C,J. Estrogen activates protein kinase C in neurons: role in neuroprotection. J Neurochem 2003;84(6):1340–8.

[38]   Perry NS, Houghton PJ, Sampson J, Theobald AE, Hart S, Lis-Balchin M, et al. *In-vitro* activity of *S. lavandulaefolia* (Spanish sage) relevant to treatment of Alzheimer's disease. J Pharm Pharmacol 2001;53(10):1347–56.

[39]   Su Y,W, Chao S,H, Lee M,H, Ou T,Y, Tsai Y,C. Inhibitory effects of citronellol and geraniol on nitric oxide and prostaglandin E_2 production. in macrophages. Planta Med 2010;76:1666–71.

[40]   Park HM, Lee JH, Yaoyao J, Jun HJ, Lee SJ. Limonene, a natural cyclic terpene, is an agonistic ligand for adenosine A(2A) receptors. Biochem Biophys Res Commun 2011;404(1):345–8.

[41]   Gonçalves JC, de Oliveira SF, Benedito RB, de Sousa DP, de Almeida RN, de Araújo DA. Antinociceptive activity of (−)-carvone: evidence of association with decreased peripheral nerve excitability. Biol Pharm Bull 2008;31:1017.

[42]   de Almeida RN, de Sousa DP, Nóbrega FF, Claudino FdeS, Araújo DA, Leite JR, et al. Anticonvulsant effect of a natural compound alpha, beta-epoxy-carvone and its action on the nerve excitability. Neurosci Lett 2008;443(1):51–5.

[43]   Lima DF, Brandão MS, Moura JB, et al. Antinociceptive activity of the mono-terpene α-phellandrene in rodents: possible mechanisms of action. J Pharm Pharmacol 2012;64:283–92.

[44]   Kawata J, Kameda M, Miyazawa M. Cyclooxygenase-2 inhibitory effects of mono-terpenoids with a p-menthane skeleton. Int J Essen Oil Therapeut 2008;2:145–8.

[45]   Hassan SB, Gali-Muhtasib H, Göransson H, Larsson R. Alpha terpineol: a poten-tial anticancer agent which acts through suppressing NF-kappaB signalling. Anticancer Res 2010;30(6):1911–9.

[46]   Watt E,E, Betts B,A, Kotey F,O, et al. Menthol shares general anesthetic activity and sites of action on the GABA(A) receptor with the intravenous agent, propofol. Eur J Pharmacol 2008;590:120–6.

[47]   Bourgou S, Pichette A, Marzouk B, Legault J. Bioactivities of black cumin essen-tial oil and its main terpenes from Tunisia. S Afric J Botan 2010;76:210–6.

[48]   Savelev S,U, Okello E,J, Perry E,K. Butyryl- and acetyl-cholinesterase inhibitory activities in essential oils of Salvia species and their constituents. Phytother Res 2004;18(4):315–24.

[49]   Vogt-Eisele AK, Weber K, Sherkheli MA, Vielhaber G, Panten J, Gisselmann G, et al. Monoterpenoid agonists of TRPV3. Br J Pharmacol 2007;151(4):530–40.

[50]   Yamamoto S, Wajima T, Hara Y, Nishida M, Mori Y. Transient receptor potential channels in Alzheimer's disease. Biochim Biophys Acta 2007; 1772(8):958–67.

[51]   Landa P, Kokoska L, Pribylova M, Vanek T, Marsik P. *In vitro* anti-inflammatory activity of carvacrol: inhibitory effect on COX-2 catalyzed prostaglandin E(2) bio-synthesis. Arch Pharm Res 2009;32(1):75–8.

[52]   Hotta M, Nakata R, Katsukawa M, Hori K, Takahashi S, Inoue H. Carvacrol, a component of thyme oil, activates PPARalpha and gamma and suppresses COX-2 expression. J Lipid Res 2010;51(1):132–9.

[53]   Guimarães AG, Xavier MA, Santana MT, et al. Carvacrol attenuates mechani-cal hypernociception and inflammatory response. Naunyn-Schmiedeberg's Arch Pharmacol 2012;385:253–63.

[54]   Magyar J, Szentandrássy N, Bányász T, Fülöp L, Varró A, Nánási PP. Effects of thymol on calcium and potassium currents in canine and human ventricular car-diomyocytes. Br J Pharma col 2002;136(2):330–8.

[55]   García DA, Bujons J, Vale C, Suñol C. Allosteric positive interaction of thy-mol with the GABAA receptor in primary cultures of mouse cortical neurons. Neuropharmacol 2006;50(1):25–35.

[56]   Abdel Fattah AM, Matsumoto K, Watanabe H. Antinociceptive effects of Nigella sativa oil and its major component, thymoquinone, in mice. Eur J Pharmacol 2000;400:89–97.

[57]   Al-Majed AA, Daba MH, Asiri YA, Al-Shabanah OA, Mostafa AA, El-Kashef HA. Thymoquinone-induced relaxation of guinea-pig isolated trachea. Res Commun Mol Pathol Pharmacol 2001;110(5–6):333–45.

[58]   Al-Majed AA, Al-Omar FA, Nagi MN. Neuroprotective effects of thymoquinone against transient forebrain ischemia in the rat hippocampus. Eur J Pharmacol 2006;543(1–3):40–7.

[59] Yi T, Cho SG, Yi Z, Pang X, Rodriguez M, Wang Y, et al. Thymoquinone inhibits tumor angiogenesis and tumor growth through suppressing AKT and extracellular signal-regulated kinase signaling pathways. Mol Cancer Ther 2008;7(7):1789–96.

[60] Sethi G, Ahn K,S, Aggarwal B,B. Targeting nuclear factor-kappa B activation pathway by thymoquinone: role in suppression of antiapoptotic gene products and enhancement of apoptosis. Mol Cancer Res 2008;6(6):1059–70.

[61] Gilhotra N, Dhingra D. Thymoquinone produced antianxiety-like effects in mice through modulation of GABA and NO levels. Pharmacol Rep 2011;63(3):660–9.

[62] Radad K, Moldzio R, Taha M, Rausch WD. Thymoquinone protects dopaminergic neurons against MPP+ and rotenone. Phytother Res 2009;23(5):696–700.

[63] Ullah I, Ullah N, Naseer MI, Lee HY, Kim MO. Neuroprotection with metformin and thymoquinone against ethanol-induced apoptotic neurodegeneration in prenatal rat cortical neurons. BMC Neurosci 2012;13:11.

[64] Swandulla D, Schafer K, Lux H,D. Calcium channel current inactivation is selectively modulated by menthol. Neurosci Lett 1986;68(1):23–8.

[65] Juergens UR, Stöber M, Vetter H. The anti-inflammatory activity of L-menthol compared to mint oil in human monocytes *in vitro:* a novel perspective for its therapeutic use in inflammatory diseases. Eur J Med Res 1998;3:539–45.

[66] Hall AC, Turcotte C,M, Betts B,A, Yeung W,Y, Agyeman A,S, Burk L,A. Modulation of human GABA A and glycine receptor currents by menthol and related monoterpenoids. Eur J Pharmacol 2004;506(1):9–16.

[67] Galeotti N, Di Cesare Mannelli L, Mazzanti G, Bartolini A, Ghelardini C. Menthol: a natural analgesic compound. Neurosci Lett 2002;322:145–8.

[68] Silva MI, Silva MA, de Aquino Neto MR, Moura BA, de Sousa HL, de Lavor EP, et al. Effects of isopulegol on pentylenetetrazol-induced convulsions in mice: possible involvement of GABAergic system and antioxidant activity. Fitoterapia 2009;80(8):506–13.

[69] Perry N,S, Houghton P,J, Theobald A, Jenner P, Perry E,K. *In-vitro* inhibition of human erythrocyte acetylcholinesterase by *salvia lavandulaefolia* essential oil and constituent terpenes. J Pharm Pharmacol 2000;52(7):895–902.

[70] Liapi C, Anifantis G, Chinou I, Kourounakis A,P, Theodosopoulos S, Galanopoulou P. Antinociceptive properties of 1,8-cineole and b-pinene, from the essential oil of *Eucalyptus camaldulensis* leaves, in rodents. Planta Med 2007;73:1247–54.

[71] Granger R,E, Campbell E,L, Johnston G,A. (+)- and (−)-borneol: efficacious positive modulators of GABA action at human recombinant alpha1beta2gamma2L GABA(A) receptors. Biochem Pharmacol 2005;69(7):1101–11.

[72] Liu R, Zhang L, Lan X, Li L, Zhang TT, Sun JH, et al. Protection by borneol on cortical neurons against oxygen-glucose deprivation/reperfusion: involvement of anti-oxidation and anti-inflammation through nuclear transcription factor kappaB signaling pathway. Neurosci 2011;176:408–19.

[73] Park TJ, Seo HK, Kang BJ, Kim KT. Non competitive inhibition by camphor of nicotinic acetylcholine receptors. Biochem Pharmacol 2001;61:787–93.

[74] Xu H, Blair N,T, Clapham D,E. Camphor activates and strongly desensitizes the transient receptor potential vanilloid subtype 1 channel in a vanilloid-independent mechanism. J Neurosci 2005;25(39):8924–37.

[75] Juergens UR, Engelen T, Racke K, Stober M, Gillissen A, Vetter H. Inhibitory activity of 1,8-cineol (eucalyptol) on cytokine production in cultured human lymphocytes and monocytes. Pulm Pharmacol Ther 2004;17:281–7.

Sesquiterpenes

2.2.1 *Illicium jiadifengpi* B.N. Chang

History The plant was first described by Ben Neng Chang in *Acta Botanica Yunnanica* published in 1982.

Family Schisandraceae Blume, 1830

Common Names Jia di feng pi (Chinese)

Habitat and Description It is a toxic tree that grows in China to a height of 20 m. The leaves are simple, exstipulate, and gathered in groups of 3–5 along the stem. The petiole is 1.5–3 cm long. The leaf blade is elliptic, 5–15 cm × 2–5 cm, tapering at the base, acuminate at the apex, and with 5–10 pairs of secondary nerves. The inflorescence is terminal. The perianth includes 30–50 tepals, which are 1.5 cm long and linear. The androecium consists of 30 stamens, which are minute. The gynoecium comprises 12 carpels, which are free and 0.5 cm long. The fruit presents 12 follicles, which are 1.5 cm long (Figure 2.2).

Medicinal Uses The seeds are crushed and apply to festered wounds to kill maggots.

Phytopharmacology The plant elaborates series of seco-prezizaane sesquiterpenes, including majucin, neomajucin, 2-oxoneomajucin, 1,2-dehydroneomajucin, 2,3-dehydroneomajucin, (2S)-hydroxy-3,4-dehydroneomajucin, (1S)-2-oxo-3,4-dehydroneomajucin, (1R)-2-oxo-3,4-dehydroneomajucin, jiadifenin,[76] jiadifenolide, jiadifenoxolane A and B, neomajucin, majucin, 1,2-epoxyneomajucin, 2-oxo-3,4-dehydroxyneomajucin, 2,3-dehydroxyneomajucin,[77] (2R)-hydroxy-norneomajucin, jiadifenone, 1,2-epoxyneomajucin, and pseudomajucin.[78]

Proposed Research Pharmacological study of (2R)-hydroxy-norneomajucin for the treatment of neurodegenerative diseases.

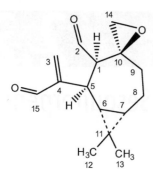

■ **CS 2.32** Plagiochilal B.

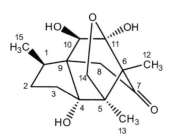

■ **CS 2.33** 11-*O*-debenzoyltashironin.

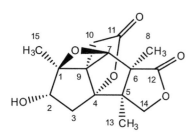

■ **CS 2.34** Merrilactone A.

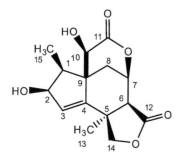

■ **CS 2.35** (2S)-hydroxy-3,4-dehydroneomajucin.

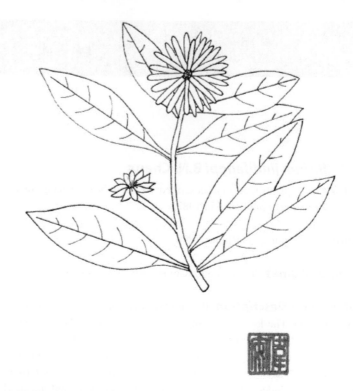

■ **FIGURE 2.2** *Illicium jiadifengpi* B.N. Chang.

Rationale A unique and intriguing feature of a number of sesquiterpenes is their remarkable ability to evoke neuritogenesis. For instance, the *seco*-aromadendrane-type sesquiterpene plagiochilal B (CS 2.32), isolated from *Plagiochila fruticosa* Mitt. (family Plagiochilaceae K. Müller & Herzog) induced the sprouting of neurites at a dose of 10^{-5} M.[79] The anisolactone sesquiterpenes 11-*O*-debenzoyltashironin (CS 2.33) and merrilactone A (CS 2.34) from *Illicium merrillianum* A.C. Sm. (family Schisandraceae Blume) induced the growth of neurites from cortical neurons at a dose of 10μmol/L[80,81] and 10μM, respectively. The seco-prezizaane-type sesquiterpene (2S)-hydroxy-3,4-dehydroneomajucin (CS 2.35) from *Illicium jiadifengpi* B.N. Chang (family Schisandraceae Blume) evoked the growth of neurites from cortical neurons at a dose of 10μM.[76] Likewise, the seco-prezizaane sesquiterpenes jiadifenolide (CS 2.36) and jiadifenoxolane A (CS 2.37) from *Illicium jiadifengpi* B.N. Chang (family Schisandraceae Blume) evoked the growth of neurites from cortical neurons at a dose of 10μM,[77] and the seco-prezizaane sesquiterpene jiadifenin (CS 2.38) from

■ **CS 2.36** Jiadifenolide.

■ **CS 2.37** Jiadifenoxolane A.

■ **CS 2.38** Jiadifenin.

■ **CS 2.39** (2R)-hydroxy-norneomajucin.

■ **CS 2.40** (4R,5R,7R)-1(10)-spirovetiven-11-ol-2-one.

■ **CS 2.41** Sesquiterpene 2.

the same plant evoked the development of neurites from cortical neurons by 21% and 28% at doses of 0.3 μM and 0.5 μM, respectively.[76] Furthermore, the seco-prezizaane sesquiterpene (2R)-hydroxy-norneomajucin (CS 2.39) evoked the growth of neurites from cortical neurons at a dose of 10 μmol/L.[78] The spirovetinane sesquiterpene (4R,5R,7R)-110-spirovetiven-11-ol-2-one (CS 2.40) from agarwood significantly induced the production of brain-derived neurotrophic factor (BDNF) in rat cortical neurons at dose of 100 μg/mL.[82] The sesquiterpene 2 (CS 2.41) and the bisabolane sesquiterpene 4 (CS 2.42) isolated from *Valeriana fauriei* Briq. (family Valerianaceae Batsch) at a dose of 100 μM enhanced the neuritogenic effect of 2 ng/mL of nerve growth factor (NGF) by 52% and 48%, respectively.[83] The eudesmane sesquiterpenes (7S,8R,8′R,4″S,5″S,7″S,10″S)-9-*O*-(11-hydroxyeudesman-4-yl)-dyhydrosesamin (CS 2.43) from *Chamaecyparis obtusa* (Siebold & Zucc.) Endl. (family Cupressaceae Gray) prompted the growth of neurites from rat pheochromocytoma (PC12) cells by 42.6% at a dose of 1.2 μg/mL in the presence of 2 ng/mL of NGF.[84] The mushroom *Russula japonica* Hongo (family Russulaceae Lotsy) produces the illudoid sesquiterpenes russujaponol I,

■ **CS 2.42** Sesquiterpene 4.

■ **CS 2.43** (7S,8R,8′R,4″S,5″S,7″S,10″S)-9-O-(11-hydroxyeudesman-4-yl)-dyhydrosesamin.

■ **CS 2.44** Russujaponol I.

■ **CS 2.45** Russujaponol J.

■ **CS 2.46** Russujaponol K.

russujaponol J, and russujaponol K (CS 2.44–2.46), which induced the growth of neurites from cortical neurons at a dose of $1\,\mu$M.[85] This evidence raises a question regarding how neurons are regulated to produce neurites upon exposure to sesquiterpenes with different chemical frameworks.

Indeed, the mechanisms by which sesquiterpenes induce neurites are presently unclear. In the brain, brain-derived neurotrophic factor (BDNF) and NGF account both for normal function and synaptic arborescence of neurons by activating the cAMP response element binding protein (CREB).[86] NGF activates the receptor tropomyosin receptor kinase A receptor (TrkA) expressed by cerebral cholinergic neurons in basal forebrain[86] and induces the simultaneous activation of Ras and phosphatidylinositol 3-kinase.[87] The activation of Ras triggers Raf, hence phosphorylation and stimulation of extracellular signal-regulated kinase (ERK1/2) via mitogen-activated protein kinase 1/2 (MEK1/2)[88] and CREB.[89] Furthermore, Ras induces Rac, which activates mitogen-activated protein kinase 1/4 (MEK1/4), which evokes mitogen-activated protein kinase (MAPK) kinase 4/7 (MKK4/7), thus inducing c-Jun N-terminal kinase (JNK), which in turns activates numerous transcription factors.

Of note, the activation of muscarinic receptors subtype 1 (M_1) stimulates phospholipase C (PLC), which cleaves inositol 4,5-bisphosphate (PIP2) into inositol triphosphate (IP3) and diacylglycerol (DAG), resulting in the release of Ca^{2+} from endoplasmic reticulum (ER) and activation of protein kinase C (PKC), which stimulates Ras.[90,91] In addition, phosphoinositide 3-Kinase (PI3K) converts phosphatidylinositol 3,4-bisphoshate (PIP2) to phosphatidylinositol 3,4,5-trisphoshate (PIP3), which activates 3-phosphoinositide-dependent kinase 1 (PDK-1) and therefore protein kinase B (Akt) Rac, resulting in neuronal survival[92,93] via JNK. NGF also has the ability to bind to the p75^{NTR} receptor that evokes tumor necrosis factor receptor-associated factor 6 (TRAF6), hence, kappa light polypeptide gene enhancer in B-cells inhibitor (IκB) kinase (IKK) and thus nuclear factor of kappa light polypeptide gene enhancer in B-cells inhibitor α (IκBα) phosphorylation and degradation, translocation of nuclear factor kappa-light-chain-enhancer of activated B cells (NF-κB), and therefore dendritic growth.[94] Furthermore, TRAF6 commands the activation of MAPK.

BDNF binds to and activates tropomyosin receptor kinase B receptor (TrkB), which induces PLC, Ras, and PI3K. PLC evokes the release of Ca^{2+} from the endoplasmic reticulum (ER), hence the activation of the Ca^{2+}-calmodulin dependent kinase (CaMKII) and consequently CREB.[95] Along the same lines, CREB is activated by Ras, Raf, ERK1/2, PI3K, and Akt.[96] Likewise, NGF and BDNF bind to p75^{NTR}, hence activation of the NF-κB TRAF6.[97]

One potential explanation for the neurotrophic properties of sesquiterpenes might be some positive interference with the aforementioned pathways and especially the induction of CREB. In effect, the eudesmane sesquiterpene β-eudesmol (CS 2.47) at a dose of $150 \mu M$ commanded the growth of neurites from PC12 cells via induction of PI3K, hence the efflux of Ca^{2+} from the endoplasmic reticulum (ER) resulting in CaM/PKA activation of ERK1/2 and consequently CREB.[98] It is therefore tempting to speculate that sesquiterpenes represent a vast pool of unexploited neurotrophic

■ CS 2.47 β-eudesmol.

agents to fight Alzheimer's disease (AD), Parkinson's disease (PD), and amyotrophic lateral sclerosis (ALS).

Within the past decade, a considerable body of evidence has accumulated in support of the view that sesquiterpenes inhibit the secretion of nitric oxide (NO), which is involved in neurodegeneration. As a matter of fact, the binding of lipopolysaccharide (LPS) to toll-like receptor 4 (TLR4) in glial cells stimulates myeloid differentiation primary response gene 88 (MyD88), hence activation of TRAF6 and consequently activation of NF-κB and activator protein-1 (AP-1), resulting in NO and prostaglandin synthesis via inducible nitric oxide synthetase (iNOS) and cyclo-oxygenase-2 (COX-2) induction.[99]*Laurus nobilis* L. (family Lauraceae Juss.) engineers the guaiane sesquiterpenes eremanthine (CS 2.48) and zaluzanin C (CS 2.49), the eudesmane sesquiterpenes magnolialide (CS 2.50) and santamarine (CS 2.51) and spirafolide (CS 2.52), which inhibited the generation of NO by macrophages exposed to LPS with IC_{50} values equal to $2\,\mu M$, $1.7\,\mu M$, $1.5\,\mu M$, $2.8\,\mu M$, and $3.8\,\mu M$, respectively.[100] Along the same lines, *Laurus nobilis* L. (family Lauraceae Juss.) produces the eudesmane sesquiterpenes $1\beta,15$-dihydroxy-5α-H,7α -H-eudesma-3,11(13)-dien-12,6α-olide (CS 2.53), santamarine, reynosin (CS 2.54), and the

■ **CS 2.48** Eremanthine.

■ **CS 2.49** Zaluzanin C.

■ **CS 2.50** Magnolialide.

■ **CS 2.51** Santamarine.

■ **CS 2.52** Spirafolide.

■ **CS 2.53** 1β,15-dihydroxy-5α-H,7α-H-eudesma-3,11(13)-dien-12,6α-olide.

■ **CS 2.54** Reynosin.

germacrane sesquiterpene costunolide (CS 2.55), which suppressed the production of NO by macrophages challenged with LPS with IC_{50} values equal to $0.8\,\mu M$, $15.5\,\mu M$, $3.7\,\mu M$, and $13.8\,\mu M$, respectively.[101] Strikingly, the aforementioned sesquiterpenes bear a lactone group which reacts with nucleophilic cysteine residues of IKK, thus inhibiting NF-κB and therefore reducing the production of nitric oxide synthetase (NOS). This contention is supported by the fact that $5\,\mu M$ of the germacrane sesquiterpene parthenolide (CS 2.56) and $20\,\mu M$ of the eudesmane sesquiterpene isohelenin (CS 2.57) nullified the activation of NF-κB in malignant cells exposed to tumor necrosis factor-α (TNF-α) by preventing IκBα degradation via the lactone ring.[102] The European medicinal *Valeriana officinalis* L. (family Valerianaceae Batsch) produces the valerenanes sesquiterpenes acetylvalerenolic acid (CS 2.58) and valerenal (CS 2.59), which mitigated the activity of NF-κB activity to 4% and 25% at doses of $100\,\mu g/mL$,[103] confirming further that a lactone ring is a prerequisite for NF-κB inhibition.

By inhibiting NF-κB, sesquiterpene lactones hamper the production of chemokines, cytokines, and COX-2. In effect, secopseudoguaianolide sesquiterpene inulicin (CS 2.60) from *Inula japonica* Thunb. (family Asteraceae

■ **CS 2.55** Costunolide.

■ **CS 2.56** Parthenolide.

■ **CS 2.57** Isohelenin.

■ **CS 2.58** Acetylvalerenolic acid.

■ **CS 2.59** Valerenal.

■ **CS 2.60** Inulicin.

■ **CS 2.61** 7α,11α-dihydroxy-8-drimen-12,11-olide.

■ **CS 2.62** 7α,11α-dihydroxy-4(13),8-coloratadien-12,11-olide.

Bercht. & J. Presl) given at a dose of 26 mg/kg to rodents prevented the neurotoxic effects of β-amyloid (Aβ$_{25-35}$) peptide by lowering the hippocampal-level expressions of COX-2 and iNOS.[104] The drimane sesquiterpenes 7α,11α-dihydroxy-8-drimen-12,11-olide (CS 2.61) and the coloratane sesquiterpene 7α,11α-dihydroxy-4(13),8-coloratadien-12,11-olide (CS 2.62) from *Discopodium penninervium* Hochst. (family Solanaceae Juss.) inhibited the production of leukotriene B4 (LB4) by neutrophiles by 11.5% and 36.6%, respectively, at a dose of 50 μM.[105] The germacrane sesquiterpene eupatolide (CS 2.63) inhibited the production of NO and prostaglandin E2 (PGE2) in macrophages (RAW 264.7) challenged with LPS with an IC$_{50}$ value of 2.2 μM and 3.1 μM, respectively, by promoting the degradation of TRAF6 by proteasome,[106] providing evidence that sesquiterpenes lactones are able to inhibit pathways upstream NF-κB. This is corroborated with the eudesmane sesquiterpene alantolactone (CS 2.64) from a member of the genus *Aucklandia* Falc. (family Asteraceae Bercht. & J. Presl), which at a dose of 10 μM impeded the production of NO, PGE2, and TNF-α by RAW 264.7 challenged with LPS by reducing the expression of iNOS and COX-2 by inhibiting AP-1 and preventing the phosphorylation of IKK and therefore degradating IκBα[107]*Tithonia diversifolia* (Hemsl.) A. Gray (Asteraceae Bercht. & J. Presl) produces the germacrane sesquiterpenes tirotundin (CS 2.65) and tagitinin A (CS 2.66), which

■ **CS 2.63** Eupatolide.

■ **CS 2.64** Alantolactone.

■ **CS 2.65** Tirotundin.

■ **CS 2.66** Tagitinin A.

■ **CS 2.67** Tussilagone.

■ **CS 2.68** 7α-(4-methylsenecioyloxy)oplopa-3(14)E,8(10)-dien-2-one.

■ **CS 2.69** 1β-angeloyloxy-7α-(4-methylsenecioyloxy)oplopa-3(14)Z,8(10)-dien-2-one.

are peroxisome proliferator-activated receptor-γ (PPAR-γ) agonists by direct binding with IC_{50} values equal to $27\,\mu M$ and $55\,\mu M$.[108] Note that activating the PPAR-γ results in increased levels of phosphorylated ERK1/2 and IκBα and therefore inhibition of nuclear translocation of NF-κB and therefore reduction of COX-2.[109]

Oddly enough, certain sesquiterpenes are able to hinder neuroinflammation via NF-κB inhibition although lacking lactone rings. This is the case of tussilagone (CS 2.67) from *Tussilago farfara* L. (family Asteraceae Bercht. & J. Presl), which inhibited the generation of NO and PGE2 by microglial (BV-2) cells activated by LPS with IC_{50} values equal to $8.6\,\mu M$ and $14.1\,\mu M$, by diminishing the levels of iNOS and COX-2 as a result of NF-κB inactivation.[110] Similarly, 7α-(4-methylsenecioyloxy)oplopa-3(14)E,8(10)-dien-2-one (CS 2.68), 1β-angeloyloxy-7α -(4-methylsenecioyloxy)oplopa3(14) Z,8(10)-dien-2-one (CS 2.69), and 7α-(3′-ethyl-cis-crotonoyloxy)-14-hydroxy-1β -(2′-methylbutyryloxy)-notonipetranone (CS 2.70) from *Tussilago farfara* L. (family Asteraceae Bercht. & J. Presl) inhibited the

■ **CS 2.70** 7α-(3'-ethyl-cis-crotonoyloxy)-14-hydroxy-1β-(2'-methylbutyryloxy)-notonipetranone.

■ **CS 2.71** Zerumbone.

■ **CS 2.72** Petasine.

production of NO by RAW 264.7 challenged with LPS with IC_{50} values equal to $10.8 \mu M$, $13.8 \mu M$, and $15.6 \mu M$, respectively.[111] In addition, the humulane sesquiterpene zerumbone (CS 2.71) from *Zingiber zerumbet* (L.) Roscoe ex Sm. (family Zingiberaceae Martinov) reduced the expression of CXC chemokine receptor 4 (CXCR4) on malignant cells via inhibition of NF-κB.[112] It is of interest to note that the stimulation of CXC chemokine receptor 4 (CXCR4) by the chemokine CXCL12 induces neuroapoptosis in the *substantia nigra*.[113] At a dose of $20 \mu mol/L$ zerumbone prevented the activation of NF-κB in RAW 264.7 challenged with receptor activator of nuclear factor-kappaB ligand (RANKL) by inhibiting IκBα degradation.[114]

There is an accumulating body of evidence to suggest that sesquiterpenes have the propensity to prolong neuronal survival by hampering the apoptotic machinery. In effect, a member of the genus *Petasites* Mill. (family Asteraceae Bercht. & J. Presl) elaborates the eremophilane sesquiterpene petasine (CS 2.72), which inhibited, at a dose of $16 \mu g/mL$, the increase of cytoplasmic Ca^{2+} into eosinophils and neutrophils exposed to platelet activating factor (PAF).[115] Platelet activating factor (PAF) binds to metabotropic receptors expressed by glial cells, which results in the activation of PLC, the production of inositol-1, 4, 5-trisphosphate (IP3), endoplasmic reticulum (ER) release of Ca^{2+} in the cytoplasm, and as a consequence neuroinflammation.[116] The sesquiterpene trilactone bilobalide (CS 2.73)

■ **CS 2.73** Bilobalide.

■ **CS 2.74** Isoatriplicolide tiglate.

■ **CS 2.75** Anislactone B.

■ **CS 2.76** Pseudomajucin.

■ **CS 2.77** Lindenenyl acetate.

from *Ginkgo biloba* L. (family Ginkgoaceae Engl.) at a dose of 50 μM pro-tected PC12 cells against reactive oxygen species (ROS) by decreasing the levels of p53, c-Myc, pro-apoptotic Bcl-2-associated X protein (Bax); and preventing the activation of caspase 3,[117] suggesting the possible stimula-tion of Akt via MDM2. In fact, bilobalide at a dose of 10 μM protected human SH-SY5Y neuroblastoma cells against β-amyloid (Aβ$_{1-42}$) peptide by activating PI3K, Akt.[118] The activation of Akt may therefore account for the effect of isoatriplicolide tiglate (CS 2.74) from *Paulownia tomen-tosa* (Thunb.) Steud. (family Paulowniaceae Nakai), which at doses of 1 μM to 10 μM protected cortical neurons against glutamate by 40% to 80%, respectively.[119] Likewise, the anislactone sesquiterpene anislactone B (CS 2.75) at a dose of 1 μM and the seco-prezizaane sesquiterpene pseu-domajucin (CS 2.76) at a dose of 10 μM from *Illicium oligandrum* Merr. & Chun (family Schisandraceae Blume) prompted the survival by 12.5% and 2.8%, respectively, of human SH-SY5Y neuroblastoma cells poisoned with H_2O_2[120] via probable activation of PI3K. This contention was further confirmed when the selinane sesquiterpene lindenenyl acetate (CS 2.77) from *Lindera strychnifolia* (Siebold & Zucc.) Fern.-Vill. (family Lauraceae

■ **CS 2.78** Bakkenolide-VI.

■ **CS 2.79** Commiterpene A.

■ **CS 2.80** Commiterpene B.

■ **CS 2.81** Commiterpene C.

■ **CS 2.82** Myrrhterpenoid K.

■ **CS 2.83** Myrrhterpenoid N.

Juss.) hindered the cytotoxic effect of glutamate against mouse hippocampal cells (HT-22) at a dose of $40\,\mu M$, by activating PI3K, Akt, ERK1/2, and therefore inducing the nuclear translocation of transcription factor Nrf2 and anti-oxidant protein heme oxygenase-1 (HO-1).[121]

The bakkane-type sesquiterpene bakkenolide-VI (CS 2.78) from *Petasites tatewakianus* Kitam. (family Asteraceae Bercht. & J. Presl) at a dose of $100\,\mu g/mL$ sustained the viability of cortical neurons by 90% against the oxidative damages incurred by Fe^{2+} ascorbic acid.[122]*Commiphora myrrha* (T. Nees) Engl. (family Burseraceae Kunth) produces the cadinane sesquiterpene commiterpenes A, B, and C (CS 2.79–2.81)[123]; the guaiane sesquiterpene myrrhterpenoid K (CS 2.82); and the eudesmane sesquiterpene myrrhterpenoid N (CS 2.83), which abolished the apoptotic effect of 1-methyl-4-phenylpyridinium ion (MPP$^+$) against dopaminergic neuroblastoma human SH-SY5Y cells at a dose of $30\,\mu M$[124] probably through PI3K activation. As a matter of fact, 1-methyl-4-phenylpyridinium ion

(MPP$^+$) inhibits the phosphorylation of Akt, allowing the translocation of Bax into mitochondria; the release of cytochrome c; and the activation of caspases 9 and 3, leading to neuroapoptosis.[125] In addition, the germacrane sesquiterpene costunolide from *Laurus nobilis* L. (family Lauraceae Juss.) reduced the levels of nuclear receptor related-1 (Nurr1), vesicular mono-amine transporter 2 (VMAT2), and dopamine transporter (DAT) by 52% and decreased α-synuclein in human SH-SY5Y neuroblastoma cells at a dose of 600 μM.[126] From the same plant, spirafolide protected human SH-SY5Y neuroblastoma cells against dopamine-induced apoptosis with an EC$_{50}$ equal to 5.7 μM with a reduced level in ROS.[127]

Evidence points to the fact that sesquiterpenes, which are lipophilic, have the ability to bind to cerebral receptors. This is the case notably for the seco-prezizaane sesquiterpene anisatin (CS 2.84) from *Illicium anisatum* L. (family Schisandraceae Blume), which blocked γ-amino butyric acid subtype A (GABA$_A$) with an IC$_{50}$ value equal to 0.4 μM.[128] Another sesquiterpene with affinity for γ-amino butyric acid receptors subtype A (GABA$_A$) is bilobalide from *Ginkgo biloba* L. (family Ginkgoaceae Engl.), which antagonized γ-amino butyric acid receptors subtype A (GABA$_A$) expressed by oocytes with an IC$_{50}$ value equal to 4.6 μM,[129] hence stimulation of hippocampal pyramidal neuronal activity leading to improved cognition.[129,130] Furthermore, the guaiane sesquiterpene isocurcumenol (CS 2.85) from *Cyperus rotundus* L. (family Cyperaceae Juss.) was agonist for γ-amino butyric acid receptors subtype A (GABA$_A$) expressed by oocytes with an IC$_{50}$ value equal to 4.6 μM.[129,130] In addition, isocurcumenol blocked the binding of [^{3}H]-Ro15-1788 by 58.3% at a dose of 10^{-4} M, and, in the presence of γ-amino butyric acid (GABA), improved the binding of [^{3}H]flunitrazepam by 106.6% at a dose of 10^{-6} M, implying facilitation of GABA-ergic activity.[131] The drimane sesquiterpene drimanial (CS 2.86)

■ **CS 2.84** Anisatin.

■ **CS 2.85** Isocurcumenol.

■ **CS 2.86** Drimanial.

■ **CS 2.87** Polygodial.

■ **CS 2.88** S-Isopetasin.

from *Drimys winteri* J.R. Forst. & G. Forst. (family Winteraceae R. Br. ex Lindl.) alleviated the pain induced by formalin and glutamate in rodent by blocking metabotropic glutamate receptors (mGluRs) as shown with lower binding of [³H]glutamate to cortical membranes with an IC_{50} value equal to $4.3\,\mu$M.[132] The stimulation of metabotropic glutamate receptor I (mGluRI) results in phosphoinositide hydrolysis, and stimulation of metabotropic glutamate receptor subtype II and III (mGluRII and III) reduces the cytoplasmic levels of cyclic adenosine monophosphate (cAMP).[133] Interestingly, the stimulation of metabotropic glutamate receptors II and III (mGluRII and III) protects neurons,[134,135] whereas the stimulation of the metabotropic glutamate receptor subtype I (mGluRI) favors the catabolism of amyloid precursor protein (APP).[136] In addition, the drimane-type sesquiterpenes polygodial (CS 2.87) and drimanial evoked the influx of Ca^{2+} into spinal synaptosomes by 70% and 47%, respectively, as a result of transient receptor potential vanilloid subtype 1 (TRPV1) activation at doses of $5\,\mu$M.[132] Indeed, activation of the TRPV1 results in increased levels of free cytoplasmic Ca^{2+}, activation of protein kinase A (PKA), activation of the arachidonate cascade, mitochondrial insult, release of cytochrome c, activation of caspase 9 and 3, and neuroapoptosis.[137] Because drimanial is a metabotropic glutamate receptor antagonist and a TRPV1 receptor agonist, one could reasonably speculate that hemisynthetic derivates of this sesquiterpene may very well produce leads for the treatment of neurodegenerative diseases.

Both GABA-ergic and glutamatergic interneurons synapse and modulate the activity of cholinergic neurons.[138] These interneurons express presynaptic muscarinic receptors subtype 1 and 3 (M_1,M_3), the stimulation of which weakens the release of GABA and glutamate.[138] Therefore, muscarinic receptor subtype 1 and 3 (M1, M3) agonists may have some potential as cholinergic potentiators of possible usefulness for the treatment of AD. In this respect, the eremophilane sesquiterpenes S-isopetasin (CS 2.88) from

Petasites formosanus Kitam. (family Asteraceae Bercht. & J. Presl) relaxed tracheal ring preparations exposed to acetylcholine at a dose of $300 \mu M$ by antagonizing muscarinic receptor subtype 3 (M3),[139] and it is reasonable to anticipate that *S*-isopetasin synthetic derivatives may produce muscarinic receptor subtype 1and 3 (M_1, M_3) agonists for the treatment of AD.

The pathophysiology of AD,[140] PD,[141] and demyelinating pathologies such as multiple sclerosis (MS) and ALS[142] involve the mobilization and extreme activation of glial cells which release massive amounts of TNF-α, Interleukin-1β (IL-1β), Interleukin-6 (IL-6), NO, and ROS, destroying neurons.[143] Intriguingly, activated glial cells undergo profound changes, including increased expression of cannabinoid receptor subtype 2 (CB$_2$) at their surface[144] and activation which dampens the release of cytokines and chemokines.[145] Therefore, cannabinoid receptor subtype 2 (CB$_2$) agonists may very well fight neurodegeneration, and such agonists may be found among the numerous sesquiterpenes engineered by medicinal plants. In fact, the caryophyllene (*E*)-β-caryophyllene (CS 2.89) found in the essential oil of *Cannabis sativa* L. (family Cannabaceae Martinov) is an agonist to the cannabinoid receptor subtype 2 (CB$_2$) with a K_i=155 nM.[146] The zizaane sesquiterpene khusimol (CS 2.90) from *Vetiveria zizanioides* (L.) Nash (family Poaceae Barnhart) hindered the binding of [^{3}H]-vasopressin to hepatic membranes vasopressin V1a receptors with an IC$_{50}$ value equal to $125 \mu M$ and *K*i of $41 \mu M$.[147] Vasopressin V1a receptor antagonists induce dementia.[148] In rodent vasopressin V1a receptors are found in the *hippocampus* and dorsolateral *septum*, and stimulation of these receptors by vasopressin improves memory by presumably modulating the noradrenergic and dopamine activity.[149,150] Although the precise molecular mode of action of V1a agonists on cognition is yet elusive, Son *et al.*[151] showed that 250 nM of V1a agonist boosted the influx of extracellular Ca^{2+} into cortical neurons to 140%, commanding the opening of L-type Ca^{2+} channels.[151] At a dose of $5 \mu M$, PKC inhibitor (BIS I) and $1 \mu M$ of PLC inhibitor (U-73122) repressed Ca^{2+} entry into cortical neurons by 90.2% and 97.8%, respectively, suggesting the activation of both PKC and PLC in the aforementioned opening.[151] In this respect, PLC is known to convert phosphatidylinositol 4,5-bisphosphate (PIP2) into diacylglycerol (DAG), which activates PKC, hence the phosphorylation of receptors for activated C kinases (RACK) which evoke L-type Ca^{2+} channels opening.[152]

In this context, one must recall that muscarinic receptor subtype 1 (M1) is coupled with Gq/11, hence activating PLC and converting membrane phospholipid phosphatidylinositol 4,5-bisphosphate (PIP2) into inositol-1,4,5-trisphosphate (IP3) and diacylglycerol (DAG).[153] DAG activates

■ CS 2.89 (*E*)-β-caryophyllene.

■ CS 2.90 Khusimol.

■ **CS 2.91** Germacrone.

■ **CS 2.92** Zederone.

■ **CS 2.93** Dehydrocurdione.

■ **CS 2.94** Curcumenol.

■ **CS 2.95** Zedoarondiol.

PKC.[153] Along the same lines, inositol-1, 4, 5-trisphosphate (IP3) binds to and compels the aperture of IP3-receptor Ca $^{2+}$ channels expressed by the endoplasmic reticulum (ER), which results in increased levels of cytoplasmic Ca $^{2+}$, which notably activates calmodulin, resulting in the closure of K^+ channels and abrogation of K^+ efflux, depolarization in neurons, and enhancement of cognition.[153] Likewise, the stimulation of serotoninergic receptor subtype 2A (5-HT2$_A$) compels PLC and PKC[154] and the germacrane sesquiterpene parthenolide from *Tanacetum parthenium* (L.) Sch. Bip. (family Asteraceae Bercht. & J. Presl) serotoninergic receptor subtype 2A (5-HT2$_A$), as shown by a reduction of [³H]-ketanserin for binding with a Ki value equal to $250\,\mu M$.[155] The serotoninergic receptor subtype 2A (5-HT2$_A$) antagonist M100907 at a dose of 0.1 mg/kg favored the release of dopamine in the prefrontal cortex of rodents induced by 0.05 mg/kg of the serotoninergic receptor subtype 1A (5-HT1$_A$) agonist R(1)-8-OH-DPAT,[154] implying that serotoninergic receptor subtype 2A (5-HT2$_A$) antagonist may have some value for the treatment of PD. Other examples of sesquiterpenes of neuropharmacological interest are found in a member of the genus *Curcuma* L. (family Zingiberaceae Martinov), which elaborates the germacrane sesquiterpenes germacrone (CS 2.91), zederone (CS 2.92), and dehydrocurdione (CS 2.93), which inhibited the enzymatic activity of 5α-reductase by 65.7%, 46%, and 45.2%, respectively, at doses of 1 mg/mL.[156] In the same experiment, the guaiane sesquiterpenes curcumenol (CS 2.94), zedoarondiol (CS 2.95), and isocurcumenol at a dose of 1 mg/mL inhibited the enzymatic activity of 5α-reductase by 17.4%, 16.4%, and 4.7%, respectively, at doses of 1 mg/mL, 0.5 mg/mL, and 1 mg/mL, respectively.[156] In neurophysiological conditions, 5-α-reductase catalyzes the synthesis of dihydrotestosterone from testosterone, which is beneficial against AD. As a matter of fact, testosterone at a dose of 200 nM induced the secretion of a secreted form of APP (sAPPα) from neurons by 20% and reduced the secretion of β-amyloid (Aβ$_{1–40}$) peptide by 30%.[157] Others sesquiterpenes that may be of value in developing derivatives to delay the onset of AD are the bisabolane sesquiterpenes

■ **CS 2.96** (+)-(S)-*ar*-curcumene.

■ **CS 2.97** (+)-(S)-dihydro-*ar*-curcumene.

■ **CS 2.98** (+)-(7S,9S)-*ar*-turmerol.

(+)-(S)-*ar*-curcumene (CS 2.96), (+)-(S)-dihydro-*ar*-curcumene (CS 2.97), (+)-(7S,9S)-*ar*-turmerol (CS 2.98), and (+)-(7S,9R)-*ar*-turmerol (CS 2.99) from a member of the genus *Peltophorum* (Vogel) Benth. (family Fabaceae Lindl.), which inhibited the enzymatic activity of acetylcholinesterase (AChE) with IC_{50} values equal to $14.6 \mu M$, $35.6 \mu M$, $35.6 \mu M$, and $42.5 \mu M$, respectively.[158]

■ **CS 2.99** (+)-(7S,9R)-*ar*-turmerol.

REFERENCES

[76] Yokoyama R, Huang J,M, Yang C,S, Fukuyama Y. New seco-prezizaane-type sesquiterpenes, jiadifenin with neurotrophic activity and 1,2-dehydroneomajucin from *Illicium jiadifengpi*. J Nat Prod 2002;65(4):527–31.

[77] Kubo M, Okada C, Huang J,M, Harada K, Hioki H, Fukuyama Y. Novel pentacyclic seco-prezizaane-type sesquiterpenoids with neurotrophic properties from *Illicium jiadifengpi*. Org Lett 2009;11(22):5190–3.

[78] Kubo M, Kobayashi K, Huang J,M, Harada K, Fukuyama Y. The first examples of seco-prezizaane-type norsesquiterpenoids with neurotrophic activity from *Illicium jiadifengpi*. Tetrahedron Lett 2012;53(10):1231–5.

[79] Fukuyama Y, Asakawa Y. Neurotrophic secoaromadendrane-type sesquiterpenes from the liverwort *Plagiochila fruticosa*. Phytochem 1991;30(12):4061–5.

[80] Huang J,M, Yokoyama R, Yang C,S, Fukuyama Y. Structure and neurotrophic activity of seco-prezizaane-type sesquiterpenes from *Illicium merrillianum*. J Nat Prod 2001;64(4):428–31.

[81] Huang JM, Yokoyama R, Yang CS, Fukuyama Y. Merrilactone A, a novel neurotrophic sesquiterpene dilactone from *Illicium merrillianum*. Tetrahedron Lett 2000;41(32):6111–4.

[82] Ueda J,Y, Imamura L, Tezuka Y, Tran QL, Tsuda M, Kadota S. New sesquiterpene from Vietnamese agarwood and its induction effect on brain-derived neurotrophic factor mRNA expression *in vitro*. Bioorg Med Chem 2006;14(10):3571–4.

[83] Guo Y, Xu J, Li Y, Watanabe R, Oshima Y, Yamakuni T, et al. Iridoids and sesquiterpenoids with NGF-potentiating activity from the rhizomes and roots of *Valeriana fauriei*. Chem Pharm Bull 2006;54(1):123–5.

[84] Kuroyanagi M, Ikeda R, Gao HY, Muto N, Otaki K, Sano T, et al. Neurite outgrowth-promoting active constituents of the Japanese cypress (*Chamaecyparis obtusa*). Chem Pharm Bull 2008;56(1):60–3.

[85] Yoshikawa K, Matsumoto Y, Hama H, Tanaka M, Zhai H, Fukuyama Y, et al. Russujaponols G-L, Illudoid sesquiterpenes, and their neurite outgrowth promoting activity from the fruit body of *Russula japonica*. Chem Pharm Bull 2009;57(3):311–4.

[86] Sofroniew M,V, Copper J,D. Neurotrophic mechanisms and neuronal degeneration. Sem Neurosci 1993;5:283–94.

[87] Niewiadomska G, Mietelska-Porowska A, Mazurkiewicz M. The cholinergic system, nerve growth factor and the cytoskeleton. Behav Brain Res 2011;221(2):515–26.

[88] English J, Pearson G, Wilsbacher J, Swantek J, Karandikar M, Xu S, et al. New insights into the control of MAP kinase pathways. Exp Cell Res 1999;253:255–70.

[89] Riccio A, Ahn S, Davenport C,M, Blendy J,A, Ginty D,D. Mediation by a CREB family transcription factor of NGF-dependent survival of sympathetic neurons. Science 1999;286(5448):2358–61.

[90] Lee SB, Rhee SG. Significance of PIP2 hydrolysis and regulation of phospholipase C isozymes. Curr Opin Cell Biol 1995;7:183–9.

[91] Wooten M,W, Seibenhener M,L, Zhou G, Vandenplas M,L, Tan T,H. Overexpression of atypical PKC in PC12 cells enhances NGF-responsiveness and survival through an NF-kappaB dependent pathway. Cell Health Differ 1999;6:753–64.

[92] Crowder R,J, Freeman R,S. Phosphatidylinositol 3-kinase and Akt protein kinase are necessary and sufficient for the survival of nerve growth factor-dependent sympathetic neurons. J Neurosci 1998;18:2933–43.

[93] Yasui H, Katoh H, Yamaguchi Y, Aoki J, Fujita H, Mori K. Differential responses to nerve growth factor and epidermal growth factor in neurite outgrowth of PC12 cells are determined by Rac1 activation systems. J Biol Chem 2001;276(18):15298–15305.

[94] Wooten M,W, Seibenhener M,L, Mamidipudi V, Diaz-Meco M,T, Barker P,A, Moscat J. The atypical protein kinase C-interacting protein p62 is a scaffold for NF-kappaB activation by nerve growth factor. J Biol Chem 2001;276:7709–12.

[95] West AE, Griffith EC, Greenberg ME. Regulation of transcription factors by neuronal activity. Nat Rev Neurosci 2002;3:921–31.

[96] Yamada M, Ohnishi H, Sano S, Nakatani A, Ikeuchi T, Hatanaka H. Insulin receptor substrate (IRS)-1 and IRS-2 are tyrosine-phosphorylated and associated with phosphatidylinositol 3-kinase in response to brain-derived neurotrophic factor in cultured cerebral cortical neurons. Biol Chem 1997;272:30334–30339.

[97] Chao M,V. Neurotrophins and their receptors: a convergence point for many signalling pathways. Nat Rev Neurosci 2003;4:299–309.

[98] Obara Y, Aoki T, Kusano M, Ohizumi Y. Beta-eudesmol induces neurite outgrowth in rat pheochromocytoma cells accompanied by an activation of mitogen-activated protein kinase. J Pharmacol Exp Ther 2002;301(3):803–11.

[99] Jalleh R, Koh K, Choi B, Liu E, Maddison J, Hutchinson M,R. Role of microglia and toll-like receptor 4 in the pathophysiology of delirium. Med Hypotheses 2012;79(6):735–9.

[100] Matsuda H, Kagerura T, Toguchida I, Ueda H, Morikawa T, Yoshikawa M. Inhibitory effects of sesquiterpenes from bay leaf on nitric oxide production in

lipopolysaccharide-activated macrophages: structure requirement and role of heat shock protein induction. Life Sci 2000;66:2151–7.

[101] De Marino S, Borbone N, Zollo F, Ianaro A, Di Meglio P, Iorizzi M. New sesquiterpene lactones from *Laurus nobilis* leaves as inhibitors of nitric oxide production. Planta Med 2005;71:706–10.

[102] Hehner SP, Heinrich M, Bork PM, Vogt M, Ratter F, Lehmann V, et al. Sesquiterpene lactones specifically inhibit activation of NF-kappaB by preventing the degradation of I kappa B-alpha and I kappa B-alpha. J Biol Chem 1998;273(3):1288–97.

[103] Jacobo-Herrera N,J, Vartiainen N, Bremner P, Gibbons S, Koistinaho J, Heinrich M. NF-kappaB modulators from *Valeriana officinalis*. Phytother Res 2006;20(10):917–9.

[104] Wang YJ, Chai XQ, Wang WS, Han M, Wen JK. Inulicin inhibits expression of COX-2 and iNOS in the hippocampus of Alzheimer disease rats. Chin J New Drugs 2008;17(15):1318–21.

[105] Wube A,A, Wenzig E,M, Gibbons S, Asres K, Bauer R, Bucar F. Constituents of the stem bark of *Discopodium penninervium* and their LTB4 and COX-1 and -2 inhibitory activities. Phytochem 2008;69(4):982–7.

[106] Lee J, Tae N, Lee JJ, Kim T, Lee JH. Eupatolide inhibits lipopolysaccharide-induced COX-2 and iNOS expression in RAW264.7 cells by inducing proteasomal degradation of TRAF6. Eur J Pharmacol 2010;636(1–3):173–80.

[107] Chun J, Choi RJ, Khan S, Lee D,S, Kim Y,C, Nam Y,J, et al. Alantolactone suppresses inducible nitric oxide synthase and cyclooxygenase-2 expression by down-regulating NF-κB, MAPK and AP-1 via the MyD88 signaling pathway in LPS-activated RAW 264.7 cells. Int Immunopharmacol 2012;14(4):375–83.

[108] Lin H,R. Sesquiterpene lactones from *Tithonia diversifolia* act as peroxisome proliferator-activated receptor agonists. Bioorg Med Chem Lett 2012;22(8):2954–8.

[109] Zhang HL, Xu M, Wei C, Qin AP, Liu CF, Hong LZ, et al. Neuroprotective effects of pioglitazone in a rat model of permanent focal cerebral ischemia are associated with peroxisome proliferator-activated receptor gamma-mediated suppression of nuclear factor-κB signaling pathway. Neurosci 2011;176:381–95.

[110] Lim HJ, Lee H,S, Ryu J,H. Suppression of inducible nitric oxide synthase and cyclooxygenase-2 expression by tussilagone from *Farfarae flos* in BV-2 microglial cells. Arch Pharm Res 2008;31(5):645–52.

[111] Li W, Huang X, Yang XW. New sesquiterpenoids from the dried flower buds of Tussilago farfara and their inhibition on NO production in LPS-induced RAW264.7 cells. Fitoterapia 2012;83:318–22.

[112] Sung B, Jhurani S, Kwang SA, Mastuo Y, Yi T, Guha S, et al. Zerumbone down-regulates chemokine receptor CXCR4 expression leading to inhibition of CXCL12-induced invasion of breast and pancreatic tumor cells. Cancer Res 2008;68(21):8938–44.

[113] Shimoji M, Pagan F, Healton EB, Mocchetti I. CXCR4 and CXCL12 expression is increased in the nigro-striatal system of Parkinson's disease. Neurotox Res 2009;16(3):318–28.

[114] Sung B, Murakami A, Oyajobi BO, Aggarwal BB. Zerumbone abolishes RaNKL-induced NF-kappaB activation, inhibits osteoclastogenesis, and suppresses human breast cancer-induced bone loss in athymic nude mice. Cancer Res 2009;69(4):1477–84.

[115] Thomet OAR, Wiesmann UN, Schapowal A, Bizer C, Simon H,U. Role of petasin in the potential anti-inflammatory activity of a plant extract of *petasites hybridus*. Biochem Pharmacol 2001;61(8):1041–7.

[116] Verkhratsky A, Kettenmann H. Calcium signalling in glial cells. Trends Neurosci 1996;19(8):346–52.

[117] Zhou LJ, Zhu X,Z. Reactive oxygen species-induced apoptosis in PC12 cells and protective effect of bilobalide. J Pharmacol Exp Ther 2000;293(3):982–8.

[118] Shi C, Wu F, Yew DT, Xu J, Zhu Y. Bilobalide prevents apoptosis through activation of the PI3K/Akt pathway in SH-SY5Y cells. Apoptosis15 2010;6:715–27.

[119] Kim S,K, Cho S,B, Moon H,I. Neuroprotective effects of a sesquiterpene lactone and flavanones from *Paulownia tomentosa* Steud. against glutamate-induced neurotoxicity in primary cultured rat cortical cells. Phytother Res 2010;24(12):1898–900.

[120] Zhu Q, Tang C,P, Ke C,Q, Wang W, Zhang H,Y, Ye Y. Sesquiterpenoids and phenylpropanoids from pericarps of *Illicium oligandrum*. J Nat Prod 2009;72(2):238–42.

[121] Li B, Jeong G,S, Kang D,G, Lee H,S, Kim Y,C. Cytoprotective effects of lindenenyl acetate isolated from *Lindera strychnifolia* on mouse hippocampal (HT-22) cells. Eur J Pharmacol 2009;614(1–3):58–65.

[122] Sun Z,L, Gao G,L, Luo J,Y, Zhang X,L, Zhang M, Feng J. A new neuroprotective bakkenolide from the rhizome of *Peatasites tatewakianus*. Fitoterapia 2011;82(3):401–4.

[123] Xu J, Guo Y, Zhao P, Guo P, Ma Y, Xie C, et al. Four new sesquiterpenes from *Commiphora myrrha* and their neuroprotective effects. Fitoterapia 2012;83(4):801–5.

[124] Xu J, Guo Y, Zhao P, Xie C, Jin D,Q, Hou W, et al. Neuroprotective cadinane sesquiterpenes from the resinous exudates of *Commiphora myrrha*. Fitoterapia 2011;82(8):1198–201.

[125] Tasaki Y, Omura T, Yamada T, Ohkubo T, Suno M, Iida S, et al. Meloxicam protects cell damage from 1-methyl-4-phenyl pyridinium toxicity via the phosphatidylinositol 3-kinase/Akt pathway in human dopaminergic neuroblastoma SH-SY5Y cells. Brain Res 2010;1344:25–33.

[126] Ham A, Lee SJ, Shin J, Kim K,H, Mar W. Regulatory effects of costunolide on dopamine metabolism-associated genes inhibit dopamine-induced apoptosis in human dopaminergic SH-SY5Y cells. Neurosci Lett 2012;507(2):101–5.

[127] Ham A, Kim B, Koo U, Nam K,W, Lee S,J, Kim KH, et al. Spirafolide from bay leaf (*Laurus nobilis*) prevents dopamine-induced apoptosis by decreasing reactive oxygen species production in human neuroblastoma SH-SY5Y cells. Arch Pharm Res 2010;33(12):1953–8.

[128] Kakemoto E, Okuyama E, Nagata K, Ozoe Y. Interaction of anisatin with rat brain gamma-aminobutyric acid(a) receptors: allosteric modulation by competitive antagonists. Biochem Pharmacol 1999;58(4):617–21.

[129] Huang SH, Duke RK, Chebib M, Sasaki K, Wada K, Johnston GAR. Bilobalide, a sesquiterpene trilactone from *Ginkgo biloba*, is an antagonist at recombinant alpha1beta2gamma2L GABA(A) receptors. Eur J Pharmacol 2003;464(1):1–8.

[130] Sasaki K, Oota I, Wada K, Inomata K, Ohshika H, Haga M. Effects of bilobalide, a sesquiterpene in *Ginkgo biloba* leaves, on population spikes in rat hippocampal slices. Comp Biochem Physiol, Part C Pharmacol Toxicol Endocrinol 1999;124C:315–21.

[131] Ha JH, Lee KY, Choi HC, Cho J, Kang BS, Lim JC, et al. Modulation of radioligand binding to the GABA(A)-benzodiazepine receptor complex by a new component from *Cyperus rotundus*. Biol Pharm Bull 2002;25(1):128–30.

[132] André E, Campi B, Trevisani M, Ferreira J, Malheiros A, Yunes RA, et al. Pharmacological characterisation of the plant sesquiterpenes polygodial and drimanial as vanilloid receptor agonists. Biochem Pharmacol 2006;71(8):1248–54.

[133] Conn PJ, Pin JF. Pharmacology and functions of metabotropic glutamate receptors. Annu Rev Pharmacol Toxicol 1997;37:205–37.

[134] Copani A, Bruno V, Battaglia G, Leanza G, Pellitteri R, Russo A, et al. Activation of metabotropic glutamate receptors protects cultured neurons against apoptosis induced by beta-amyloid peptide. Mol Pharmacol 1995;47:890–7.

[135] Maiese K, Greenberg R, Boccone L, Swiriduk M. Activation of the metabotropic glutamate receptor is neuroprotective during nitric oxide toxicity in primary hippocampal neurons of rats. Neurosci Lett 1995;194:173–6.

[136] Lee R,K,K, Wurtman R,J, Cox A,J, Nitsch R,M. Amyloid precursor protein processing is stimulated by metabotropic glutamate receptors. Proc Natl Acad Sci USA 1995;92:8083–7.

[137] Maccarrone M, Lorenzon T, Bari M, Melino G, Finazzi-Agro A. Anandamide induces apoptosis in human cells via vanilloid receptors. Evidence for a protective role of cannabinoid receptors. J Biol Chem 2000;275:31938–31945.

[138] Sugita S, Uchimura N, Jiang Z,G, North RA. Distinct muscarinic receptors inhibit release of gamma-aminobutyric acid and excitatory amino acids in mammalian brain. Proc Natl Acad Sci USA 1991;88(6):2608–11.

[139] Lin L,H, Huang T,J, Wang S,H, Lin YL, Wu S,N, Ko W,C. Bronchodilatory effects of S-isopetasin, an antimuscarinic sesquiterpene of *Petasites formosanus*, on obstructive airway hyperresponsiveness. Eur J Pharmacol 2008;584(2–3):398–404.

[140] McGeer P,L, Rogers J. Anti-inflammatory agents as a therapeutic approach to Alzheimer's disease. Neurology 1992;42:447–9.

[141] McGeer P,L, Yasojima K, McGeer E,G. Inflammation in Parkinson's disease. Adv Neuro 2001;86:83–9.

[142] Martino G, Adorini L, Rieckmann P, Hillert J, Kallmann B, Comi G, et al. Inflammation in multiple sclerosis: the good, the bad, and the complex. Lancet Neurol 2002;1:499–509.

[143] Liu B, Hong J,S. Role of microglia in inflammation-mediated neurodegenerative diseases: mechanisms and strategies for therapeutic intervention. J Pharmacol Exp Ther 2003;304:1–7.

[144] Benito C, Tolon R,M, Nunez E, Pazos M,R, Romero J. Neuroinflammation and the glial endocannabinoid system Koffalvi A, editor. *Cannabinoids and the Brain*. : Springer Verlag, New York, USA; 2007. p. 314–34.

[145] Benito C, Tolón R,M, Pazos M,R, Nunez E, Castillo A,I, Romero J. Cannabinoid CB2 receptors in human brain inflammation. Br J Pharmacol 2008;153(2):277–85.

[146] Gertsch J, Leonti M, Raduner S, Racz I, Jian-Zhong Chen JZ, Xie XQ, et al. Beta-caryophyllene is a dietary cannabinoid. Proc Natl Acad Sci U S A. 2008;105(26):9099–104.

[147] Rao RC, Gal C,S,L, Granger I, Gleye J, Augereau J,M, Bessibes C. Khusimol, a non-peptide ligand for vasopressin V(1a) receptors. J Nat Prod 1994;57(10):1329–35.

[148] Everts HGJ, Koolhaas JM. Differential modulation of lateral septal vasopressin receptor blockade in spatial-learning, social recognition, and anxiety-related behaviors in rats. Behav Brain Res 1999;99:7–16.

[149] Winslow J,T, Insel T,R. Neuroendocrine basis of social recognition. Curr Opin Neurobiol 2004;14(2):248–53.

[150] Versteeg D,H, De Kloet E,R, Greidanus T,V, De Wied D. Vasopressin modulates the activity of catecholamine containing neurons in specific brain regions. Neurosci Lett 1979;11(1):69–73.

[151] Son M,C, Brinton R,D. Regulation and mechanism of L-type calcium channel activation via V1a vasopressin receptor activation in cultured cortical neurons. Neurobiol Learn Mem 2001;76(3):388–402.

[152] Kamp T,J, Hell J,W. Regulation of cardiac L-type calcium channels by protein kinase A and protein kinase C. Circ Res 2000;87(12):1095–102.

[153] Berstein G, Blank JL, Smrcka AV, Higashijima T, Sternweis PC, Exton JH, et al. Reconstitution of agonist stimulated phosphatidylinositol 4,5-bisphosphate hydrolysis using purified m1 muscarinic receptor, Gq/11, and phospholipase C-beta. J Biol Chem 1992;267:8081–8.

[154] Ichikawa J, Ishii H, Bonaccorso S, Fowler W,L, O'Laughlin I,A, Meltzer H,Y. 5-HT(2A) and D(2) receptor blockade increases cortical DA release via 5-HT(1A) receptor activation: a possible mechanism of atypical antipsychotic-induced cortical dopamine release. J Neurochem 2001;76(5):1521–31.

[155] Weber JT, O'Connor M-F, Hayataka K, Colson N, Medora R, Russo EB, et al. Activity of Parthenolide at 5HT(2A) receptors. J Nat Prod 1997;60(6):651–3.

[156] Suphrom N, Pumthong G, Khorana N, Waranuch N, Limpeanchob N, Ingkaninan K. Anti-androgenic effect of sesquiterpenes isolated from the rhizomes of *Curcuma aeruginosa* Roxb. Fitoterapia 2012;83(5):864–71.

[157] Gouras G,K, Xu H, Gross R,S, Greenfield J,P, Hai B, Wang R, et al. Testosterone reduces neuronal secretion of Alzheimer's beta-amyloid peptides. Proc Natl Acad Sci U S A 2000;97(3):1202–5.

[158] Fujiwaraj M, Yagi N, Miyazawa M. Acetylcholinesterase inhibitory activity of volatile oil from *Peltophorum dasyrachis* Kurz ex Bakar (yellow batai) and bisabolane-type sesquiterpenoids. J Agric Food Chem 2010;58(5):2824–9.

Topic **2.3**

Diterpenes

2.3.1 *Ajuga ciliata* **Bunge**

History The plant was first described by Alexander Andrejewitsch Bunge in *Mémoires Presentes a l'Académie Impériale des Sciences de St.-Pétersbourg par Divers Savants et lus dans ses Assemblées* published in 1833.

Family Lamiaceae Martinov, 1820

Common Names jin gu cao (Chinese)

Habitat and Description It is a perennial herb which grows to 50 cm tall in the grassy and watery spots of China. The stem is erect, quadrangular, and purplish. The leaves are simple, decussate, and exstipulate. The petiole is purplish, ciliate, and 1 cm long. The blade is elliptic, dull light green, 4 cm–10 cm × 3 cm–6 cm, incised, ciliate, and presents a conspicuous midrib that is channeled and purplish at the base and develops 4–7 pairs of secondary nerves. The inflorescence is terminal, 10 cm long, and spike-like. The calyx is campanulate, 0.7 cm long, and develops 5 teeth, which are ciliate. The corolla is 1 cm long, purple, and bilabiate; the upper lip is short, straight, and emarginated; and the lower lip is deeply trilobed. The androecium consists of 2 pairs of stamens that are didynamous and exerted. The fruit consists of 4 nutlets packed in the persistent calyx (Figure 2.3).

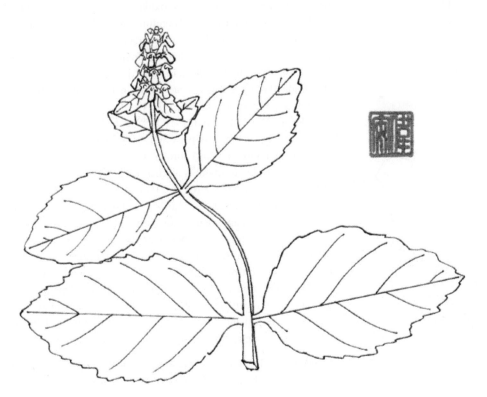

■ **FIGURE 2.3** *Ajuga ciliata* Bunge.

Medicinal Uses In China, the plant is used to treat inflammation.

Phytopharmacology The plant abounds with clerodane diterpenes such as ajugaciliatin A, B, C, D, E, F, G, and H[159]; deacetylajugarin IV; and ajugalide D[160,161]; and produces iridoid glucosides, including reptoside.[162]

Proposed Research Pharmacological study of ajugaciliatin B for the treatment of Parkinson's disease (PD).

Rationale An interesting point of note is that the loss of motor neurons in amyotrophic lateral sclerosis (ALS) results from glutamate activation of N-Methyl-D-aspartate (NMDA), followed by massive influx of Ca^{2+}, resulting in the activation of calpain, mitogen-activated protein kinase (MAPK) p38, c-Jun N-terminal kinase (JNK), inducible nitric oxide synthetase (iNOS), and mitochondrial insults followed by caspase 3 activation, reticulum endoplasmic (RE) stress, burst in reactive oxygen species (ROS) and apoptosis as the antioxidant activity of superoxide dismutase (SOD) is compromised.[163] Therefore, natural products and especially diterpenes with the ability to counteract the deleterious effect of glutamate are of immense neuropharmacological value. In this light, the labdane 15-methoxypinusolidic acid (CS 2.100) isolated from *Biota orientalis* (L.) Endl. (family Cupressaceae Gray) increased the viability of cortical neurons exposed to glutamate by 72.4% at a dose of $10\,\mu M$[164] by reducing cytoplasmic Ca^{2+}, nitric oxide (NO),[165] and ROS and synergistically maintaining the enzymatic activity of SOD.[166] Furthermore, 15-methoxypinusolidic acid at a dose of $50\,\mu M$ abrogated the production of NO by BV-2 glial cells challenged with lipopolysaccharides[167] and reduced the glial levels of iNOS, cyclo-oxygenase-2 (COX-2), tumor necrosis factor-α (TNF-α), and Interleukin-6 (IL-6) and, interestingly, without interfering with nuclear factor kappa-light-chain-enhancer of activated B cells (NF-κB),[167] implying that this labdane not only rends neurons resistant to glutamate challenge but inhibits the activation of glial cells, which in fact secrete glutamate.[168] One could suggest that the aforementioned reduction of COX-2 may result from the activation of the transcriptional activity of peroxisome proliferator-activated receptor-γ (PPAR-γ)[169] as 15-methoxypinusolidic acid at a dose of $200\,\mu M$ inhibited the transcriptional activity of PPAR-γ in preadipocytes,[170] suggesting the involvement of activator protein-1 (AP-1).

Pinusolide (CS 2.101) prevented the apoptosis of cortical neurons exposed to the protein kinase C (PKC) inhibitor staurosporine by blocking Ca^{2+}

■ **CS 2.100** 15-methoxypinusolidic acid.

■ **CS 2.101** Pinusolide.

■ **CS 2.102** 19-hydroxylabda-8(17)-en-16,15-olide.

■ **CS 2.103** 3β,19-dihydroxylabda-8(17), 11E-dien-16,15-olide.

■ **CS 2.104** 19-hydroxylabda-8(17),13-dien-16, 15-olide.

increase and ROS insults, and sustaining the enzymatic activity of SOD.[166] Furthermore, pinusolide inhibited the production of leukotriene C_4 (LTC_4) by bone marrow mast cells with a reduction of Ca^{2+} and hypophosphorylation of JNK.[171] JNK sustained activation and the subsequent transcription of COX-2 via the AP-1 account for the death of dopaminergic neurons exposed to 1-methyl-4-phenyl-1,2,3,6-tetrahydropyridine (MPTP),[172] suggesting that pinusolide and congener may be useful to treat PD. Furthermore, pinusolide inhibited the binding of platelet activating factor (PAF) to platelets with an IC_{50} value equal to 2.3×10^{-5} M,[173] thus hindering the aggregation of platelets, the activation of β and γ-secretases,[174] and the chimeric cleavage of the amyloid precursor protein (APP) into the β-amyloid ($A\beta_{1-42}$) peptide, which accounts for Alzheimer's disease (AD). The mushroom *Antrodia camphorata* (M. Chang & CH Su) Sheng H. Wu, Ryvarden & T.T (family Polyporaceae Fr. ex Corda, 1839) contains the labdane diterpenes 19-hydroxylabda-8(17)-en-16,15-olide (CS 2.102), 3β,19-dihydroxylabda-8(17),11E-dien-16,15-olide (CS 2.103), 19-hydroxylabda-8(17),13-dien-16,15-olide (CS 2.104), and 14-deoxy-11,12-didehydroandrographolide (CS 2.105), which protected cortical neurons against β-amyloid ($A\beta_{25-35}$) peptide insults by 28.1%, 29.5%, 28.9%, and 30.8%, respectively, at a dose of 10 μM.[175] Other neuroprotective labdane diterpenes are leojaponin (CS 2.106) from *Leonurus japonicus* Houtt. (family Lamiaceae Martinov), which boosted the viability of cortical neurons of rodents exposed to glutamate by 68.9% at a dose of 10 μM.[176] In addition, *Leonurus heterophyllus* Sweet (family Lamiaceae Martinov) shelters the labdane diterpenes heteronin F, leoheteronin A, leoheteronin D, and leopersin G

■ **CS 2.105** 14-deoxy-11, 12-didehydroandrographolide.

■ **CS 2.106** Leojaponin.

■ **CS 2.107** Heteronin F.

■ **CS 2.108** Leoheteronin A.

■ **CS 2.109** Leoheteronin D.

■ **CS 2.110** Leopersin G.

■ **CS 2.111** 6α,7β-dihydroxy-labda-8(17),12(E),14-triene.

■ **CS 2.112** 6-oxo-2α-hydroxy-labda-7,12(E),14-triene.

(CS 2.107–2.110), which inhibited the enzymatic activity of acetylcholinesterase (AChE) with IC$_{50}$ values equal to 16.1 μM, 11.6 μM, 18.4 μM, and 12.9 μM, respectively.[177] 6α,7β-Dihydroxy-labda-8(17),12(E),14-triene (CS 2.111) and 6-oxo-2α -hydroxy-labda-7,12(E),14-triene (CS 2.112) from *Fritillaria ebeiensis* G.D. Yu & G.Q.Ji (family Liliaceae Juss.) increased the life span of human SH-SY5Y neuroblastoma cells poisoned with 1-methyl-4-phenylpyridinium ion (MPP$^+$) by 118.3% and 115.8% at a dose of 30 μM[178] via probable mitigation of oxidative insults.

The migration of methyl groups from C4 and C10 of the labdane framework to C5 and C9 produces the clerodane diterpenes, which have the interesting, albeit yet unexplained, propensity to evoke the growth of neurites. For instance, ptychonal hemiacetal (CS 2.113) and ptychonal

■ **CS 2.113** Ptychonal hemiacetal.

■ **CS 2.114** Ptychonal.

■ **CS 2.115** 6α,7α-dihydroxyannonene.

■ **CS 2.116** 7α,20-dihydroxyannonene.

■ **CS 2.117** 15,16-epoxy-15α-methoxy-ent-clerod-3-en-18-oic acid.

■ **CS 2.118** 13-epi-15,16-epoxy-15β-methoxy-ent-clerod-3-en-18-oic acid.

(CS 2.114) from *Ptychopetalum olacoides* Benth. (family Olacaceae R. Br.) at a dose of 10 μM boosted the growth of neurite of rat pheochromocytoma (PC12) cells exposed to 20 ng/mL of nerve growth factor (NGF).[179] Likewise, the clerodane diterpenes 6α,7α-dihydroxyannonene (CS 2.115) and 7α,20-dihydroxyannonene (CS 2.116) from the same plant evoked the sprouting of neurites from PC12 cells in the presence of 20 ng/mL of NGF at doses of 50 μM and 30 μM, respectively.[180] The ent-clerodane 15,16-epoxy-15α-methoxy-ent-clerod-3-en-18-oic acid (CS 2.117) and 13-epi-15,16-epoxy-15β-methoxy-ent-clerod-3-en-18-oic acid (CS 2.118) isolated from *Baccharis gaudichaudiana* DC. (family Asteraceae Bercht. & J. Presl) at a dose of 100 μmol promoted the growth of neurite by PC12 cells exposed to 2 ng/mL of NGF to 49% and 53%, respectively.[181] From

■ **CS 2.119** 15,16-epoxy-7α,18-dihydroxy-15-methoxy-ent-clerod-3-ene.

■ **CS 2.120** 13,14-dihydro-marrubiagenine.

■ **CS 2.121** (12S,2'S)-12,19-diacetoxy-18-chloro-4α,6α-dihydroxy-1β-(2-methylbutanoyloxy)-neo-clerod-13-en-15,16-olide.

the same plant, 15,16-epoxy-7α,18-dihydroxy-15-methoxy-ent-clerod-3-ene (CS 2.119) and 13,14-dihydro-marrubiagenine (CS 2.120) promoted the growth of PC12D cells in the presence of 2 ng/mL of NGF by 68% and 28%, respectively, at a dose of 100 μmol.[182]

Polyoxygenated clerodane diterpenes have the compelling activity to protect dopaminergic neurons against apoptosis. For instance, (12S,2'S)-12,19-diacetoxy-18-chloro-4α,6α-dihydroxy-1β-(2–methylbutanoyloxy)-neo-clerod-13-en-15,16-olide (CS 2.121), (12S)-6α,18,19-triacetoxy-4α,12-dihydroxy-1β-tigloyloxy-neo-clerod-13-en-15,16-olide (CS 2.122),

■ **CS 2.122** (12S)-6α,18,19-triacetoxy-4α,12-dihydroxy-1β-tigloyloxy-neo-clerod-13-en-15,16-olide.

■ **CS 2.123** Deacetylajugarin IV.

■ **CS 2.124** Ajugaciliatin B.

■ **CS 2.125** Ajugamarin A2 chlorohydrin.

deacetylajugarin IV (CS 2.123), and ajugalide D from *Ajuga ciliata* Bunge (family Lamiaceae Martinov) safeguarded SH-SY5Y cells against 1-methyl-4-phenylpyridinium ion (MPP$^+$) by 80%, 70%, 70%, and 70%, respectively, at a dose of 30 μM.[160] From the same plant, the clerodane diterpenes ajugaciliatin B (CS 2.124), ajugamarin A2 chlorohydrin (CS 2.125),

■ **CS 2.126** Ajugamarin A1 chlorohydrin.

■ **CS 2.127** Ajugaciliatin G.

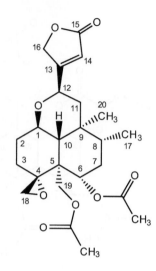

■ **CS 2.128** Ajugaciliatin H.

■ **CS 2.129** Ajugamarin H.

■ **CS 2.130** Ajugatakasin A.

ajugamarin A1 chlorohydrin (CS 2.126), ajugaciliatin G (CS 2.127), ajugaciliatin H (CS 2.128), ajugamarin H (CS 2.129), ajugatakasin A (CS 2.130), ajugamarin A1 (CS 2.131), and ajuganipponin A (CS 2.132) at a dose of 30 μM enhanced the survival of human SH–SY5Y neuroblastoma cells exposed to 1-methyl-4-phenylpyridinium ion (MPP+) by 88.2%, 83.7%, 73.8%, 81.5%, 78.2%, 87.5%, 78.2%, 81.9%, and 76.6%, respectively.[159]

■ **CS 2.131** Ajugamarin A1.

■ **CS 2.132** Ajuganipponin A.

■ **CS 2.133** (12S)-6α-acetoxy-4α,18-epoxy-12-hydroxy-19-tigloyloxy-neoclerod-13-en-15,16-olide.

■ **CS 2.134** Ajugalide D.

The pathophysiology of PD involves the progressive and inexorable loss of dopaminergic neurons in the *substantia nigra* and *loci* of activation of glial cells notably by α-synuclein.[183] In this light, it is of interest to mention that the neo-clerodane diterpenes (12S)-6α-acetoxy-4α,18-epoxy-12-hydroxy-19-tigloyloxy-neoclerod-13-en-15,16-olide (CS 2.133) and ajugalide D (CS 2.134) inhibited the generation of NO production in microglial BV-2 cells challenged

■ **CS 2.135** Salvinorin A.

■ **CS 2.136** Ballotenic acid.

■ **CS 2.137** Ballodiolic acid.

with LPS with IC_{50} values equal to $28.6 \mu M$ and $43.5 \mu M$, respectively.[184] Considering evidence that clerodane protects dopaminergic neurons from intrinsic and extrinsic insults, one could reasonably contemplate the possibility of identifying leads for the treatment of PD from this group of terpenes.

Salvia divinorum Epling & Játiva (family Lamiaceae Martinov) produces the *ent*-clerodane diterpene salvinorin A (CS 2.135), which is a robust opioid receptor subtype κ agonist with an IC_{50} value equal to 1 nM.[185] Indeed, salvinorin A at a dose of 100 nM increased the release of noradrenaline and decreased the exocytosis of serotonine from hippocampal synaptosomes induced by K^+.[186] These effects were abrogated by the opioid receptor subtype κ antagonist norbinaltorphimine, further confirming that salvinorin A binds to presynaptic opioid receptor subtype κ.[186] Salvinorin A at a dose of $1 \mu M$ evoked the dilation of the cerebral artery by stimulation of opioid receptor subtype κ, nitric oxide synthetase (NOS), and adenosine triphosphate-sensitive potassium (K_{ATP}) channels.[187] Note that the opioid receptor subtype κ agonists command the activation of $G\alpha$ protein, inhibit voltage-activated Ca^{2+} channel opening, and activate phospholipase C (PLC); hence, an increase of inositol 1,4,5-trisphosphate induces PKC and extracellular signal-regulated kinase (ERK1/2) and JNK,[188,189] raising the possibility that opioid receptor subtype κ agonists like salvinorin A and semisynthetic derivatives may favor the growth and survival of neurons.

The clerodane diterpenes ballotenic acid (CS 2.136) and ballodiolic acid (CS 2.137) from *Ballota limbata* Benth. (family Lamiaceae Martinov) inhibited the enzymatic activity of lipoxygenase (LOX) with IC_{50} values equal to $99.6 \mu M$ and $38.3 \mu M$, respectively,[190] raising the possibility of

■ **CS 2.138** Ballatenolide A.

■ **CS 2.139** 15-methoxy-16-oxo-15,16H-hardwickiic acid.

■ **CS 2.140** Limbatolide A.

■ **CS 2.141** Limbatolide B.

■ **CS 2.142** Limbatolide C.

finding 12-lipoxygenase (12-LOX) inhibitors among this class of terpenes. From the same plant, the clerodane diterpenes ballatenolide A (CS 2.138) and 15-methoxy-16-oxo-15,16H-hardwickiic acid (CS 2.139) inhibited the enzymatic activity of AChE with IC_{50} values equal to $50\,\mu M$ and $52\,\mu M$.[191] Likewise, the clerodane diterpenes limbatolide A, limbatolide B, and limbatolide C (CS 2.140–2.142) inhibited the enzymatic activity of AChE with IC_{50} values equal to $38.5\,\mu M$, $47.2\,\mu M$, and $103.7\,\mu M$, respectively.[192] Furthermore, the clerodane diterpene 5α,8α-2-oxokolavenic acid from *Detarium microcarpum* Guill. & Perr. (family Fabaceae Lindl.) inhibited the enzymatic activity of AChE at a dose of $0.1\,\mu g$.[193]

■ **CS 2.143** Carnosic acid.

■ **CS 2.144** Carnosol.

The formation of a covalent bond between C14 and C8 of the labdane skeleton followed by the migration of C16 and C17 methyls in C15 yields the abietane diterpenes, such as carnosic acid (CS 2.143) and carnosol (CS 2.144), which are engineered by *Rosmarinus officinalis* L. (family Lamiaceae Martinov). Carnosic acid and carnosol are antioxidants which scavenge, for instance, α-diphenyl-β-picrylhydrazyl radicals with IC_{50} values equal to $0.6\,\mu$M and $0.5\,\mu$M, respectively,[194] and preserve mouse hippocampal cell (HT-22) neurons against glutamate insults at a dose of $3\,\mu$M[195] via the probable induction of PPAR-γ as shown in Cos7 cells with EC_{50} values equal to $41.2\,\mu$M and $19.6\,\mu$M, respectively.[196] Activating the PPAR-γ results in NF-κB inhibition and therefore reduction of both COX-2 and iNOS.[197]

Carnosic acid shielded dopaminergic neurons (SN4741) against dieldrin-induced apoptosis by contravening the activation of caspase 3 and 12, and JNK at a dose of $10\,\mu$M.[198] Likewise, PC12 cells exposed to $6.8\,\mu$g/mL of carnosic acid had increased levels of acetylcholine production, and strikingly, evoked the growth of neurites as a result of ERK1/2.[199] In addition, carnosic acid at a dose of $20\,\mu$mol/mL induced the nuclear translocation of transcription factor (Nrf2) and therefore the production of NADPH: quinone oxidoreductase 1 in clone 9 cells via phosphorylation of MAPK p38.[200] Note that NAD(P)H: quinone oxidoreductase 1 is antioxidant and protected neurons against dopamine, 6-hydroxydopamine, and hydrogen peroxidase insults.[201] Along the same lines, carnosic acid at a dose of $1\,\mu$M sustained the viability of SH-SY5Y neurons against 6-hydroxydopamine (6-OHDA) and 1-methyl-4-phenylpyridinium ion (MPP$^+$) by 45% and 32%, respectively, by inducing Nrf2 and increasing γ-GCL protein and the subsequent synthesis of GSH, hence a decrease in

ROS, inhibition of MAPK p38 and JNK, and abrogation of apoptosis.[202] The activation of transcription factor NF-E2–related factor-2 (Nrf2) by nucleophiles comprises its release from KEAP, nuclear translocation, and binding to antioxidant response elements (AREs) from which the transcription of heme oxygenase-1 (HO-1) and NAD(P)H: quinone oxidoreductase 1 (NOQ1) protect neurons and glial cells against oxidative insults incurred by β-amyloid ($A\beta_{1-42}$) peptide.[203] Thus, carnosic acid may very well be of value in delaying the progression of PD, ALS, and AD. In this regard, carnosic acid at a dose of $30\,\mu M$ hampered the production of β-amyloid ($A\beta_{1-42}$) peptide by human SH_SY5Y neuroblastoma cells to 39% and at the same time increased a secreted form of APP (sAPPα) as a result of α-secretase induction.[204] Along the same lines, carnosic acid boosted the synthesis of NGF by T98G cells at a dose of $100\,\mu M$[205] via the induction of Nrf2[206] and inhibited the aggregation of platelets exposed to collagen, arachidonic acid, and thrombin with IC_{50} values equal to $39\,\mu M$, $34\,\mu M$, and $48\,\mu M$, respectively.[207]

Carnosol is not only antioxidant *per se* but inhibits the secretion of cytokines in glial cells. Indeed, this aromatic abietane inhibited the generation of NO by macrophages (RAW 264.7) challenged with LPS with an IC_{50} value equal to $9.4\,\mu M$.[193] At a dose of $60\,\mu M$, it reduced the production of prostaglandin E2 (PGE2), and COX-2 induced in HER-2/neu-transformed human mammary epithelial cells (184B5/HER) by PKC activator phorbol 12-myristate 13-acetate (PMA) via the inhibition of PKC, ERK1/2, MAPK p38, and JNK, and consequently AP-1.[208] The inhibition of AP-1 was further demonstrated in malignant cells, whereas $10\,\mu M$ of carnosol inhibited this protein.[209] The activation of NF-κB involves nuclear factor of kappa light polypeptide gene enhancer in B-cells inhibitor α (IκBα) activation by protein kinase B (Akt) and results in an increase in iNOS.[193] In this light, $10\,\mu M$ of carnosol induced HO-1 in PC12 cells deprived of serum via the activation of phosphoinositide 3-kinase (PI3K) and thus Akt, transcription factor NF-E2-related factor-2 (Nrf2), and ARE sequences.[210]

The cardinal role of transcription factor NF-E2-related factor-2 (Nrf2) in carnosol antioxidant activity was further substantiated as $5\,\mu$mol/L of this diterpene boosted GSH by 160% in malignant cells via the induction of glutamate-cysteine ligase (GCL), activation of transcription factor NF-E2-related factor-2 (Nrf2), and binding to ARE sequences.[211] In the same experiment, the aforementioned increase of GSH resulted in NF-κB inactivation.[211] There is evidence indicating that *Salvia miltiorrhiza* Bunge (family Lamiaceae Martinov) produces a series of neuroprotective abietane

■ **CS 2.145** Cryptotanshinone.

■ **CS 2.146** 15,16-dihydrotanshinone I.

■ **CS 2.147** Tanshinone IIA.

diterpenes such as cryptotanshinone (CS 2.145). Cryptotanshinone and 15,16-dihydrotanshinone I (CS 2.146) abated the production of NO by BV-2 cells challenged with LPS at a dose of $10\,\mu$M.[212] In the same experiment, 15,16-dihydrotanshinone I repressed the expression of iNOS, the secretion of Interleukin-1-β (IL-β), and TNF-α,[212] implying the possible hindrance of NF-κB, hence β-secretase inhibition.[213] As a matter of fact, cryptotanshinone fought dementia in APP/PS1 rodent at a dose of 15 mg/Kg/day by mitigating the deposition of amyloid plaques, decreasing β-amyloid (Aβ_{1-42}) peptide by 50%, and increasing the enzymatic activity of α-secretase and thus a secreted form of APP (sAPPα).[214] Along the same lines, cortical neurons expressing APPs exposed to cryptotanshinone at a dose of $10\,\mu$M had their ability to produce β-amyloid (Aβ_{1-42}) peptide hindered by 45% together with an increase of a secreted form of APP (sAPPα) on account of PI3K induction, activation of Akt, hence α-secretase expression.[215]

Likewise, C2C12 myotubes treated with $20\,\mu$M of cryptotanshinone experienced PI3K and Akt activation and therefore activation of the mammalian target of rapamycin (mTOR),[216] suggesting the possible use of this diterpene to enhance neurite outgrowth. Furthermore, cryptotanshinone at a dose of $5\,\mu$M increased the survival of cortical neurons exposed to glutamate by 82.2% on account of PI3K and Akt activation and the consequent inhibition of pro-apoptotic Bcl-2-associated X protein (Bax).[217] In addition, cryptotanshinone inhibited the enzymatic activity of AChE with an IC$_{50}$ value equal to $4\,\mu$M, and at a dose of 5 mg/kg protected rodents against the dementia induced by scopolamine.[218] Likewise, tanshinone protected rodent against the cholinergic depletion in the *hippocampus* induced by β-amyloid (Aβ_{1-40}) peptide at a dose of 50 mg/kg with a decrease of iNOS in the hippocampus.[219] Considering the evidence, cryptotanshinone and derivatives are able to rescue neurons from excitotoxic challenges and to potentially command the growth of neurites; the present results suggest that this terpene may be of value for the treatment of ALS, AD, and PD.

Along the same lines, *Salvia miltiorrhiza* Bunge (family Lamiaceae Martinov) engineers tanshinone IIA (CS 2.147), which at a dose of $10\,\mu$mol/L protected PC12 cells against ethanol with a decrease of ROS and abrogation of apoptosis via pro-apoptotic protein (p53),[220] suggesting antioxidant effects. Indeed, tanshinone IIA at a dose of $1\,\mu$M nullified the neurotoxic effects of β-amyloid (Aβ_{25-35}) peptide toward cortical neurons by boosting the enzymatic activity of SOD, resulting in a decrease in ROS and therefore protection of mitochondrial integrity and release of

■ **CS 2.148** Tanshinone IIB.

■ **CS 2.149** Kahweol.

cytochrome c and inactivation of caspase 3.[221] Along the same lines, $10\,\mu M$ of tanshinone IIA preserved cortical neurons against H_2O_2 by reducing cytoplasmic Ca^{2+} and inactivating pro-apoptotic Bax.[222] In the same experiment, tanshinone II at a dose of $10\,\mu M$ maintained synaptic transmission in hippocampal preparations exposed to H_2O_2.[222] Another possibly antioxidant and neuroprotective abietane diterpene sheltered by this plant is tanshinone IIB (CS 2.148), which protected rodents against experimental stroke, as shown by a decrease in focal infarct volume at a dose of 25 mg/kg,[223] and defended neurons against staurosporine-induced apoptosis with an IC_{50} equal to $75.4\,\mu M$ by inactivating pro-apoptotic Bax, increasing Bcl-2, and inhibiting caspase 3 activation at a dose of $100\,\mu M$.[224]

The abietane diterpene kahweol (CS 2.149) isolated from a member of the genus *Coffea* L. (family Rubiaceae Juss.) nullified the production of NO in RAW 264.7 exposed to LPS at a dose of $20\,\mu M$ via a mechanism involving the blockade of NF-κB and therefore the production of iNOS.[225] Along the same lines, endothelial cells exposed to kahweol at a dose of $10\,\mu M$ resisted the effects of TNF-α *id sunt* to impose the expression of vascular cell adhesion molecule-1 (VCAM-1) and intercellular cell adhesion molecule-1 (ICAM-1) by activating protein tyrosine kinase JAK2, PI3K, and therefore Akt, kappa light polypeptide gene enhancer in B-cells inhibitor (IκB) kinase (IKK), and NF-κB,[226] suggesting the use of this diterpene to counteract neuroinflammation. Furthermore, kahweol at a dose of $10\,\mu M$ abrogated the deleterious effects of 6-hydroxydopamine (6-OHDA) against human SH-SY5Y neuroblastoma cells by combating ROS via PI3K, Akt, and MAPK p38 activation, Nrf2 induction, and subsequently HO-1 production.[227] In light of these observations, one could reasonably assume that kahweol and semisynthetic derivatives may have some usefulness for the treatment of PD.

Likewise, *Tripterygium wilfordii* Hook. f. (family Celastraceae R.Br.) produces the abietane diterpene tripchlorolide (CS 2.150), which at a dose of 10^{-10} M provoked the sprouting of neurites from mesencephalic dopaminergic neurons by 43%[228] and protected the aforementioned neurons against 1-methyl-4-phenylpyridinium ion (MPP$^+$) insults at a dose of 10^{-8} M together with an increase of brain-derived neurotrophic factor (BDNF) production.[228] Furthermore, tripchlorolide at a dose of 1 μg/kg protected dopaminergic neurons in the *substantia nigra* of C57BL/6 mice poisoned with 1-methyl-4-phenyl-1,2,3,6-tetrahydropyridine (MPTP) by 80% and sustained the levels of dopamine.[229] In the same experiment, triptochloride at a dose of 1 μg/kg improved the locomotor activity of C57BL/6 mice poisoned with 1-methyl-4-phenyl-1,2,3,6-tetrahydropyridine (MPTP).[229] The precise molecular mechanism involved here is yet elusive, but emerging evidence suggests the inactivation of glial cells. Indeed, tripchlorolide at a dose of 10 nM protected primary cortical neurons against LPS insults by 40% and 38%, respectively, by inactivating glial cells and therefore abating the secretion of TNF-α, IL-1β, NO, and PGE2.[231] In microglial BV-2 cells, tripchlorolide at a dose of 10 nM halved the LPS-induced burst of intracellular ROS and reduced the levels of iNOS and COX-2 by 74% and 56%, respectively,[230] via the probable inhibition of NF-κB. In effect, microglial BV-2 cells challenged with β-amyloid (Aβ_{1-42}) peptide and at the same time exposed to tripchlorolide at a dose of 10 nM had their capacities to secrete TNF-α, IL-1β, PGE2, and NO reduced by 51.5%, 56.4%, 27.6%, and 37.3%, respectively, as a result of iNOS and COX-2 inhibition as well as JNK and NF-κB inactivation.[231] Note that tripchlorolide at a dose of 10 nM protected Neuro-2A cells and primary cortical neurons against β-amyloid (Aβ_{1-42}) peptide insults by 28% and 35%, respectively, as a result of microglial inactivation.[231] The abietane diterpene triptolide (CS 2.151) from the same plant protected dopaminergic neurons against LPS insults by 46% at a dose of 10 nM as a result of microglial inactivation, leading to a decrease in TNF-α

■ **CS 2.150** Tripchlorolide.　　　　■ **CS 2.151** Triptolid.

and NO production.[232] Triptolide at a dose of $5\,\mu g/kg$ protected rodents against the intrathecal injection of LPS in the *substancia nigra*, which compelled the degeneration of dopaminergic neurons following the activation of glial cells.[233] The secretion of PGE2 by microglial cells challenged with LPS was nullified by 50nM of triptolide as a result of JNK and NF-κB inactivation and subsequent reduction of COX-2.[234] Along the same lines, triptolide at a dose of 100ng/mL inhibited the phosphorylation of Akt, MAPK p38, ERK1/2, and consequently NF-κB, COX-2, and PGE2 in PC12 cells challenged with LPS.[235]

An interesting point of note is that triptolide at a dose of 10^{-8} M imposed the synthesis and secretion of NGF by 400% and two-fold, respectively, from astrocytes.[236] Furthermore, triptolide boosted the synthesis of survival motor neuron (SMN) protein in fibroblast obtained from patients with spinal muscular atrophy (SMA) at a dose of 1 pM.[237] Additionally, triptolide at dose of 1 mg/kg increased the production of survival motor neuron (SMN) protein in muscles and spinal cord of SMA-like rodents which likewise exhibited increased viability.[238] Triptolide at a dose of $1\,\mu M$ prevented the physical activation of astrocytes via hypophosphorylation of STAT3 and JAK2.[237] In the same experiment, rodents with spinal cord injuries displayed some levels of locomotor improvement upon triptolide administration at a dose of 0.1 mg/kg,[237] preliminarily suggesting that triptochloride, triptolide, and semisynthetic derivatives may very well offer leads for the treatment of ALS and spinal cord injuries.

The cyclization of the labdane skeleton by the formation of a covalent bond between C14 and C8, C13 and C16, followed by the attachment of C17 methyl to C16, and the formation of an extra cycle via C15 and C16 bonding yields the kaurane diterpenes which offer robust protection against neuro-inflammation. In this regard, it is interesting to mention that the *ent*-kaurane diterpenes longikaurin C, longikaurin B, longikaurin D (CS 2.152–2.154),

■ **CS 2.152** Longikaurin C.

■ **CS 2.153** Longikaurin B.

■ **CS 2.154** Longikaurin D.

and lasiokaurin (CS 2.155) from *Isodon japonicus* (Burm. f.) H. Hara (family Lamiaceae Martinov) inhibited the production of NO by macrophages (RAW264.7) exposed to LPS with IC_{50} values equal to $1\,\mu$M, $0.9\,\mu$M, $2\,\mu$M, and $1.6\,\mu$M, respectively.[238] More pertinently, kamebanin (CS 2.156), kamebacetal A (CS 2.157), kamebakaurin, and excisanin A (CS 2.158) from the same plant dampened the generation of NO and PGE2 from macrophages (RAW264.7) challenged with LPS with IC_{50} values equal to $0.06\,\mu$M, $0.5\,\mu$M, $0.1\,\mu$M, and $0.3\,\mu$M, respectively, and $0.8\,\mu$M, $2.8\,\mu$M, $2.6\,\mu$M, and $5.3\,\mu$M, respectively, as a result of NF-κB inhibition[239] on probable account of PPAR-γ induction. Correspondingly, 18-hydroxy-1α,6α-diacetoxy-6,7-seco-ent-kaur-16-en-15-one-7,20-olide (CS 2.159) isolated from the same plant at a dose of $20\,\mu$g/mL nullified the deleterious effects of β-amyloid (Aβ_{25-35}) peptide against PC12 cells via inhibition of NF-κB, hence a reduction of COX-2, and hindered apoptosis by decreasing pro-apoptotic Bax, protecting mitochondrial integrity, and inactivating caspase 3.[240] In microglial BV-2 cells exposed to LPS, 18-hydroxy-1α,6α -diacetoxy-6,7-seco- ent-kaur-16-en-15-one-7,20-olide at a dose of $10\,\mu$g/mL evoked the hypophosphorylation of

■ **CS 2.155** Lasiokaurin.

■ **CS 2.156** Kamebanin.

■ **CS 2.157** Kamebacetal A.

■ **CS 2.158** Excisanin A.

■ **CS 2.159** 18-hydroxy-1α,6α-diacetoxy-6,7-seco-ent-kaur-16-en-15-one-7,20-olide.

ERK1/2, MAPK p38, and JNK; inhibited NF-κB; reduced the expression of iNOS and COX-2; and hindered the secretion of NO, PGE2, IL-1β, IL-6, and TNF-α.[241]

From the same plant, the kaurane diterpene kamebakaurin (CS 2.160) at a dose of 5 μM hypophosphorylated MAPK p38 and JNK, inhibited NF-κB, abated the activation of iNOS and COX-2, and therefore the secretion of NO and PGE2, TNF-α, IL-1β, and IL-6 by microglial BV-2 cells challenged with LPS.[242] Likewise, the kaurane diterpene effusanin C (CS 2.161) at a dose of 3 μg/mL attenuated the secretion of NO, TNF-α, and IL-1β from macrophages (RAW265.7) challenged with LPS as a consequence of MAPK p38, JAK, and NF-κB inhibition.[243] Note that *Fritillaria ebeiensis* G.D. Yu & G.Q.Ji (family Liliaceae Juss.) shelters the *ent*-kaurane diterpenes ent-3β-butanoyloxykaur-15-ene-17-ol (CS 2.162), ent-kaur-15-ene-17-ol[160] (CS 2.163), ent-kaur-15- ene-3β,17-diol (CS 2.164),

■ **CS 2.160** Kamebakaurin.

■ **CS 2.161** Effusanin C.

■ **CS 2.162** ent-3β-butanoyloxykaur-15-ene-17-ol.

■ **CS 2.163** *ent*-kaur-15-ene-17-ol.

■ **CS 2.164** *ent*-kaur-15-ene-3β,17-diol.

ent-3β-acetoxykaur-15-ene-17-ol (CS 2.165), and ent-kauran-16β,17-diol (CS 2.166), which sustained the viability of human SH-SY5Y neuroblastoma cells against 1-methyl-4-phenylpyridinium ion (MPP$^+$) by 90%, 90%, 100%, 95%, and 80% respectively, at a dose of 30 μM.[244] Given the fact that neuroinflammation accounts for the pathophysiology of AD, PD, and ALS, it is tempting to speculate that kaurane diterpenes may create leads for the treatment of neurodegenerative diseases.

Mushrooms belonging to the genus *Sarcodon* Quél. ex P. Karst (family Bankeraceae Donk) synthesize polyoxygenated cyathane diterpenes with compelling neurotrophic activities. For instance, *Sarcodon scabrosus* (Fr.) P. Karst. elaborates scabronine A (CS 2.167), which at a dose of 100 pM imposed the synthesis of NGF in astrocytoma cells by 746 pg/mL.[245] Likewise, scabronines B, C, E, and F (CS 2.168–2.171) from the same

■ **CS 2.165** *ent*-3β-acetoxykaur-15-ene-17-ol.

■ **CS 2.166** *ent*-kauran-16β,17-diol.

■ **CS 2.167** Scabronine A.

■ **CS 2.168** Scabronine B.

species increased the production of NGF by astroglial cells 31.1 pg/mL, 12.3 pg/mL, 21.4 pg/mL, and 4.7 pg/mL at doses of 100 μg/mL, 100 μg/mL, 33 μg/mL, and 100 μg/mL, respectively.[246] Of note, the cyathane diterpenes scabronines A and G (CS 2.172) commanded the growth of neurites in PC12 cells via the production of NGF by human astrocytoma cells at a dose of 100 μM.[247] Scabronine G methyl ester (CS 2.173) at a dose of 100 μM induced the activation of PKC, hence NF-κB and the synthesis of NGF.[248] The cyathane diterpenes sarcodonins G and A (CS 2.174–2.175) from *Sarcodon scabrosus* (Fr.) P. Karst. at a dose of 25 μM enhanced the sprouting of neurites from PC12 cells exposed to 20 ng/mL of NGF by 24.9% and 20.0%, respectively.[249]

■ **CS 2.169** Scabronine C.

■ **CS 2.170** Scabronine E.

■ **CS 2.171** Scabronine F.

■ **CS 2.172** Scabronine G.

■ **CS 2.173** Scabronine G methyl ester.

■ **CS 2.174** Sarcodonin G.

■ **CS 2.175** Sarcodonin A.

■ **CS 2.176** Cyrneine A.

■ **CS 2.177** Cyrneine C.

■ **CS 2.178** Glaucopine C.

■ **CS 2.179** Tricholomalide A.

■ **CS 2.180** Tricholomalide B.

The cyathane diterpenes cyrneines A (CS 2.176) and B from *Sarcodon cyrneus* Maas Gest induced the sprouting of dendrites in PC12 cells via a mechanism involving the activation of Rac, AP-1, and NF-κB at a dose of 100 μM.[250,251] The cyathane diterpenes cyrneine C (CS 2.177) and glaucopine C (CS 2.178) from the same mushroom exhibited mild neuritogenic effects at dose of 200 μM in PC12 cells.[252] Other examples of fungal neurotrophic diterpenes are the trichoaurantiane diterpenes tricholomalides A, B, and C (CS 2.179–2.181) isolated from members of the genus *Tricholoma* (Fries) Staude (family Tricholomataceae R. Heim ex Pouzar), which induced the sprouting of neurites from PC12 cells at a dose of 100 μM.[253] The close neurotrophic similarity to NGF and the prospect of dendritogenesis *in vivo* give rise to the exciting concept that cyathane diterpenes may be a source of leads for the treatment of neurodegenerative diseases and spinal cord injuries.

Other neurogenic diterpenes are the daphnanes synaptolepis factor K7 (CS 2.182) and kirkinine B (CS 2.183) from *Synaptolepis kirkii* Oliv. (family Thymelaeaceae Juss.), which elicited neurotrophic properties with EC$_{50}$

■ **CS 2.181** Tricholomalide C.

■ **CS 2.182** Synaptolepis factor K7.

■ **CS 2.183** Kirkinine B.

■ **CS 2.184** Kirkinine.

values equal to 8.8×10^{-9} M and 4.5×10^{-8} M, respectively, in spinal neurons of chicks[254] via probable activation of PKC. In fact, synaptolepis factor K7 induced the activation of PKC and the subsequent phosphorylation of ERK1/2 and cAMP response element binding protein (CREB) in human SH-SY5Y neuroblastoma cells at a dose of 0.3μM.[255] From the same plant, kirkinine (CS 2.184) boosted the viability of neurons by 57% at a dose of 70 nM.[256] The vibsane diterpenes neovibsanin A, neovibsanin B, neovibsanin

■ **CS 2.185** Neovibsanin A.

■ **CS 2.186** Neovibsanine B.

■ **CS 2.187** Neovibsanin L.

■ **CS 2.188** (8Z)-neovibsanin M.

L, and (8Z)-neovibsanin M (CS 2.185–2.188) from *Viburnum sieboldii* Miq. (family Adoxaceae E. Mey.) induced the growth of neurites in PC12 cells at a dose of 40 μM in the presence of 20 ng/mL of NGF,[257] suggesting the activation of ERK1/2. At a concentration of 30 μM, the myrsinol diterpenes euphorbiaproliferin H, euphorbiaproliferin I, and euphorbiaproliferin J

■ **CS 2.189** Euphorbiaproliferin H.

■ **CS 2.190** Euphorbiaproliferin I.

■ **CS 2.191** Euphorbiaproliferin J.

(CS 2.189–2.191) isolated from *Euphorbia prolifera* Buch.-Ham. ex D. Don (family Euphorbiaceae Juss.) protected human SH-SY5Y neuroblastoma cells against 1-methyl-4-phenylpyridinium ion (MPP⁺) by 99.4 %, 101.3 %, and 104.2 %, respectively.[258]

REFERENCES

[159] Guo P, Li Y, Xu J, Liu C, Ma Y, Guo Y. Bioactive neo-clerodane diterpenoids from the whole plants of *Ajuga ciliata* Bunge. J Nat Prod 2011;74(7):1575–83.

[160] Guo P, Li Y, Xu J, Guo Y, Jin D,Q, Gao J, et al. neo-Clerodane diterpenes from *Ajuga ciliata* Bunge and their neuroprotective activities. Fitoterapia 2011;82(7):1123–7.

[161] Shimomura H, Sashida Y, Ogawa K. neo-Clerodane diterpenes from *Ajuga ciliata* var. *villosior*. Chem Pharm Bull 1989;37:988–92.

[162] Shimomura H, Sashida Y, Ogawa K. Iridoid glucosides and phenylpropanoid glycosides in *Ajuga* species of Japan. Phytochem 1987;26(7):1981–3.

[163] Pasinelli P, Brown RH. Molecular biology of amyotrophic lateral sclerosis: Insights from genetics. Nat Rev Neurosci 2006;7(9):710–23.

[164] Koo K,A, Sung S,H, Kim Y,C. A new neuroprotective pinusolide derivative from the leaves of *Biota orientalis*. Chem Pharm Bull 2002;50(6):834–6.

[165] Koo KA, Kim SH, Lee MK, Kim YC. 15-Methoxypinusolidic acid from *Biota orientalis* attenuates glutamate-induced neurotoxicity in primary cultured rat cortical cells. Toxicol in Vitro 2006;20(6):936–41.

[166] Koo KA, Lee MK, Kim SH, Jeong EJ, Kim SY, Oh TH, et al. Pinusolide and 15-methoxypinusolidic acid attenuate the neurotoxic effect of staurosporine in primary cultures of rat cortical cells. Br J Pharmacol 2007;150(1):65–71.

[167] Choi Y, Moon A, Kim YC. A pinusolide derivative, 15-methoxypinusolidic acid from *Biota orientalis* inhibits inducible nitric oxide synthase in microglial cells: implication for a potential anti-inflammatory effect. Int Immunopharmacol 2008;8(4):548–55.

[168] Piani D, Spranger M, Frei K, Schaffner A, Fontana A. Macrophage-induced cytotoxicity of N-methyl-D-aspartate receptor positive neurons involves excitatory amino acids rather than reactive oxygen intermediates and cytokines. Eur J Immunol 1992;22(9):2429–36.

[169] Yang WL, Frucht H. Activation of the PPAR pathway induces apoptosis and COX-2 inhibition in HT-29 human colon cancer cells. Carcinogenesis 2001;22(9):1379–83.

[170] Lee YS, Sung SH, Hong JH, Hwang ES. Suppression of adipocyte differentiation by 15-methoxypinusolidic acid through inhibition of PPARγ activity. Arch PharmRes 2010;33(7):1035–41.

[171] Jin Y, Yang HO, Son JK, Chang HW. Pinusolide isolated from *Biota orientalis* inhibits 5-lipoxygenase dependent leukotriene C4 generation by blocking c-Jun N-terminal kinase pathway in mast cells. Biol Pharm Bull 2012;35(8):1374–8.

[172] Hunot S, Vila M, Teismann P, Davis RJ, Hirsch E,C, Przedborski S, et al. JNK-mediated induction of cyclooxygenase 2 is required for neurodegeneration in a mouse model of Parkinson's disease. Proc Natl Acad Sci U S A. 2004;101(2):665–70.

[173] Kim KA, Moon TC, Lee SW, Chung KC, Han BH, Chang HW. Pinusolide from the leaves of *Biota orientalis* as potent platelet activating factor antagonist. Planta Med 1999;65(1):39–42.

[174] Gianni D, Zambrano N, Bimonte M, Minopoli G, Mercken L, Talamo F, et al. Platelet-derived growth factor induces the beta-gamma-secretase-mediated cleavage of Alzheimer's amyloid precursor protein through a Src-Rac-dependent pathway. J Biol Chem 2003;278(11):9290–2.

[175] Chen CC, Shiao YJ, Lin RD, Shao YY, Lai MN, Lin CC, et al. Neuroprotective diterpenes from the fruiting body of *Antrodia camphorata*. J Nat Prod 2006;69(4):689–91.

[176]   Moon HI. Three diterpenes from *Leonurus japonicus* Houtt protect primary cultured rat cortical cells from glutamate-induced toxicity. Phytother Res 2010;24(8):1256–9.

[177]   Hung TM, Luan TC, Vinh BT, Cuong TD, Min BS. Labdane-type diterpenoids from *Leonurus heterophyllus* and their cholinesterase inhibitory activity. Phytother Res 2011;25(4):611–4.

[178]   Xu J, Liu C, Guo P, Guo Y, Jin D,Q, Song X, et al. Neuroprotective labdane diterpenes from *Fritillaria ebeiensis*. Fitoterapia 2011;82(5):772–6.

[179]   Tang W, Hioki H, Harada K, Kubo M, Fukuyama Y. Clerodane diterpenoids with NGF-potentiating activity from *Ptychopetalum olacoides*. J Nat Prod 2008;71(10):1760–3.

[180]   Tang W, Kubo M, Harada K, Hioki H, Fukuyama Y. Novel NGF-potentiating diterpenoids from a Brazilian medicinal plant, *Ptychopetalum olacoides*. Bioorg Med Chem Lett 2009;19(3):882–6.

[181]   Guo Y, Li Y, Xu J, Watanabe R, Oshima Y, Yamakuni T, et al. Bioactive ent-clerodane diterpenoids from the aerial parts of *Baccharis gaudichaudiana*. J Nat Prod 2006;69(2):274–6.

[182]   Guo Y, Li Y, Xu J, Li N, Yamakuni T, Ohizumi Y. Clerodane diterpenoids and flavonoids with NGF-potentiating activity from the aerial parts of *Baccharis gaudichaudiana*. Chem Pharm Bull 2007;55(10):1532–4.

[183]   Zhang W, Wang T, Pei Z, Miller DS, Wu X, Block ML, et al. Aggregated alpha-synuclein activates microglia: a process leading to disease progression in Parkinson's disease. FASEB J 2005;19(6):533–42.

[184]   Guo P, Li Y, Jin D,Q, Xu J, He Y, Zhang L, et al. neo-Clerodane diterpenes from *Ajuga ciliata* and their inhibitory activities on LPS-induced NO production. Phytochem Lett 2012;5(3):563–6.

[185]   Roth B,L, Baner K, Westkaemper R, Siebert D, Rice K,C, Steinberg S, et al. Salvinorin A: a potent naturally occurring nonnitrogenous k opioid selective agonist. Proc Nat Acad Sci USA 2002;99(18):11934–11939.

[186]   Grilli M, Neri E, Zappettini S, Massa F, Bisio A, Romussi G, et al. Salvinorin A exerts opposite presynaptic controls on neurotransmitter exocytosis from mouse brain nerve terminals. Neuropharmacol 2009;57(5–6):523–30.

[187]   Su D, Riley J, Kiessling W,J, Armstead W,M, Liu R. Salvinorin A produces cerebrovasodilation through activation of nitric oxide synthase, κ receptor, and adenosine triphosphate-sensitive potassium channel. Anesthesiology 2011;114(2):374–9.

[188]   Bohn L,M, Belcheva M,M, Coscia C,J. Mitogenic signaling via endogenous kappa-opioid receptors in C6 glioma cells: evidence for the involvement of protein kinase C and the mitogen-activated protein kinase signaling cascade. J Neurochem 2000;74(2):564–73.

[189]   Bruchas M,R, Chavkin C. Kinase cascades and ligand-directed signaling at the kappa opioid receptor. Psychopharmacol 2010;210(2):137–47.

[190]   Ahmad VU, Farooq U, Hussain J, Ullah F, Nawaz SA, Choudhary MI. Two new diterpenoids from *Ballota limbata*. Chem Pharm Bull 2004;52(4):441–3.

[191]   Ahmad VU, Farooq U, Abbaskhan A, Hussain J, Abbasi MA, Nawaz SA, et al. Four new Diterpenoids from *Ballota limbata*. Helvetica Chimica Acta 2004;87(3):682–9.

[192]   Ahmad V,U, Khan A, Farooq U, Kousar F, Khan SS, Nawaz SA, et al. Three new cholinesterase-inhibiting cis-clerodane diterpenoids from *Otostegia limbata*. Chem Pharm Bull 2005;53(4):378–81.

[193] Cavin A,L, Hay A,E, Marston A, Stoeckli-Evans H, Scopelliti R, Diallo D, et al. Bioactive diterpenes from the fruits of *Detarium microcarpum*. J Nat Prod 2006;69(5):768–73.

[194] Lo AH, Liang YC, Lin-Shiau SY, Ho CT, Lin JK. Carnosol, an antioxidant in rosemary, suppresses inducible nitric oxide synthase through down-regulating nuclear factor-kappaB in mouse macrophages. Carcinogenesis 2002;23(6):983–91.

[195] Satoh T, Izumi M, Inukai Y, Tsutsumi Y, Nakayama N, Kosaka K, et al. Carnosic acid protects neuronal HT-22 cells through activation of the antioxidant-responsive element in free carboxylic acid- and catechol hydroxyl moieties-dependent manners. Neurosci Lett 2008;434(3):260–5.

[196] Rau O, Wurglics M, Paulke A, Zitzkowski J, Meindl N, Bock A, et al. Carnosic acid and carnosol, phenolic diterpene compounds of the labiate herbs rosemary and sage, are activators of the human peroxisome proliferator-activated receptor gamma. Planta Med 2006;72:881–7.

[197] Collino M, Aragno M, Mastrocola R, Gallicchio M, Rosa A,C, Dianzani C, et al. Modulation of the oxidative stress and inflammatory response by PPAR-gamma agonists in the hippocampus of rats exposed to cerebral ischemia/reperfusion. Eur J Pharmacol 2006;530(1–2):70–80.

[198] Park J,A, Kim S, Lee S,Y, Kim C,S, Kimdo K, Kim S,J, et al. Beneficial effects of carnosic acid on dieldrin-induced dopaminergic neuronal cell death. Neuroreport 2008;19(13):1301–4.

[199] El Omri A, Han J, Yamada P, Kawada K, Abdrabbah MB, Isoda H. *Rosmarinus officinalis* polyphenols activate cholinergic activities in PC12 cells through phosphorylation of ERK1/2. J Ethnopharmacol 2010;131(2):451–8.

[200] Tsai CW, Lin CY, Wang YJ. Carnosic acid induces the NAD(P)H: quinone oxidoreductase 1 expression in rat clone 9 cells through the p38/nuclear factor erythroid-2 related factor 2 pathway. J Nutr 2011;141(12):2119–25.

[201] Jia Z, Zhu H, Misra H,P, Li Y. Potent induction of total cellular GSH and NQO1 as well as mitochondrial GSH by 3H-1,2-dithiole-3-thione in SH-SY5Y neuroblastoma cells and primary human neurons: protection against neurocytotoxicity elicited by dopamine, 6-hydroxydopamine, 4-hydroxy-2-nonenal, or hydrogen peroxide. Brain Res 2008;1197:159–69.

[202] Chen JH, Ou HP, Lin CY, Lin FJ, Wu CR, Chang SW, et al. Carnosic acid prevents 6-hydroxydopamine-induced cell death in SH-SY5Y cells via mediation of glutathione synthesis. Chem Res Toxicol 2012;25(9):1893–901.

[203] Kanninen K, Malm TM, Jyrkkänen HK, Goldsteins G, Keksa-Goldsteine V, Tanila H, et al. Nuclear factor erythroid 2-related factor 2 protects against beta amyloid. Mol Cell Neurosci 2008;39(3):302–13.

[204] Meng P, Yoshida H, Matsumiya T, Imaizumi T, Tanji K, Xing F, et al. Carnosic acid suppresses the production of amyloid-β 1–42 by inducing the metalloprotease gene TACE/ADAM17 in SH-SY5Y human neuroblastoma cells. Neurosci Res 2013;75(2):94–102.

[205] Kosaka K, Yokoi T. Carnosic acid, a component of rosemary (*Rosmarinus officinalis* L.), promotes synthesis of nerve growth factor in T98G human glioblastoma cells. Biol Pharm Bull 2003;26(11):1620–2.

[206] Mimura J, Kosaka K, Maruyama A, Satoh T, Harada N, Yoshida H, et al. Nrf2 regulates NGF mRNA induction by carnosic acid in T98G glioblastoma cells and normal human astrocytes. J Biochem 2011;150(2):209–17.

[207] Lee JJ, Jin YR, Lee JH, Yu JY, Han XH, Oh KW, et al. Antiplatelet activity of carnosic acid, a phenolic diterpene from *Rosmarinus officinalis*. Planta Med 2007;73(2):121–7.

[208] Subbaramaiah K, Cole P,A, Dannenberg A,J. Retinoids and carnosol suppress cyclooxygenase-2 transcription by CREB-binding protein/p300-dependent and -independent mechanisms. Cancer Res 2002;62(9):2522–30.

[209] Huang S,C, Ho C,T, Lin-Shiau S,Y, Lin J,K. Carnosol inhibits the invasion of B16/F10 mouse melanoma cells by suppressing metalloproteinase-9 through down-regulating nuclear factor-kappa B and c-Jun. Biochem Pharmacol 2005;69(2):221–32.

[210] Martin D, Rojo AI, Salinas M, Diaz R, Gallardo G, Alam J, et al. Regulation of heme oxygenase-1 expression through the phosphatidylinositol 3-kinase/Akt pathway and the Nrf2 transcription factor in response to the antioxidant phytochemical carnosol. J Biol Chem 2004;279:8919–29.

[211] Chen CC, Chen HL, Hsieh CW, Yang YL, Wung BS. Upregulation of NF-E2-related factor-2-dependent glutathione by carnosol provokes a cytoprotective response and enhances cell survival. Acta Pharmacol Sinica 2011;32(1):62–9.

[212] Lee P, Hur J, Lee J, Kim J, Jeong J, Kang I, et al. 15,16-Dihydrotanshinone I suppresses the activation of BV-2 cell, a murine microglia cell line, by lipopolysaccharide. Neurochem Int 2006;48(1):60–6.

[213] Paris D, Beaulieu-Abdelahad D, Bachmeier C, Reed J, Ait-Ghezala G, Bishop A, et al. Anatabine lowers Alzheimer's Aβ production *in vitro* and *in vivo*. Eur J Pharmacol 2011;670(2–3):384–91.

[214] Mei Z, Zhang F, Tao L, Zheng W, Cao Y, Wang Z, et al. Cryptotanshinone, a compound from *Salvia miltiorrhiza* modulates amyloid precursor protein metabolism and attenuates beta-amyloid deposition through upregulating alpha-secretase *in vivo* and *in vitro*. Neurosci Lett 2009;452(2):90–5.

[215] Mei Z, Situ B, Tan X, Zheng S, Zhang F, Yan P, et al. Cryptotanshinione upregulates alpha-secretase by activation PI3K pathway in cortical neurons. Brain Res 2010;1348:165–73.

[216] Kim EJ, Jung SN, Son KH, Kim SR, Ha TY, Park MG, et al. Antidiabetes and antiobesity effects of cryptotanshinone via activation of AMP-activated protein kinase. Mol Pharmacol 2007;72:62–72.

[217] Zhang F, Zheng W, Pi R, Mei Z, Bao Y, Gao J, et al. Cryptotanshinone protects primary rat cortical neurons from glutamate-induced neurotoxicity via the activation of the phosphatidylinositol 3-kinase/Akt signaling pathway. Exp Brain Res 2009;193:109–18.

[218] Wong K,K, Ho M,T, Lin H,Q, Lau K,F, Rudd J,A, Chung R,C, et al. Cryptotanshinone, an acetylcholinesterase (AChE) inhibitor from *Salvia miltiorrhiza*, ameliorates scopolamine-induced amnesia in morris water maze task. Planta Med 2010;76(3):228–34.

[219] Li LX, Dai JP, Ru LQ, Yin GF, Zhao B. Effects of tanshinone on neuropathological changes induced by amyloid beta-peptide1–40 injection in rat hippocampus. Acta Pharmacologica Sinica 2004;25(7):861–8.

[220] Meng XF, Zou XJ, Peng B, Shi J, Guan XM, Zhang C. Inhibition of ethanol-induced toxicity by tanshinone IIA in PC12 cells. Acta Pharmacologica Sinica 2006;27(6):659–64.

[221] Liu T, Jin H, Sun Q,R, Xu J,H, Hu H,T. The neuroprotective effects of tanshinone IIA on β-amyloid-induced toxicity in rat cortical neurons. Neuropharmacol 2010;59(7–8):595–604.

[222] Wang W, Zheng L,L, Wang F, Hu Z,L, Wu W,N, Gu J, et al. Tanshinone IIA attenuates neuronal damage and the impairment of long-term potentiation induced by hydrogen peroxide. *J Ethnopharmacol*1 2011;34(1):147–55.

[223] Yu X,Y, Lin S,G, Zhou Z,W, Chen X, Liang J, Duan W, et al. Tanshinone IIB, a primary active constituent from *Salvia miltiorrhza*, exhibits neuro-protective activity in experimentally stroked rats. Neurosci Lett 2007;417(3):261–5.

[224] Yu XQ, Xue CC, Zhou ZW, Li CG, Zhou SF. Tanshinone IIB, a primary active constituent from *Salvia miltiorrhiza*, exerts neuroprotective effect via inhibition of neuronal apoptosis *in vitro*. Phytother Res 2008;22(6):846–50.

[225] Kim J,Y, Jung K,S, Lee K,J, Na H,K, Chun H,K, Kho Y,H, et al. The coffee diterpene kahweol suppress the inducible nitric oxide synthase expression in macrophages. Cancer Lett 2004;213(2):147–54.

[226] Kim HG, Kim JY, Hwang YP, Lee KJ, Lee KY, Kim DH, et al. The coffee diterpene kahweol inhibits tumor necrosis factor-alpha–induced expression of cell adhesion molecules in human endothelial cells. Toxicol Appl Pharmacol 2006;217(3):332–41.

[227] Hwang Y,P, Jeong H,G. The coffee diterpene kahweol induces heme oxygenase-1 via the PI3K and p38/Nrf2 pathway to protect human dopaminergic neurons from 6-hydroxydopamine-derived oxidative stress. FEBS Lett 2008;582(17):2655–62.

[228] Li F,Q, Cheng X,X, Liang X,B, Wang X,H, Xue B, He Q,H, et al. Neurotrophic and neuroprotective effects of tripchlorolide, an extract of Chinese herb *Tripterygium wilfordii* Hook F, on dopaminergic neurons. Exp Neurol 2003;179(1):28–37.

[229] Hong Z, Wang G, Gu J, Pan J, Bai L, Zhang S, et al. Tripchlorolide protects against MPTP-induced neurotoxicity in C57BL/6 mice. Eur J Neurosci 2007;26(6):1500–8.

[230] Pan X, Chen X, Zhu Y, Zhang J, Huang T, Chen L, et al. Neuroprotective role of tripchlorolide on inflammatory neurotoxicity induced by lipopolysaccharide-activated microglia. Biochem Pharmacol 2008;76(3):362–72.

[231] Pan X,D, Chen X,C, Zhu Y,G, Chen L,M, Zhang J, Huang T,W, et al. Tripchlorolide protects neuronal cells from microglia-mediated beta-amyloid neurotoxicity through inhibiting NF-kappaB and JNK signaling. Glia 2009; 57(11):1227–38.

[232] Li FQ, Lu XZ, Liang XB, Zhou HF, Xue B, Liu XY, et al. Triptolide, a Chinese herbal extract, protects dopaminergic neurons from inflammation-medi-ated damage through inhibition of microglial activation. J Neuroimmunol 2004;148(1–2):24–31.

[233] Zhou H,F, Liu X,Y, Niu D,B, Li F,Q, He Q,H, Wang X,M. Triptolide protects dopaminergic neurons from inflammation-mediated damage induced by lipo-polysaccharide intranigral injection. Neurobiol Disease 2005;18(3):441–9.

[234] Gong Y, Xue B, Jiao J, Jing L, Wang X. Triptolide inhibits COX-2 expression and PGE2 release by suppressing the activity of NF-kappaB and JNK in LPS-treated microglia. J Neurochem 2008;107(3):779–88.

[235] Geng Y, Fang M, Wang J, Yu H, Hu Z, Yew DT, et al. Triptolide down-regulates COX-2 expression and PGE2 release by suppressing the activity of NF-κB and MAP kinases in lipopolysaccharide-treated PC12 cells. Phytother Res 2012;26(3):337–43.

[236] Xue B, Jiao J, Zhang L, Li K,R, Gong Y,T, Xie J,X, et al. Triptolide upregulates NGF synthesis in rat astrocyte cultures. Neurochem Res 2007;32(7):1113–9.

[237] Hsu Y,Y, Jong Y,J, Tsai H,H, Tseng Y,T, An L,M, Lo Y,C. Triptolide increases transcript and protein levels of survival motor neurons in human SMA fibroblasts and improves survival in SMA-like mice. Br J Pharmacol 2012;166(3):1114–26.

[238] Hong SS, Lee SA, Han XH, Hwang JS, Lee C, Lee D, et al. ent-Kaurane diterpenoids from *Isodon japonicus*. J Nat Prod 2008;71(6):1055–8.

[239] Kim HS, Lim JY, Sul D, Hwang BY, Won TJ, Hwang KW, et al. Neuroprotective effects of the new diterpene, CBNU06 against beta-amyloid-induced toxicity through the inhibition of NF-kappaB signaling pathway in PC12 cells. Eur J Pharmacol 2009;622(1–3):25–31.

[240] Hwang BY, Lee JH, Koo TH, Kim HS, Hong YS, Ro JS, et al. Kaurane diterpenes from *Isodon japonicus* inhibit nitric oxide and prostaglandin E2 production and NF-kappaB activation in LPS-stimulated macrophage RAW264.7 cells. Planta Med 2001;67(5):406–10.

[241] Lim JY, Won TJ, Hwang BY, Kim HR, Hwang KW, Sul D, et al. The new diterpene isodojaponin D inhibited LPS-induced microglial activation through NF-kappaB and MAPK signaling pathways. Eur J Pharmacol 2010;642(1–3):10–18.

[242] Kim B,W, Koppula S, Kim I,S, Lim H,W, Hong S,M, Han S,D, et al. Anti-neuroinflammatory activity of kamebakaurin from *Isodon japonicus* via inhibition of c-Jun NH₂-terminal kinase and p38 mitogen-activated protein kinase pathway in activated microglial cells. J Pharmacol Sci 2011;116(3):296–308.

[243] Kim J,Y, Kim H,S, Kim Y,J, Lee H,K, Kim J,S, Kang J,S, et al. Effusanin C inhibits inflammatory responses via blocking NF-κB and MAPK signaling in monocytes. Int Immunopharmacol 2013;15(1):84–8.

[244] Xu J, Guo P, Liu C, Sun Z, Gui L, Guo Y, et al. Neuroprotective kaurane diterpenes from Fritillaria ebeiensis. Biosci Biotechnol Biochem 2011;75(7):1386–8.

[245] Ohta T, Kita T, Kobayashi N, Obara Y, Nakahata N, Ohizumi Y, et al. Scabronine A, a novel diterpenoid having potent inductive activity of the nerve growth factor synthesis, isolated from the mushroom, *Sarcodon scabrosus*. Tetrahedron Lett 1998;39(34):6229–32.

[246] Kita T, Takaya Y, Oshima Y, Ohta T, Aizawa K, Hirano T, et al. Scabronines B, C, D, E and F, novel diterpenoids showing stimulating activity of nerve growth factor-synthesis, from the mushroom *Sarcodon scabrosus*. Tetrahedron 1998;54(39):11877–11886.

[247] Obara Y, Nakahata N, Kita T, Takaya Y, Kobayashi H, Hosoi S, et al. Stimulation of neurotrophic factor secretion from 1321N1 human astrocytoma cells by novel diterpenoids, scabronines A and G. Eur J Pharmacol 1999;370(1):79–84.

[248] Obara Y, Kobayashi H, Ohta T, Ohizumi Y, Nakahata N. Scabronine G-methylester enhances secretion of neurotrophic factors mediated by an activation of protein kinase C-zeta. Mol Pharmacol 2001;59(5):1287–97.

[249] Shi X,W, Liu L, Gao J,M, Zhang A,L. Cyathane diterpenes from Chinese mushroom *Sarcodon scabrosus* and their neurite outgrowth-promoting activity. Eur J Med Chem 2011;46(7):3112–7.

[250] Obara Y, Hoshino T, Marcotullio M,C, Pagiotti R, Nakahata N. A novel cyathane diterpene, cyrneine A, induces neurite outgrowth in a Rac1-dependent mechanism in PC12s cells. Life Sci 2007;80(18):1669–77.

[251] Marcotullio MC, Pagiott R, Maltese F, Obara Y, Hoshino T, Nakahata N, et al. Neurite outgrowth activity of cyathane diterpenes from *Sarcodon cyrneus*, cyrneines A and B. Planta Med 2006;72(9):819–23.

[252] Marcotullio M,C, Pagiotti R, Maltese F, Oball–Mond Mwankie GN, Hoshino T, Obara Y, et al. Cyathane diterpenes from *Sarcodon cyrneus* and evaluation of their activities of neuritegenesis and nerve growth factor production. Bioorg Med Chem 2007;15(8):2878–82.

[253] Tsukamoto S, Macabalang AD, Nakatani K, Obara Y, Nakahata N, Ohta T. Tricholomalides A–C, new neurotrophic diterpenes from the mushroom *Tricholoma* sp. J Nat Prod 2003;66(12):1578–81.

[254] He W, Cik M, Van Puyvelde L, Van Dun J, Appendino G, Lesage A, et al. Neurotrophic and antileukemic daphnane diterpenoids from *Synaptolepis kirkii*. Bioorg Med Chem 2002;10(10):3245–55.

[255] Van Kolen K, Bruinzeel W, He W, De Kimpe N, Van Puyvelde L, Cik M, et al. Investigation of signalling cascades induced by neurotrophic synaptolepis factor K7 reveals a critical role for novel PKCε. Eur J Pharmacol 2013;701(1):73–81.

[256] He W, Cik M, Lesage A, Van der Linden I, De Kimpe N, Appendino G, et al. Kirkinine, a new daphnane orthoester with potent neurotrophic activity from *Synaptolepis kirkii*. J Nat Prod 2000;63(9):1185–7.

[257] Kubo M, Kishimoto Y, Harada K, Hioki H, Fukuyama Y. NGF-potentiating vibsane-type diterpenoids from *Viburnum sieboldii*. Bioorg Med Chem Lett 2010;20(8):2566–71.

[258] Xu J, Guo Y, Xie C, Li Y, Gao J, Zhang T, et al. Bioactive myrsinol diterpenoids from the roots of *Euphorbia prolifera*. J Nat Prod 2011;74(10):2224–30.

Topic **2.4**

Triterpenes

2.4.1 *Boswellia serrata* Roxb. ex Colebr.

History The plant was first described by William Roxburgh in *Asiatic Researches* published in 1807.

Family Burseraceae Kunth, 1824

Common Names Guggula (Sanskrit)

Habitat and Description It is a magnificent resinous tree growing in India. The leaves are imparipinnate, spirally arranged, apical, and exstipulate. The rachis is 40 cm long and presents 7–15 pairs of folioles, which are sessile, 1–10 cm × 0.5–4 cm, elliptic, and serrate. The inflorescence is an axillary raceme. The calyx is tubular, with 5 lobes. The corolla consists of 5 petals, which are light pink, lanceolate, and 0.8 cm long. The androecium comprises 10 stamens, which are inserted around a conspicuous nectary disc. The fruit is fleshy, light dull green, and irregularly trigonal, and it encloses a woody stone that conceals 3 seeds (Figure 2.4).

Medicinal Uses In India, the plant is used to treat inflammation.

Phytopharmacology The plant builds a series of triterpenes, including 11-keto-β-boswellic acid, 3-α-acetyl-11-keto-β-boswellic acid, α-boswellic acid, β-boswellic acid, 3-acetyl-α-boswellic acid, and 3-acetyl-β-boswellic acid[259]

Proposed Research Pharmacological study of 11-keto-β-boswellic acid for the treatment of Parkinson's disease (PD).

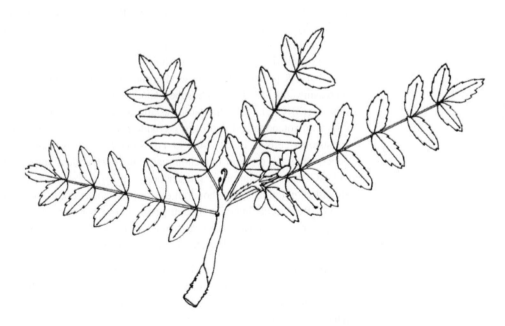

■ **FIGURE 2.4** *Boswellia serrata* Roxb. ex Colebr.

■ **CS 2.192** Asiatic acid.

Rationale Several lines of evidence point to the fact ursane triterpenes are neuroprotective. One such triterpene is asiatic acid (CS 2.192) from *Centella asiatica* (L.) Urb. (family Apiaceae Lindl.), which at dose of $1\,\mu$M protected neurons against the deleterious effects of β-amyloid (Aβ_{25-35}) peptide by 97% at a dose of $1\,\mu$M.[260,261] Furthermore, asiatic acid at a dose of $1\,\mu$mol/L protected primary cortical neurons against C2-ceramide by reducing reactive oxygen species (ROS), repressing pro-apoptotic Bcl-2–associated X protein (Bax), preserving the integrity of mitochondria, and increasing the levels of phosphorylated extracellular signal-regulated kinase (ERK1/2).[262] Furthermore, asiatic acid abolished the deleterious effects of rotenone and hydrogen peroxide against human SH-SY5Y neuroblastoma cells at doses equal to 1 nM and 10 nM, respectively, with a concomitant decrease in mitochondrial voltage-dependent anion channels (VDAC) and reservation of mitochondrial integrity.[263] In this context, it should be noted that mitochondrial dysfunction and subsequent accumulation of ROS in cholinergic, dopaminergic, and motor neurons account for the progressive apparition of Alzheimer's disease (AD), Parkinson's disease (PD), and amyotrophic lateral sclerosis (ALS).[264]

Asiatic acid at a dose of 75 mg/kg enhanced the cognition of rodents subjected to experimental stroke with reduced infarcts volumes.[265] In the same experiment, asiatic acid at a dose of $1\,\mu$g/mL maintained mitochondrial integrity and therefore reduced the release of cytochrome c.[265] Strikingly, asiatic acid at a dose of $10\,\mu$M inhibited β-secretase and induced α-secretase in primary cortical neurons.[266] Furthermore, asiatic acid at a dose of $1\,\mu$M commanded the growth of neurites in human SH-SY5Y neuroblastoma cells via the activation of MAP kinase (MEK1/2).[267] The neuroprotective potencies of asiatic acid also may involve glial cells, as

■ **CS 2.193** AS-2.

■ **CS 2.194** AS-2-9-006.

■ **CS 2.195** AS-9-006.

■ **CS 2.196** Madecassic acid.

at a dose of $120\,\mu$M this ursane mitigated the secretion of Interleukin-6 (IL-6), Interleukin-1β (IL-1β), and tumor necrosis factor-α (TNF-α) by macrophages (RAW 264.7) challenged with lipopolysaccharide (LPS) by inhibiting the phosphorylation of Raf, ERK1/2, mitogen-activated protein kinase (MAPK) p38, and c-Jun N-terminal kinase (JNK), and therefore the phosphorylation of IκBα.[268]

Asiatic acid derivatives AS-2, AS-2-9-006, and AS-9-006 (CS 2.193–2.195) increased the enzymatic activity of choline acetyltransferase (ChAT) in S-20Y cells by 32.6 pmol/mg, 33.5 pmol/mg, and 35.8 pmol/mg, respectively,[265] and at a dose of 1 mg/kg protected rodents against the dementia induced by scopolamine.[265] Along the same lines, madecassic acid (CS 2.196) from *Centella asiatica* (L.) Urb. (family Apiaceae Lindl.)

■ **CS 2.197** Ursolic acid.

at a dose of $1\,\mu$M protected B103 cells against β-amyloid (Aβ_{25-35}) peptide by 56%.[261] AD is characterized by the deposition of β-amyloid peptide into amyloid plaques and the neurofibrillary tangles which include glycoproteinic complexes known as advanced glycation end-products (AGEs), the formation of which is favored by diabetes and oxidation.[269,270]

The chronic deposition of AGEs such as N^{ε}-(carboxymethyl) lysine (CML) in cerebral arteries reduces blood perfusion, enhancing further Alzheimerian dementia.[270] Therefore, agents able to inhibit the generation of AGEs may very well delay the course of AD, and, in fact, the ursane triterpene ursolic acid (CS 2.197) from *Pyrus malus* L. (family Rosaceae Juss.) at a dose of 0.2% of diet reduced the occurrence of N^{ε}-(carboxymethyl) lysine (CML) in the kidneys of rodents with a decrease of aldose reductase activity.[271,272] Of further interest is the fact that ursolic acid preserved rat pheochromocytoma (PC12) cells against β-amyloid (Aβ_{25-35}) peptide insults at a dose of $5\,\mu$M by halving the production of ROS.[273] In addition, ursolic acid potentiated the neuritogenic property of $2\,$ng/mL of nerve growth factor (NGF) toward PC12 cells at a dose of $3\,\mu$M[274] and inhibited the enzymatic activity of cyclo-oxygenase-2 (COX-2) with IC$_{50}$ values equal to $130\,\mu$M.[275] At a dose of $15\,\mu$M, ursolic acid rended hippocampal neurons resistant against kainate by lowering the cytoplasmic levels of ROS and preserving mitochondrial integrity.[276] In addition, ursolic acid at a dose of $20\,$mg/kg protected rodents against the locomotor deficiencies imposed by intraperitoneal injections of LPS by blocking nuclear factor kappa-light-chain-enhancer of activated B cells (NF-κB) and therefore the expression of inducible nitric oxide synthetase (iNOS), COX-2, and the synthesis of Interleukin-1β (IL-1β), Interleukin-2 (IL-2), Interleukin-6 (IL-6) and TNF-α.[277] Along the same lines, ursolic

■ **CS 2.198** Acetyl-11-keto-β-boswellic acid.

■ **CS 2.199** 11-keto-β-boswellic acid.

acid evoked the expression of Nrf2 in rodents with cerebral ischemia and decreased toll-like receptor 4 (TLR4) and NF-κB expression.[278] NF-κB is a cytoplasmic protein which translocates into the nucleus of microglia where it imposes the transcription of cytokines such as IL-1, IL-2, TNF-α, chemokines like Interleukin-8 (IL-8), adhesion molecules like vascular cell adhesion molecule (VCAM), and inducible enzymes, including COX-2 and iNOS.[279] Therefore, ursanes may very well produce leads to prevent or delay AD.

Members of the genus *Boswellia* Roxb. ex Colebr. (family Burseraceae Kunth) accumulate in their aromatic resins series of ursane triterpenes such as acetyl-11-keto-β-boswellic acid (CS 2.198) and 11-keto-β- boswellic acid, which inhibited the enzymatic activity of 5-lipoxygenase with IC_{50} values equal to 1.5 μM and 2.8 μM by virtue of their 11-keto moiety.[280] 5-Lipoxygenase (5-LOX) catalyzes the oxidation of fatty acids and neuro-degeneration in the elderly as a consequence of hyperglucocorticoidemia,[281] and 11-keto-β-boswellic acids (CS 2.199) may be used to hamper the progression of neurodegenerative diseases. Along the same lines, acetyl-11-keto-β-boswellic acid at a dose of 50 μM inhibited HER-2, NF-κB, and subsequently CXCR4 in malignant cells.[282] β-Boswellic acid (CS 2.200), which lacks a carbonyl moiety in C11, did not inhibit 5-LOX,[280] but at a dose of 30 nM evoked the growth neurites from fetal hippocampal neurons by boosting the polymerization of tubulin.[283]

Migration of C19 methyl of the ursane framework to C20 yielded oleanane triterpenes such as oleanolic acid (CS 2.201), which abound in olives, namely the fruits of *Olea europaea* L. (family Oleaceae Hoffmanns. &

■ **CS 2.200** β-boswellic acid.

■ **CS 2.201** Oleanolic acid.

■ **CS 2.202** Erythrodiol.

Link), which inhibited the enzymatic activity of COX-2 with IC_{50}, a value equal to $295\,\mu M$.[275] Strikingly, oleanolic acid and the oleanane triterpene erythrodiol (CS 2.202) at a dose of $50\,mg/kg$ attenuated paralysis incurred by the injection of myelin oligodendrocyte glycoprotein $(MOG)_{35-55}$ to C57BL/6 mice with decreased infiltration of leukocytes in the spinal cord; reduced amounts of myelin oligodendrocyte glycoprotein $(MOG)_{35-55}$ -specific IgM and IgG, TNF-α, interferon-γ (INF-γ), and IL-6; and sustained blood brain barrier integrity.[284] In addition, $15\,\mu M$ of oleanolic acid or erythrodiol abated the expression of COX-2 and iNOS by BV-2 microglia challenged with LPS.[284] Furthermore, oleanolic acid potentiated the neuritogenic property of $2\,ng/mL$ of NGF toward PC12 cells at a dose of $10\,\mu M$[274] and at a dose of 0.2% of

■ **CS 2.203** Maslinic acid.

■ **CS 2.204** α-spinasterol.

diet reduced N^ε-(carboxymethyl) lysine (CML) and boosted glyoxalase in the kidneys of rodents.[271] Along the same lines, oleanolic acid at a dose of $200\,\mu M$ induced the expression of heme oxidase (HO-1) as a result of Nrf2-ARE binding in malignant cells,[285] making it a first-line candidate for the treatment of demyelinating pathologies such as multiple sclerosis (MS) and amyotrophic lateral sclerosis (ALS).

Maslinic acid (CS 2.203) from *Olea europaea* L. (family Oleaceae Hoffmanns. & Link) at a dose of $10\,\mu M$ sustained the viability of cortical neurons deprived of oxygen and glucose by 77.8% with reduction of iNOS, nitric oxide (NO), pro-apoptotic Bax, and inactivation of caspases 3 and 9.[286] Under oxygen and glucose, neurons ought to face waves of glutamate,[287] which overwhelm the capacity of astrocytic sodium-dependent glutamate transporters GLAST and GLT-1.[288,289]

Maslinic acids preserved cortical neurons against the deleterious effects of glutamate by inducing the expression of glutamate transporters GLAST and GLT-1 at a dose of $10\mu M$.[290] Maslinic acid and oleanolic acid at a dose of $200\,\mu M$ induced the expression of HO-1 as a result of Nrf2-ARE binding in malignant cells.[285] Given that maslinic acid boosted the survival of hypoxic and hypoglycemic neurons and hindered glial activation, it is tempting to speculate that this oleanane may be of value for stroke prevention and neonatal neurological diseases. Note that *Vernonia tweedieana* Baker (family Aseraceae Bercht. & Presl) elaborates α-spinasterol (CS 2.204), which at a dose of $0.3\,\mu mol/kg$ protected rodents against the pain inflicted by heat by 26.6% as a result of TRVP1 antagonism.[291] Antagonists of vanilloid receptor transient receptor potential vanilloid subtype 1 (TRPV1) prevent the increase of cytoplasmic Ca^{2+}; inhibit protein kinase A (PKA), the arachidonate cascade, mitochondrial insults, the release of cytochrome c, and the activation of caspase 3; halting thus neuroapoptosis.[292]

The biological activity of oleanane triterpenes has stimulated considerable synthetic activity, which led to the isolation of 2-cyano-3,12-dioxoolean-1,9-dien-28-oic acid derivatives such as 2-cyano-3,12-dioxoolean-1,9-dien-28-oic acid methyl ester, which at a dose of 100 nM inhibited the activation of BV-2 cells challenged by LPS as shown by reduced production of Interleukin-12 (IL-12) and Interleukin-6 (IL-6) and low cytoplasmic levels of ROS.[293] In addition, 10 nM of 2-cyano-3,12-dioxoolean-1,9-dien-28-oic acid methyl ester preserved dopaminergic cells against conditioned media from LPS-treated BV-2 microglia cell 4.[293] In addition, 2-cyano-3,12-dioxoolean-1,9-dien-28-oic acid methyl ester potentiated the BV-2 microglia cell phagocytosis capacities upon β-amyloid (Aβ_{1-42}) peptide exposure.[294] 2-Cyano-3,12-dioxoolean-1,9-dien-28-oic acid amide; 2-cyano-3,12-dioxoolean-1,9-dien-28-oic acid butyl ester; 2-cyano-3,12-dioxoolean-1,9-dien-28-oic acid imidazoline; and 2-cyano-3,12-dioxoolean-1,9-dien-28-oic acid trifluoroethyl amide boosted the expression of NADPH: quinone oxidoreductase 1 (NQO1) in astrocytes.[294] In the same experiment, 100 nM of 2-cyano-3,12-dioxoolean-1,9-dien-28-oic acid amide increased NADPH: quinone oxidoreductase 1 (NQO1) in motor neurons by 39%.[295] At a dose of 300 nM, 2-cyano-3,12-dioxoolean-1,9-dien-28-oic acid trifluoroethyl amide boosted the levels of Nrf2 and consequently NADPH: quinone oxidoreductase 1 (NQO1), HO-1, and glutathione reductase (GR) in motor neurons.[295] Furthermore, G93A SOD1 transgenic mice fed with a diet containing 400 mg/kg of 2-cyano-3,12-dioxoolean-1,9-dien-28-oic acid ethyl amide or 2-cyano-3,12-dioxoolean-1,9-dien-28-oic acid trifluoroethyl amide displayed translocated Nrf2 in spinal neurons with increased levels of NADPH: quinone oxidoreductase 1 (NQO1) and HO-1.[295] Along the same lines, 2-cyano-3,12-dioxoolean-1,9-dien-28-oic acid ethyl amide and 2-cyano-3,12-dioxoolean-1,9-dien-28-oic acid trifluoroethyl amide prolonged the life span of G93A SOD1 transgenic mice by 20.6 days and 17.6 days, respectively.[295]

The quinone methide celastrol from *Tripterygium wilfordii* Hook. f. (family Celastraceae R. Br.) inhibited the peroxidation of lipids in hepatic mitochondria poisoned with adenosine 5'-diphosphate and Fe^{2+} with an IC_{50} equal to $7 \mu M$.[296] In monocytes exposed to LPS, celastrol (CS 2.205) at a dose of 70 nM and with an IC_{50} equal to 200 nM reduced the production of TNF-α, IL-1, and NO, respectively.[297] Likewise, celastrol at a dose of 100 nM preserved microglia against the activation induced by LPS and elevated the cognition of rodents at a dose of $7 \mu g/kg$.[297] Celastrol inhibited NF-κB in HEK293 cells exposed to TNF-α with an IC_{50} inferior to $1 \mu M$ by reacting with nucleophilic Cys-179 of kappa light polypeptide gene enhancer in B-cells inhibitor (IκB) kinase (IKK), thus preventing nuclear factor of kappa light polypeptide

■ **CS 2.205** Celastrol.

gene enhancer in B-cells inhibitor α (IκBα) phosphorylation and therefore the activation of NF-κB.[298,299] Along the same lines, celastrol reduced the production of β-amyloid (Aβ_{1-42}) peptide with an IC$_{50}$ value equal to 900 nM in Chinese hamster ovary (CHO) cells on account of NF-κB inactivation and consequent reduced levels of β-secretase.[300] At a dose of 1 mg/Kg, this quinone methide hindered the genesis of β-amyloid (Aβ_{1-42}) peptide by 40% in Tg PS1/APPsw mice.[300] Celastrol at a dose of 3 μM attenuated the expression of CXCR4 in malignant cells as a result of NF-κB inhibition.[301]

Strikingly, celastrol at a dose of 1 μM evoked an increase of heat shock proteins Hsp27, Hsp32, Hsp70, and Hsp70B in human SH-SY5Y neuroblastoma cells[302] via the probable inhibition of heat shock protein Hsp90.[303] Indeed, neuronal insults precipitate the release of heat shock factor-1 (HSF-1) from Hsp90, followed by the trimerization heat shock factor-1 (HSF-1), nuclear translocation, and synthesis of Hsp70 and activator Hsp40.[304] Heat shock chaperones such as Hsp40 account for the catabolism of misfolded peptides,[304] such as β-amyloid (Aβ_{1-42}) peptide, α-synucleine, and huntingtin. Huntington's disease (HD) involves the aggregation of chimeric huntingtin protein,[305] and in fact, celastrol at a dose of 10 μM abrogated the aggregation of GST-Q58-Htn and thrombin.[306] G93A SOD1 mice treated with celastrol at a dose of 8 mg/kg had a 13% life span increase with elevation of HSP70 and reduction of TNF-α, iNOS in the spine, and protection of lumbar spinal neurons.[306] Celastrol at a dose of 3 mg/kg protected rodents against the loss of dopaminergic neurons in the *substantia nigra* as a result of 1-methyl-4-phenyl-1,2,3,6-tetrahydropyridine (MPTP) poisoning, sustained the levels of 3,4-dihydroxyphenylacetic acid, increased expression of Hsp70 heat shock chaperone, and reduced TNF-α as a consequence of NF-κB inhibition.[307]

■ **CS 2.206** Lupeol.

■ **CS 2.207** β-amyrin.

■ **CS 2.208** Ulmicin A.

■ **CS 2.209** Ulmicin B.

The lupane triterpene lupeol from *Lycopersicon esculentum* Mill. (family Solanaceae Juss.) at a dose of 50 mg/kg reduced by 39% the edema produced by plantar injection of Complete Freund's Adjuvant (CFA) in rodents.[308] The anti-inflammatory activity of lupeol (CS 2.206) was further shown, as at a dose of 2 mg, this lupane inhibited the activation of phosphoinositide 3-kinase (PI3K) and subsequent phosphorylation of protein kinase B (Akt), activation of IKK, phosphorylation, and degradation of IκBα in CD-1 mouse skin cells exposed to chemical insults.[309] IKK inhibition may account for the ability of lupeol at a dose of 5 μM to enhance the viability of mouse hippocampal (HT-22) cells exposed to glutamate.[310] Note that β-amyrin (CS 2.207) is structurally close to lupeol, implying neuroprotective potentials.

■ **CS 2.210** Ulmicin C.

■ **CS 2.211** Ulmicin D.

■ **CS 2.212** Ulmicin E.

■ **CS 2.213** Betulin.

The lupane triterpenes ulmicin A, B, C, D, and E (CS 2.208–2.212) from *Ulmus davidiana* var. *japonica* (Rehder) Nakai (family Ulmaceae Mirb.) elicited 48%, 51.2%, 49.2%, 44.2%, and 41% protection of cortical cells against glutamate insults.[311] The lupanes betulin, epibetulinic acid, and betulonic acid (CS 2.213–2.215) isolated from members of the genus *Maytenus* Molina

■ **CS 2.214** Epibetulinic acid.

■ **CS 2.215** Betulonic acid.

■ **CS 2.216** Calenduladiol.

■ **CS 2.217** Gedunin.

(family Celastraceae R. Br.) inhibited the production of NO and prostaglandin E2 (PGE2) in mouse macrophages (RAW 264.7) stimulated with LPS with IC_{50} values equal to $5\,\mu M$, $0.7\,\mu M$, $0.3\,\mu M$ and $12.9\,\mu M$, $0.6\,\mu M$, $2.7\,\mu M$, respectively.[312] The lupane calenduladiol (CS 2.216) from *Chuquiraga erinacea* D. Don (family Asteraceae Bercht. & J. Presl) inhibited the enzymatic activity of acetylcholinesterase (AChE) by 31.2% at a dose of $0.5\,\mu M$.[313]

Azadirachta indica A. Juss. (family Meliaceae Juss.) engineers cytotoxic tetranortriterpene, which exhibits neuroprotective activities at low doses. As an example, gedunin (CS 2.217) at a dose of $20\,\mu M$ inhibited HSP90 in LNCaP cells,[314] suggesting subsequent synthesis of heat shock protein Hsp40, which

■ **CS 2.218** Deoxygedunin.

■ **CS 2.219** Obacunone.

■ **CS 2.220** Limonin.

■ **CS 2.221** Toosendanin.

accounts for the clearance of neurotoxic misfolded peptides.[315] From the same plant, deoxygedunin (CS 2.218) protected hippocampal neurons against glutamate and glucose oxygen deprivation via the activation of TrkB receptors and activation of Akt and ERK1/2 at a dose of $0.5\,\mu$M.[314] Along the same lines, $0.1\,\mu$M of obacunone (CS 2.219) and limonin (CS 2.220) from *Dictamnus dasycarpus* Turcz. (family Rutaceae Juss.) protected cortical cells against glutamate insults by 68.3% and 65.8%, respectively, by attenuating the cytoplasmic levels of Ca^{2+}, nitric oxide, ROS, and sustaining the enzymatic activity of superoxide dismutase (SOD).[316] The tetranortriterpene toosendanin (CS 2.221) from *Melia toosendan* Siebold & Zucc. (family Meliaceae Juss.) at a dose

■ **CS 2.222** 22R-hydroxycholesterol.

■ **CS 2.223** Disodium 2β,3α-dihydroxy-5α-cholestan-6-one disulfate.

of 10^{-7} M increased cytoplasmic Ca^{2+} and neuritogenesis in PC12 cells[317] via probable activation of calmodulin and cAMP response element binding protein (CREB).

Emerging evidence indicates that cholestanes are neurotrophic and neuro-protective. In effect, the mammalian cholestane 22R-hydroxycholesterol (CS 2.222) at a dose of $50\,\mu M$ protected PC12 cells against β-amyloid (Aβ$_{1-42}$) peptide insults by 98.1% by direct binding to amino acids.[318] Along the same lines, the synthetic cholestanes disodium 2β,3α-dihydroxy-5α-cholestan-6-one disulfate (CS 2.223) and disodium

■ **CS 2.224** 24-methyl-cholest-5,22-dien-3β-ol.

■ **CS 2.225** 20-hydroxyecdysone.

2β,3α-dihydroxy-5α-cholestane disulfate inhibited the enzymatic activity of AChE with IC_{50} values equal to 14.5 μM and 59.6 μM, respectively.[319] 24-Methyl-cholest-5,22-dien-3β-ol (CS 2.224) from the Chinese medicinal mushroom *Ganoderma lucidum* (Curtis) P. Karst. (family Ganodermataceae Donk) commanded the survival of cortical neurons deprived of both oxygen and glucose by enhancing the enzymatic activity of SOD and therefore reducing the cytoplasmic levels of ROS at a dose of 1 μg/mL.[320] Furthermore, 24-methyl-cholest-5,22-dien-3β-ol inhibited the degradation of IκBα, NF-κB at low doses of activity and the production of IL-1β and TNF-α by hippocampal neurons.[320] The molting hormone 20-hydroxyecdysone (CS 2.225), which evokes neurogenesis during

■ **CS 2.226** Sc-7.

■ **CS 2.227** Withanolide A.

the metamorphosis of *Drosophila melanogaster*,[321] protected PC12 cells against $CoCl_2$ by 73.6% at a dose of $200\,\mu M$ with a decrease in ROS, thus protecting mitochondrial integrity and negating apoptosis.[322] Note that the incapacity of SOD to control ROS accounts for oxidative insults and apoptosis of motoneurons, hence ALS, and in fact, the synthetic cholest-4-en-3-one, oxime mentioned earlier at a dose of $10\,\mu M$ prompted the survival and neuritogenesis of motor neurons deprived of growth factors by 74%.[323] Furthermore, cholest-4-en-3-one, oxime at a dose of 30 mg/kg imposed the regeneration of injured sciatic nerves by inhibiting mitochondrial mPTP opening.[323] The aquatic mold *Achlya heterosexualis* Whiffen-Barksdale (family Saprolegniaceae) produces the cholestane Sc-7 (CS 2.226), which at a dose of $90\,\mu M$ induced the differentiation of mouse embryonic teratocarcinoma P19 cells into neurons with increased expression of βIII-tubulin, MAP2, ChAT, and DCX.[324] Furthermore, intrathecal injection of Sc-7 in rodents at a dose of $375\,\mu M$ engaged the multiplication of progenitor cells in the subventricular zone.[324]

The oxidation of C22 and C26 on the 24-methylcholestane framework yields δ-lactone ergostanes or withanolides steroids, which occur in members of the genus *Physalis* L. (family Solanaceae Juss.). For instance, withanolide A (CS 2.227) from *Withania somnifera* (L.) Dunal (family Solanaceae Juss.) at a dose of $100\,\mu M$ reduced β-secretase and enhanced α-secretase in primary cortical neurons.[266] In addition withanolide A at a dose of $1\,\mu M$

■ **CS 2.228** Jaborosalactone A.

■ **CS 2.229** 18-hydroxywithanolide D.

■ **CS 2.230** 7-acetyl-14 deoxywithanolide U.

■ **CS 2.231** Jaborosalactone 1.

prompted the neuritogenesis and synaptogenesis of cortical neurons exposed to β-amyloid (Aβ_{25-35}) peptide.[325] In the same experiment, withanolide A at a dose of 10μmol/kg induced neuritogenesis and synaptogenesis in the *cortex* and the *hippocampus* of rodent poisoned by intrathecal injection of β-amyloid (Aβ_{25-35}) peptide.[325] Note that the withanolides jaborosalactone A, 18-hydroxywithanolide D 7-acetyl-14 deoxywithanolide U, jaborosalactone 1, withaphysalin J, and trechonolide A (CS 2.228–2.233) boosted the

■ **CS 2.232** Withaphysalin J.

■ **CS 2.233** Trechonolide A.

■ **CS 2.234** (20S,22R)-3α,6α-epoxy-4β,5β,27-trihydroxy-1-oxowitha-24-enolide.

■ **CS 2.235** (20S,22R)-4β,5β,6α,27-tetrahydroxy-1-oxo-witha-2,24-dienolide.

enzymatic activity of quinone reductase at a dose of 0.2 μM.[326] In addition, (20S,22R)-3α,6α -epoxy-4β,5β,27-trihydroxy-1-oxowitha-24-enolide (CS 2.234) and (20S,22R)-4β,5β,6α,27-tetrahydroxy-1-oxo-witha-2,24-dienolide (CS 2.235) from *Withania somnifera* (L.) Dunal (family Solanaceae Juss.) imposed the growth of neurites from human neuroblastoma SH-SY5Y at a dose of 1 μM.[327] Although the precise molecular changes that underlie these neurotrophic effects are unknown, one can reasonably speculate that witha-nolides activate CREB.

■ **CS 2.236** 24-epibrassinolide.

■ **CS 2.237** (22S,23S)-homocastasterone.

■ **CS 2.238** (22E,24S)-3β-hydroxy-5α-stigmastan-22-en-6-one.

■ **CS 2.239** (22E,24S)-2α,3α-dihydroxy-5α-stigmast-22-en-6-one.

Other phytosterols of neurological interest are the castesterones and brassinoids. An example of a brassinoid is 24-epibrassinolide (CS 2.236) from *Brassica juncea* (L.) Czern. (family Brassicaceae Burnett.), which preserved the viability of PC12 cells poisoned with 1-methyl-4-phenylpyridinium ion (MPP$^+$) by 60% with a reduction of ROS; an increase in the enzymatic activity of SOD, CAT, and GPx; and preservation of mitochondrial integrity.[328] Likewise, 24-epibrassinolide, (22S,23S)-homocastasterone (CS 2.237) and the stigmastanes (22E,24S)-3β-hydroxy-5α-stigmastan-22-en-6-one (CS 2.238), (22E,24S)-2α,3α-dihydroxy-5α-stigmast-22-en-6-one (CS 2.239), (22R, 23R)-homocastasterone

■ **CS 2.240** (22R,23R)-homocastasterone.

■ **CS 2.241** (22E,24S)-3α-hydroxy-5α-stigmastan-22-en-6-one.

■ **CS 2.242** Daedalol C.

(CS 2.240), and (22E,24S)-3α-hydroxy-5α-stigmastan-22-en-6-one (CS 2.241) prolonged the life span of PC12 cells exposed to 1-methyl-4-phenylpyridinium ion (MPP⁺) by more than 50%.[329]

The lanostane daedalol C (CS 2.242) isolated from the genus *Daedalea* Pers. ex Gray (family Polyporaceae Fr. ex Corda, 1839) inhibited the enzymatic activity of β-secretase (BACE-1) with an IC_{50} value equal to 14.2 μM.[330] Along the same lines, the lanostane n-butyl ganoderate H (CS 2.243), lucidumol B (CS 2.244), n-butyl lucidenate N (CS 2.245),

■ **CS 2.243** n-butyl ganoderate H.

■ **CS 2.244** Lucidumol B.

■ **CS 2.245** n-butyl lucidenate N.

■ **CS 2.246** n-butyl lucidenate A.

and n-butyl lucidenate A (CS 2.246) isolated from *Ganoderma lucidum* (Curtis) P. Karst. (family Ganodermataceae Donk) inhibited the enzymatic activity of AChE with IC_{50} values equal to 9.4 μM, 16.2 μM, 11.5 μM, and 12.2 μM, respectively.[331]

In rodents, squalene at a dose of 1 g/kg protected rodents against 6-hydroxy-dopamine (6-OHDA) poisoning, as shown by sustained the levels of dopamine implying that statins, which are used to lower cholesterol, might have the tendency to favor the development of neurodegenerative diseases.[332]

REFERENCES

[259] Ganzera M, Khan IA. A reversed phase high performance liquid chromatography method for the analysis of boswellic acids in *Boswellia serrata*. Planta Med 2001;67(8):778–80.

[260] Mook-Jung I, Shin JE, Yun SH, Huh K, Koh JY, Park HK, et al. Protective effects of asiaticoside derivatives against beta-amyloid neurotoxicity. J Neurosci Res 1999;58(3):417–25.

[261] Jew S,S, Yoo C,H, Lim D,Y, Kim H, Mook-Jung I, Jung M,W, et al. Structure-activity relationship study of asiatic acid derivatives against beta amyloid (A beta)-induced neurotoxicity. Bioorg Med Chem Lett 2000;10(2):119–21.

[262] Zhang X, Wu J, Dou Y, Xia B, Rong W, Rimbach G, et al. Asiatic acid protects primary neurons against C2-ceramide-induced apoptosis. Eur J Pharmacol 2012;679(1–3):51–9.

[263] Xiong Y, Ding H, Xu M, Gao J. Protective effects of asiatic acid on rotenone- or H_2O_2-induced injury in SH-SY5Y cells. Neurochem Res 2009;34(4):746–54.

[264] Lin MT, Beal MF. Mitochondrial dysfunction and oxidative stress in neurodegenerative diseases. Nature 2006;443(7113):787–95.

[265] Kim SR, Koo KA, Lee MK, Park HG, Jew SS, Cha KH, et al. Asiatic acid derivatives enhance cognitive performance partly by improving acetylcholine synthesis. Pharm Pharmacol 2004;56(10):1275–82.

[266] Patil S,P, Maki S, Khedkar S,A, Rigby A,C, Chan C. Withanolide a and asiatic acid modulate multiple targets associated with amyloid-beta precursor protein processing and amyloid-beta protein clearance. J Nat Prod 2010;73(7):1196–202.

[267] Soumyanath A, Zhong YP, Gold SA, Yu X, Koop DR, Bourdette D, et al. *Centella asiatica* accelerates nerve regeneration upon oral administration and contains multiple active fractions increasing neurite elongation *in-vitro*. J Pharm Pharmacol 2005;57(9):1221–9.

[268] Yun KJ, Kim JY, Kim JB, Lee KW, Jeong SY, Park HJ, et al. Inhibition of LPS-induced NO and PGE2 production by asiatic acid via NF-kappa B inactivation in RAW 264.7 macrophages: possible involvement of the IKK and MAPK pathways. Int Immunopharmacol 2008;8(3):431–41.

[269] Reddy VP, Obrenovich ME, Atwood CS, Perry G, Smith MA. Involvement of Maillard reactions in Alzheimer disease. Neurotox Res 2002;4:191–209.

[270] Takeda A, Wakai M, Niwa H, Dei R, Yamamoto M, Li M, et al. Neuronal and glial advanced glycation end product [Nepsilon-(carboxymethyl)lysine] in Alzheimer's disease brains. Acta Neuropathol 2001;101:27–35.

[271] Yin MC. Anti-glycative potential of triterpenes: a mini-review. Biomed 2012;2(1):2–9.

[272] Wang ZH, Hsu CC, Huang CN, Yin MC. Anti-glycative effects of oleanolic acid and ursolic acid in kidney of diabetic mice. Eur J Pharmacol 2010;628:255–60.

[273] Hong SY, Jeong WS, Jun M. Protective effects of the key compounds isolated from *Corni fructus* against β-amyloid-induced neurotoxicity in PC12 cells. Molecules 2012;17(9):10831–10845.

[274] Li Y, Ishibashi M, Satake M, Chen X, Oshima Y, Ohizumi Y. Sterol and triterpenoid constituents of *Verbena littoralis* with NGF-potentiating activity. J Nat Prod. 2003;66(5):696–8.

[275] Ringbom T, Segura L, Noreen Y, Perera P, Bohlin L. Ursolic acid from *Plantago major*, a selective inhibitor of cyclooxygenase-2 catalyzed prostaglandin biosynthesis. J Nat Prod 1998;61(10):1212–5.

[276] Shih YH, Chein YC, Wang JY, Fu YS. Ursolic acid protects hippocampal neurons against kainate-induced excitotoxicity in rat. Neurosci Lett 2004;362(2):136–40.

[277] Wang YJ, Lu J, Wu DM, Zheng ZH, Zheng YL, Wang XH, et al. Ursolic acid attenuates lipopolysaccharide-induced cognitive deficits in mouse brain through suppressing p38/NF-κB mediated inflammatory pathways. Neurobiol Learn Mem 2011;96(2):156–65.

[278] Li L, Zhang X, Cui L, Wang L, Liu H, Ji H, et al. Ursolic acid promotes the neuroprotection by activating Nrf2 pathway after cerebral ischemia in mice. Brain Res 2013;1497:32–9.

[279] Barnes PJ, Karin M. Nuclear factor-kappaB: a pivotal transcription factor in chronic inflammatory diseases. New Engl J Med 1997;336:1066–71.

[280] Sailer ER, Subramanian LR, Rall B, Hoernlein RF, Ammon HPT, Safayhi H. Acetyl-11-keto-β-boswellic acid (AKBA): Structure requirements for binding and 5-lipoxygenase inhibitory activity. Br J Pharmacol 1996;117(4):615–8.

[281] Manev H, Uz T, Sugaya K, Qu T. Putative role of neuronal 5-lipoxygenase in an aging brain. FASEB J 2000;14(10):1464–9.

[282] Park B, Sung B, Yadav VR, Cho SG, Liu M, Aggarwal BB. Acetyl-11-keto-β-boswellic acid suppresses invasion of pancreatic cancer cells through the downregulation of CXCR4 chemokine receptor expression. Int J Cancer 2011;129(1):23–33.

[283] Karima O, Riazi G, Yousefi R, Movahedi AA. The enhancement effect of beta-boswellic acid on hippocampal neurites outgrowth and branching (an *in vitro* study). Neurol Sci 2010;31(3):315–20.

[284] Martín R, Hernández M, Córdova C, Nieto ML. Natural triterpenes modulate immune-inflammatory markers of experimental autoimmune encephalomyelitis: Therapeutic implications for multiple sclerosis. Br J Pharmacol 2012;166(5):1708–23.

[285] Yap WH, Khoo KS, Ho ASH, Lim YM. Maslinic acid induces HO-1 and NOQ1 expression via activation of Nrf2 transcription factor. Biomed Preventive Nut 2012;2(1):51–8.

[286] Qian Y, Guan T, Tang X, Huang L, Huang M, Li Y, et al. Maslinic acid, a natural triterpenoid compound from *Olea europaea*, protects cortical neurons against oxygen-glucose deprivation-induced injury. Eur J Pharmacol 2011;670(1):148–53.

[287] Goldberg MP, Choi DW. Combined oxygen and glucose deprivation in cortical cell culture: calcium-dependent and calcium-independent mechanisms of neuronal injury. J Neurosci 1993;13(8):3510–24.

[288] Dienel GA, Hertz L. Astrocytic contributions to bioenergetics of cerebral ischemia. Glia 2005;50:362–88.

[289] Dunlop J. Glutamate-based therapeutic approaches: targeting the glutamate transport system. Curr Opin Pharmacol 2006;6:103–7.

[290] Qian Y, Guan T, Tang X, Huang L, Huang M, Li Y, et al. Neuroprotection of maslinic acid, a novel glycogen phosphorylase inhibitor, in type 2 diabetic rats. Chinese J Nat Med 2010;8(4):293–7.

[291] Trevisan G, Rossato M,F, Walker C,I, Klafke J,Z, Rosa F, Oliveira S,M, et al. Identification of the plant steroid α-spinasterol as a novel transient receptor potential vanilloid 1 antagonist with antinociceptive properties. J Pharmacol Exp Ther 2012;343(2):258–69.

[292] Maccarrone M, Lorenzon T, Bari M, Melino G, Finazzi-Agro A. Anandamide induces apoptosis in human cells via vanilloid receptors. Evidence for a protective role of cannabinoid receptors. J Biol Chem 2000;275:31938–31945.

[293] Tran TA, McCoy MK, Sporn MB, Tansey MG. The synthetic triterpenoid CDDO-methyl ester modulates microglial activities, inhibits TNF production, and provides dopaminergic neuroprotection. J Neuroinflammation 2008;12(5):14.

[294] Graber D,J, Park P,J, Hickey W,F, Harris B,T. Synthetic triterpenoid CDDO derivatives modulate cytoprotective or immunological properties in astrocytes neurons, and microglia. J Neuroimmune Pharmacol 2011;6(1):107–20.

[295] Neymotin A, Calingasan N,Y, Wille E, Naseri N, Petri S, Damiano M, et al. Neuroprotective effect of Nrf2/ARE activators, CDDO ethylamide and CDDO trifluoroethylamide, in a mouse model of amyotrophic lateral sclerosis. Free Radical Biology and Medicine 2011;51(1):88–96.

[296] Sassa H, Takaishi Y, Terada H. The triterpene celastrol is a very potent inhibitor of lipid peroxidation in mitochondria. Biochem Biophys Res Comm 1990;172:890–7.

[297] Allison A,C, Cacabelos R, Lombardi V,R, Alvarez X,A, Vigo C. Celastrol, a potent antioxidant and anti-inflammatory drug, as a possible treatment for Alzheimer's disease. Prog Neuropsychopharmacol Biol Psychiatry 2001;25(7):1341–57.

[298] Allison AC, Cacabelos R, Lombardi VRM, Álvarez XA, Vigo C. Central nervous system effects of celastrol, a potent antioxidant and antiinflammatory agent. CNS Drug Rev 2000;6(1):45–62.

[299] Lee J,H, Koo T,H, Yoon H, Jung H,S, Jin H,Z, Lee K, et al. Inhibition of NF-kappa B activation through targeting I kappa B kinase by celastrol, a quinone methide triterpenoid. Biochem Pharmacol 2006;72(10):1311–21.

[300] Paris D, Ganey N,J, Laporte V, Patel N,S, Beaulieu-Abdelahad D, Bachmeier C, et al. Reduction of beta-amyloid pathology by celastrol in a transgenic mouse model of Alzheimer's disease. J Neuroinflammation 2010;7:17.

[301] Yadav VR, Sung B, Prasad S, Kannappan R, Cho SG, Liu M, et al. Celastrol suppresses invasion of colon and pancreatic cancer cells through the downregulation of expression of CXCR4 chemokine receptor. J Mol Med 2010;88(12):1243–53.

[302] Chow AM, Brown IR. Induction of heat shock proteins in differentiated human and rodent neurons by celastrol. Cell Stress Chaperones 2007;12(3):237–44.

[303] Hieronymus H, Lamb J, Ross KN, Peng XP, Clement C, Rodina A, et al. Gene expression signature-based chemical genomic prediction identifies a novel class of HSP90 pathway modulators. Cancer Cell 2006;10(4):321–30.

[304] Vabulas RM, Raychaudhuri S, Hayer-Hartl M, Hartl FU. Protein folding in the cytoplasm and the heat shock response. Cold Spring Harb Perspect Biol 2010;2(12):1–18.

[305] Gusella JF, MacDonald ME. Molecular genetics: unmasking polyglutamine triggers in neurodegenerative disease. Nat Rex Neurosci 2000;1:109–15.

[306] Wang J, Gines S, MacDonald M,E, Gusella J,F. Reversal of a full-length mutant huntingtin neuronal cell phenotype by chemical inhibitors of polyglutamine-mediated aggregation. BMC Neurosci 2005;6:1–12.

[307] Cleren C, Calingasan N,Y, Chen J, Beal M,F. Celastrol protects against MPTP- and 3-nitropropionic acid-induced neurotoxicity. J Neurochem 2005;94:995–1004.

[308] Geetha T, Varalakshmi P. Anti-inflammatory activity of lupeol and lupeol linoleate in rats. J Ethnopharmacol 2001;76(1):77–80.

[309] Saleem M, Afaq F, Adhami VM, Mukhtar H. Lupeol modulates NF-kappaB and PI3K/Akt pathways and inhibits skin cancer in CD-1 mice. Oncogene 2004;23(30):5203–14.

[310] Brimson JM, Brimson SJ, Brimson CA, Rakkhitawatthana V, Tencomnao T. *Rhinacanthus nasutus* extracts prevent glutamate and amyloid-β neurotoxicity in HT-22 mouse hippocampal cells: possible active compounds include lupeol, stigmasterol and β-sitosterol. Int J Mol Sci 2012;13(4):5074–97.

[311] Lee M,K, Kim Y,C. Five novel neuroprotective triterpene esters of *Ulmus davidiana* var. *japonica*. J Nat Prod 2001;64(3):328–31.

[312] Reyes CP, Núñez MJ, Jiménez IA, Busserolles J, Alcaraz MJ, Bazzocchi IL. Activity of lupane triterpenoids from *Maytenus* species as inhibitors of nitric oxide and prostaglandin E2. Bioorg Med Chem 2006;14(5):1573–9.

[313] Gurovic M,S,V, Castro M,J, Richmond V, Faraoni M,B, Maier M,S, Murray A,P. Triterpenoids with acetylcholinesterase (AChE) inhibition from *chuquiraga erinacea* D. Don. subsp. *erinacea* (Asteraceae). Planta Med 2010;76(6):607–10.

[314] Jang S,W, Liu X, Chan C,B, France S,A, Sayeed I, Tang W, et al. Deoxygedunin, a natural product with potent neurotrophic activity in mice. PLoS One 2010;5(7):11528.

[315] Vabulas RM, Raychaudhuri S, Hayer-Hartl M, Hartl FU. Protein folding in the cytoplasm and the heat shock response. Cold Spring Harb Perspect Biol 2007;2(12):1–18.

[316] Yoon JS, Yang H, Kim SH, Sung SH, Kim YC. Limonoids from *dictamnus dasycarpus* protect against glutamate-induced toxicity in primary cultured rat cortical cells. J Mol Neurosci 2010;42(1):9–16.

[317] Tang M,Z, Wang Z,F, Shi Y,L. Toosendanin induces outgrowth of neuronal processes and apoptosis in PC12 cells. Neurosci Res 2003;45(2):225–31.

[318] Yao Z,X, Brown R,C, Teper G, Greeson J, Papadopoulos V. 2R-Hydroxycholesterol protects neuronal cells from beta-amyloid-induced cytotoxicity by binding to beta-amyloid peptide. J Neurochem 2002;83:1110–9.

[319] Richmond V, Garrido Santos G,A, Murray A,P, Maier M,S. Synthesis and acetyl-cholinesterase inhibitory activity of $2\beta,3\alpha$-disulfoxy-5α-cholestan-6-one. Steroids 2011;76(10–11):1160–5.

[320] Zhao HB, Wang SZ, He QH, Yuan L, Chen AF, Lin ZB. *Ganoderma* total sterol (GS) and GS 1 protect rat cerebral cortical neurons from hypoxia/reoxygenation injury. Life Sci 2005;76(9):1027–37.

[321] Kraft R, Levine RB, Restifo LL. The steroid hormone 20-hydroxyecdysone enhances neurite growth of *Drosophila mushroom* body neurons isolated during metamorphosis. J Neurosci 1998;18(21):8886–99.

[322] Hu J, Zhao TZ, Chu WH, Luo CX, Tang WH, Yi L, et al. Protective effects of 20-hydroxyecdysone on CoCl(2)-induced cell injury in PC12 cells. J Cell Biochem 2010;111(6):1512–21.

[323] Bordet T, Buisson B, Michaud M, Drouot C, Galéa P, Delaage P, et al. Identification and characterization of cholest-4-en-3-one, oxime (TRO19622), a novel drug candidate for amyotrophic lateral sclerosis. J Pharmacol Exp Ther 2007;322(2):709–20.

[324] Lecanu L, Hashim A,I, McCourty A, Papadopoulos V. A steroid isolated from the water mold *Achlya heterosexualis* induces neurogenesis *in vitro* and *in vivo*. Steroids 2012;77(3):224–32.

[325] Kuboyama T, Tohda C, Komatsu K. Neuritic regeneration and synaptic reconstruction induced by withanolide A. Br J Pharmacol 2005;144(7):961–71.

[326] Misico RI, Song LL, Veleiro AS, Cirigliano AM, Tettamanzi MC, Burton G, et al. Induction of quinone reductase by withanolides. J Nat Prod 2002;65(5):677–80.

[327] Zhao J, Nakamura N, Hattori M, Kuboyama T, Tohda C, Komatsu K. Withanolide derivatives from the roots of *Withania somnifera* and their neurite outgrowth activities. Chem Pharm Bull 2002;50(6):760–5.

[328] Carange J, Longpré F, Daoust B, Martinoli MG. 24-Epibrassinolide, a phytosterol from the brassinosteroid family, protects dopaminergic cells against MPP-induced oxidative stress and apoptosis. J Toxicol 2011:392859.

[329] Ismaili J, Boisvert M, Longpré F, Carange J, Gall CL, Martinoli MG, et al. Brassinosteroids and analogs as neuroprotectors: synthesis and structure-activity relationships. Steroids 2012;77(1–2):91–9.

[330] Sorribas A, Jiménez JI, Yoshida WY, Williams PG. Daedalols A-C, fungal-derived BACE1 inhibitors. Bioorg Med Chem 2011;19(22):6581–6.

[331] Lee I, Ahn B, Choi J, Hattori M, Min B, Bae K. Selective cholinesterase inhibition by lanostane triterpenes from fruiting bodies of *Ganoderma lucidum*. Bioorg Med Chem Lett 2011;21(21):6603–7.

[332] Kabuto H, Yamanushi TT, Janjua N, Takayama F, Mankura M. Effects of squalene/squalane on dopamine levels, antioxidant enzyme activity, and fatty acid composition in the striatum of Parkinson's disease mouse model. J Oleo Sci 2013;62(1):21–8.

Chapter 3

Phenolics

■ INTRODUCTION

Neurons are exposed to approximately 20% of the oxygen consumed by the body. The neuronal mitochondrial metabolism of oxygen generates reactive oxygen species (ROS), which are neutralized by glutathione (GSH), glutathione peroxidase (GPx), and superoxide dismutase (SOD). However, aging inexorably involves not only a decrease in the aforementioned antioxidative defenses but also the progressive setting of a pro-oxidative cerebral environment which favors DNA damage, which in turn stimulates the mutation of particularly sensitive genes such as those coding for secretase cleaving APP, or handling the clearance of cytoplasm of dopamine, or coding for superoxide dismutase (SOD), resulting in the irreversible settlement of Alzheimer's disease (AD), Parkinson's disease (PD), and amyotrophic lateral sclerosis (ALS), respectively. Indeed, neurodegenerative diseases engross a boost in cytoplasmic reactive oxygen species resulting in the activation of the pro-apoptotic protein (p53), hence apoptosis. Once the pro-apoptotic protein (p53) is activated by the reactive oxygen species, pro-apoptotic Bcl-2-associated X proteins (Bax) are activated. The translocation of pro-apoptotic Bcl-2-associated X protein (Bax) provokes irremediable mitochondrial insults, leakage of cytochrome c, and caspase 9 and 3 activation and apoptosis. Therefore, one might argue that phenolic compounds which are antioxidant are beneficial against neuronal oxidative insults, but phenolic compounds by definition shelter in their frameworks one or more oxygen atoms either as hydroxyl, ester, ether, or ketone moieties, and current evidence clearly demonstrates that phenolics are, at high doses, pro-oxidant. The concept of dose or "dose widow" is of critical importance in understanding the neurotoxic, neuroprotective, or even neurotrophic properties of phenolics. In regard to their neurotoxic effect, an obvious increase in reactive oxygen species is to be considered, whereas the neuroprotective potencies of phenolics involve the inhibition or activation of kinases. In effect, reactive oxygen species compel the phosphorylation and therefore activation of mitogen-activated protein (MAP) kinases, which include p38

C. Wiart: Lead Compounds from Medicinal Plants for the Treatment of Neurodegenerative Diseases.
DOI: http://dx.doi.org/10.1016/B978-0-12-398373-2.00003-0

mitogen-activated protein kinase (p38 MAPK), c-Jun N-terminal kinase (JNK), and extracellular signal-regulated kinase (ERK1/2). The activation of p38 mitogen-activated protein kinase (p38 MAPK) evokes the translocation of pro-apoptotic Bcl-2-associated X protein (Bax) and BIM, a decrease of Bcl-2, mitochondrial insults, release of cytochrome c, activation of caspases 9 and 3, and apoptosis. Furthermore, activation of p38 mitogen-activated protein kinase (p38 MAPK) by reactive oxygen species results in the inhibition of protein kinase B (Akt) and the activation of pro-apoptotic protein (p53) and caspase 8; hence, caspase 3 activation and apoptosis. Of note, flavonoids such as calycopterin from *Dracocephalum kotschyi* Boiss. (3.1.1) from *Dracocephalum kotschyi* Boiss. save neurons by blocking p38. Likewise, the activation of Jun N-terminal kinase is followed by the cleavage of BIM, translocation of pro-apoptotic Bcl-2-associated X protein (Bax), and therefore mitochondrial insults. In parallel, activation of c-Jun N-terminal kinase is followed by an increase of the death receptor (DR5), which activates caspase 8. Thus, phenolic hindering of the activation of Jun N-terminal kinase provides neuroprotection, as shown with flavonoids, which include kaempferol from *Camellia chinensis* (L.) Kuntze (3.1.1). Sustained activation of extracellular signal-regulated kinase induces p53 and apoptosis via mitochondrial insults, and flavonoids like jaceosidin rescue neurons against apoptosis by inhibiting ERK1/2. Another interesting anti-apoptotic ability of phenolics is the induction of antioxidant enzymes, including superoxide dismutase (SOD), glutathione peroxidase (GPX), and catalase (CAT) as with, for instance, butin from *Toxicodendron vernicifluum* (Stokes) F.A. Barkley (3.1.1), glabridin from *Glycyrrhiza glabra* L. (3.1.1), decursin from *Angelica gigas* Nakai (3.1.2), gomisin N from *Schisandra chinensis* (Turcz.) Baill. (3.3.1), sauchinone from *Saururus chinensis* (Lour.) Baill. (3.3.1), and sesamin from *Sesamum indicum* L. (3.3.1).

As mentioned earlier, the chronic activation of extracellular signal-regulated kinase provokes neuroapoptosis, whereas a transient activation of extracellular signal-regulated kinase within a "trophic window dose" induces neuritogenesis as what could be termed a final attempt for survival because neurons do not divide. Numerous phenolic compounds command, within a "trophic window dose," the growth of neurites via the stimulation of ERK1/2 such as acacetin from members of the genus *Populus* L. (3.1.1), luteolin from genus *Capsicum* L. (3.1.1), auraptene from *Dictamnus albus* L. (3.1.2), garcinol from *Garcinia indica* Choisy (3.1.3), obovatol from *Magnolia officinalis* Rehder & E.H. Wilson (3.1.3), and schizandrin from *Schisandra chinensis* (Turcz.) Baill. (3.1.3). Because phenolic compounds are massively catabolized in the intestine and liver and, by virtue of their hydrophilicity, are not prone to cross the blood brain barrier, it could be argued that this group of natural products is of no use for the treatment of neurodegenerative diseases. An answer would be to suggest either the use of "pro-phenolics," which would liberate phenolics in the cytoplasm of neurons or in the cerebrospinal fluid (CSF), or the injection of therapeutic phenolics directly into the cerebrospinal fluid. In brief, phenolics offer a bewildering array of chemical frameworks, thus implying a bewildering array of molecular pathways and hence the possibility of discovering new neuroprotective and/or neurotrophic proteins that could be targeted for the treatment of Alzheimer's disease, Parkinson's disease, amyotrophic lateral sclerosis, and also spinal cord injuries. In fact, the biggest challenge with possible phenolic leads will be to maintain the drug within a "protective" or "trophic window." In this chapter, several phenolics of value for the treatment of neurodegenerative diseases are identified, including calycosine, umbelliferone, phylloquinone, emodin, and dihydroguaiaretic acid.

Topic # 3.1

Benzopyrones

3.1.1 *Astragalus mongholicus* Bunge

History The plant was first described by Alexander Andrejewitsch Bunge in *Mémoires de l'Académie impériale des Sciences de St. Pétersbourg*, published in 1868.

Family Fabaceae Lindl., 1836

Synonyms *Astragalus borealimongolicus* Y.Z. Zhao, *Astragalus membranaceus* Bunge, *Astragalus membranaceus* var. *mongholicus* (Bunge) P.G. Xiao, *Astragalus mongholicus* var. *dahuricus* (DC.) Podl., *Astragalus propinquus* Schischk., *Astragalus purdomii* N.D. Simpson, *Phaca abbreviata* Ledeb., *Phaca alpina* var. *dahurica* DC., *Phaca macrostachys* Turcz.

Common Names Meng gu huang qi (Chinese)

Habitat and Description It is a gracile herb which grows to 50 cm tall in China, Mongolia, and Russia. The stem is terete and hairy. The leaves are imparipinnate and stipulate. The stipules are 1 cm long. The rachis is 15 cm long and supports 8–12 pairs of folioles, which are elliptic 0.5–2 cm × 0.3–1 cm. The inflorescence is an axillary raceme which is 4–5 cm long. The calyx is minute, tubular, and produces 5 teeth. The corolla is 1 cm long, papilionaceous, yellow, and includes 5 petals. The fruit is a hairy pod which is 2.5 cm long (Figure 3.1).

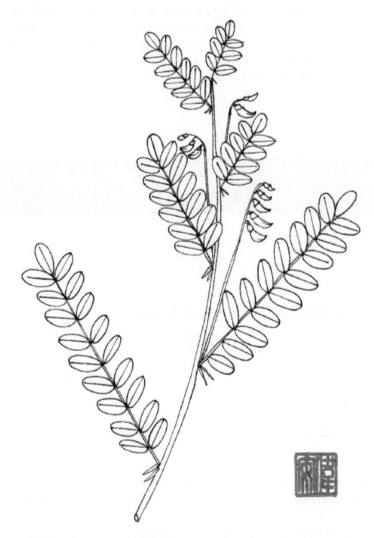

■ **FIGURE 3.1** *Astragalus mongholicus* Bunge.

Phytopharmacology The plant contains flavonoids, including the iso-flavans 7-*O*-methylisomucronulatol, isomucronulatol-7,2′-di-*O*-glucoside, 5′-hydroxyisomucronulatol-2′,5′-di-*O*-glucoside; the pterocarpan 3,9-di-*O*-methylnissolin[1]; and isoflavones, including formononetin, ononin, and calycosine[2,3]; and saponins such as saponin, iso-astragaloside IV, and astragaloside IV.[4]

Proposed Research Pharmacological study of calycosine for the treatment of neurodegenerative diseases.

Medicinal Uses In China, the plant is used to invigorate.

Rationale Chronic exposure of neurons to glutamate and/or β-amyloid peptide imposes apoptosis via the generation of cytoplasmic reactive oxygen species (ROS) which inactivate protein kinase B (Akt). In this light, it should be recalled that flavonoids, at low dose, scavenge ROS in the cytoplasm by virtue of their aromatic hydroxyl moieties and assuage neuroapoptosis and neuroinflammation, whereas at high dose these phenolics favor the generation of the aforementioned oxygenated free radicals thus inducing apoptosis or neuritogenesis. One such flavonoid is the flavone chrysin (CS 3.1) found in members of the genus *Passiflora* L. (family Passifloraceae Juss. ex Roussell), which protected dopaminergic neurons against MPP$^+$ at a dose of $40\,\mu$M.[5] In BV-2 cells challenged with lipopolysaccharide (LPS), chrysin at a dose of $10\,\mu$M hindered the production of nitric oxide (NO),[6] and at a dose of $20\,\mu$M reduced nitric oxide synthetase (iNOS), NO, and tumor necrosis factor-α (TNF-α).[7] The neuroprotective effects of chrysin were confirmed *in vivo*, whereas 30 mg/kg of this flavone mitigated dementia as well as hippocampal neurodegeneration incurred by experimental hypoperfusion in rodents.[8] The reduction of chrysin at $\Delta_{2\text{-}3}$ produces pinocembrin (CS 3.2) occurring in *Alpinia galanga* (L.) Willd. (family Zingiberaceae Martinov), which at a dose of 500 nM protected hippocampal

■ **CS 3.1** Chrysin.

■ **CS 3.2** Pinocembrin.

neurons against glutamate insults.[9] Likewise, pinocembrin at a dose of 10^{-5} mol/L protected human SH-SY5Y neuroblastoma cells against glutamate insults with a decrease of p53 and thus pro-apoptotic Bcl-2–associated X protein (Bax).[10] In rodents, pinocembrin at a dose of 30 mg/kg mitigated the locomotor impairments incurred by experimental ischemia with parallel increase in glutathione (GSH), decreased levels of inducible nitric oxide synthetase (iNOS), NO, and inactivation of caspase 3 in cortical neurons.[11]

7,8-Dihydroxyflavone (CS 3.3) evoked the growth of neurites in hippocampal neurons by activating Akt and extracellular signal-regulated kinase (ERK1/2),[9] suggesting the stimulation of Ras. Methoxylation of 7,8-dihydroxyflavone at C8 and hydroxylation in C5 produces wogonin (CS 3.4) found in *Scutellaria baicalensis* Georgi (family Lamiaceae Martinov), which at a dose of 300 μg/mL abated the deleterious effects of glutamate, NMDA, H_2O_2, and xanthine/xanthine oxidase against cortical cells.[12] Wogonin at a dose of 50 μM protected hippocampal neurons against oxygen and glucose deprivation,[13] and at a dose of 3 μM reduced iNOS and cyclo-oxygenase-2 (COX-2) in BV-2 glial cells challenged with LPS and interferon-γ (INF-γ) via inactivation of IκB, ERK1/2, MEK1/2, and Src.[14] 5-Hydroxy-7-methoxyflavone (CS 3.5), 5-hydroxy-3,7,4′-trimethoxyflavone (CS 3.6), and 5-hydroxy-3,7,3′,4′-tetramethoxyflavone (CS 3.7) from a member of the

■ **CS 3.3** 7,8-dihydroxyflavone.

■ **CS 3.4** Wogonin.

■ **CS 3.5** 5-hydroxy-7-methoxyflavone.

■ **CS 3.6** 5-hydroxy-3,7,4′-trimethoxyflavone.

family Zingiberaceae Martinov protected cortical cells against glutamate by 16,3%, 26.7%, and 63.4%, respectively, at a dose of $10 \mu M$.[15]

Hydroxylation of chrysin at C6 forms baicalein (CS 3.8), which occurs in members of the genus *Scutellaria* L. (family Lamiaceae Martinov). Baicalein protected human SH-SY5Y neuroblastoma cells and rat pheochromocytoma (PC12) cells against 6-hydroxydopamine (6-HODA) at a dose of $5 \mu g/mL$[16] and defended glioma C6 cells against H_2O_2 at a dose of $50 \mu M$ by inactivating ERK1/2 and therefore p53.[17] Oddly enough, baicalein induced heme oxidase (HO1) in glioma C6 cells void of H_2O_2,[17] suggesting that this flavonoid induces ERK1/2. In rodents subjected to experimental stroke and poisoned with MPTP, baicalein improved neurological function at a dose of 4 mg/kg and 280 mg/kg by preserving the enzymatic activity of glutathione peroxidase (GPx) in the *cortex* and *hippocampus* and *substantia nigra*.[18] Of note, 7,8-dihydroxyflavone at a dose of 500 nM evoked the growth of neurites in hippocampal neurons by activating Akt and ERK1/2.[9]

Hydroxylation of chrysin in C4′ forms apigenin, which occurs in *Petroselinum crispum* (Mill.) Nyman ex A.W. Hill. (family Apiaceae Lindl.). Apigenin at a dose of $50 \mu M$ protected SH-SY5Y cells against H_2O_2,[19] and at a dose of $10 \mu M$ reduced the production of NO by BV-2 cells challenged with LPS.[6] The saturation of apigenin at Δ_{2-3} provides naringenin (CS 3.9), which protected PC12 cells against β-amyloid ($A\beta_{25-35}$) peptide at a dose

■ **CS 3.7** 5-hydroxy-3,7,3′,4′-tetramethoxyflavone.

■ **CS 3.8** Baicalein.

■ **CS 3.9** Naringenin.

of $100\,\mu M^{20}$ and dopaminergic neurons against MPP$^+$ at a dose of $40\,\mu M$.[5] Naringenin at a dose of 80 mg/kg improved the cognition of rodents poisoned with D-galactose with increased enzymatic activity of cerebral superoxide dismutase (SOD) and glutathione S-transferase (GST),[21] implying that Δ_{2-3} is not a prerequisite for neuroprotection.

Methoxylation of apigenin (CS 3.10) in C4′ provides acacetin (CS 3.11) produced by members of the genus *Populus* L. (family Salicaceae Mirb.). Acacetin at a dose of 200 nM protected *in vitro* and *in vivo* mesencephalic cells against MPP$^+$ insults with decrease of NO, prostaglandin E2 (PGE2), and TNF-α.[22] Acacetin at a dose of $100\,\mu M$ induced the growth of neurites in PC12 cells via the activation of ERK1/2, JNK, and mitogen-activated protein kinase (MAPK) p38,[23] suggesting the generation of ROS.[24] Hydroxylation of acacetin in C5′ yields diosmetin (CS 3.12), which at a dose of 500 nM protected hippocampal neurons against glutamate insults.[9]

Hydroxylation of apigenin at C5′ forms luteolin (CS 3.13) in members of the genus *Capsicum* L. (family Solanaceae Juss.). Luteolin at a dose of $50\,\mu M$ protected human SH-SY5Y neuroblastoma cells against H_2O_2.[19] In

■ **CS 3.10** Apigenin.

■ **CS 3.11** Acacetin.

■ **CS 3.12** Diosmetin.

■ **CS 3.13** Luteolin.

PC12 cells, luteolin at a dose of $10\,\mu$M prompted the growth of neurites via the activation of ERK1/2, nuclear translocation of Nrf2, binding of Nfr2 to ARE, and transcription of heme oxygenase-1 (HO-1).[25] Luteolin at a dose of $100\,\mu$M activated ERK1/2, JNK, and MAPK p38 in PC12 cells,[23] suggesting neurotrophic potencies. Luteolin at a dose of $10\,\mu$M reduced NO by BV-2 glial cells challenged with LPS.[6] In the same experiment, apigenin at a dose of $10\,\mu$M reduced the levels of iNOS, COX-2, and PGE2 and inactivated JNK and MAPK p38 in BV-2 glial cells.[6,26]

The hydroxylation of apigenin in C3 provides kaempferol (CS 3.14), found in *Camellia chinensis* (L.) Kuntze (family Theaceae Mirb.). Kaempferol at a dose of $10\,\mu$M nullified the deleterious effects of 4-HNE against PC12 cells with inactivation of JNK and inhibition of NOX.[27] Kaempferol at a dose of $30\,\mu$M shielded SH-SY5Y cells against rotenone via inactivation of JNK and MAPK p38 and maintenance of the mitochondrial integrity.[28] Kaempferol at a dose of $10\,\mu$M reduced the production of NO by BV-2 glial cells challenged with LPS.[6] Kaempferol at a dose of $100\,\mu$M inhibited the enzymatic activity of iNOS and COX-2; reduced the production of NO, PGE2, and interleukin-1β (IL-1β); and inactivated JNK, MAPK p38, and Akt by BV-2 cells exposed to LPS.[29]

Hydroxylation of kaempferol in C2' yields morin (CS 3.15) from *Chlorophora tinctoria* (L.) Gaudich. ex Benth. (family Moraceae Gaudich.), which at a dose of $5\,\mu$mol/L sustained the viability of PC12 cells exposed to MPP$^+$ by 81.2% with a decrease in ROS.[24] Along the same line, morin at a dose of 100 nM protected cortical neurons against glutamate by 56.9% with a decrease in ROS, increased SOD activity, preservation of mitochondrial integrity, inactivation of calpain, activation of Akt, and hence inactivation of ERK1/2.[30] At a dose of 40 mg/kg, morin attenuated the locomotor disturbances incurred by MPTP by protecting the *Substantia nigra*.[24]

The hydroxylation of kaempferol in C5' yields quercetin (CS 3.16) from *Allium sativum* L. (family Amaryllidaceae J. St. Hil.), which protected neurons against MPP$^+$ and β-amyloid (Aβ_{1-42}) peptide at a dose of $40\,\mu$M[5]

■ **CS 3.14** Kaempferol.

■ **CS 3.15** Morin.

■ **CS 3.16** Quercetin.

■ **CS 3.17** (+)-dihydroquercetin.

■ **CS 3.18** 3-methoxy-quercetin.

■ **CS 3.19** Myricetin.

and $5\,\mu$M,[31] respectively. Quercetin at a dose of $100\,\mu$M protected human SH-SY5Y neuroblastoma cells against H_2O_2 with an increase in GSH, inactivation of caspase 9 and caspase 3, and an increase in Bcl-2,[32] implying the inhibition of p53 and ERK1/2. Quercetin, (+)-dihydroquercetin (CS 3.17), and 3-methoxy-quercetin (CS 3.18) protected cortical cells against the deleterious effect of H_2O_2.[33] Quercetin at a dose of $10\,\mu$M reduced the production of NO by BV-2 glial cells challenged with LPS.[6] Quercetin at a dose of 30 mg/kg mitigated locomotor impairment incurred by experimental stroke with reduction of infarct volumes and in the hippocampus with an increase of GSH, GPx, GST, SOD, and inactivation of caspase 3.[34]

The hydroxylation of quercetin in C3′ provides myricetin (CS 3.19) from *Brassica oleracea* L. (family Brassicaceae Burnett), which at a dose of 10^{-8} mol/L sustained the viability of dopaminergic neurons poisoned with MPP[+] by preserving mitochondrial integrity, inactivating mitogen-activated protein kinase (MAPK) kinase 4 (MKK4) and therefore JNK.[35] Note that myricetin has the interesting ability to bind directly to MEK-ATP binding site.[25,36,37] In addition, myricetin attaches itself to and inhibits Akt at the ATP-binding site via chemical interactions between the C3 hydroxyl and

■ **CS 3.20** Fisetin.

asparagine 292, the C7 hydroxyl and alanine 230, and the C3′ hydroxyl group with glutamine 234.[38]

The removal of C5 from quercetin creates fisetin (CS 3.20), which abounds in members of the genus *Fragaria* L. (family Rosaceae Juss.). Fisetin sustained the survival of primary neurons deprived of serum at a dose of $2\,\mu M$[39] and induced the sprouting of neurites from PC12 cells at a dose of $20\,\mu M$ by activating Ras, MEK1, and ERK1/2,[39,40] suggesting the inactivation of Akt. In BV-2 cells exposed to LPS, fisetin at a dose of $2\,\mu g/mL$ annihilated the production of NO and TNF-α, PGE2 by inactivating MAPK p38, and of nuclear factor kappa-light-chain-enhancer of activated B cells (NF-κB)[41] via possible inhibition of kappa light polypeptide gene enhancer in B-cells inhibitor (IκB) kinase (IKK).[42]

Inula britannica L. (family Asteraceae Bercht. & J. Presl) shelters the flavones patuletin (CS 3.21), nepetin (CS 3.22), and axillarin (CS 3.23), which at a dose of $50\,\mu M$ protected cortical neuronal cells against glutamate by 70.6%, 62.4%, and 59.7%, respectively.[43] Patuletin and axillarin blocked the influx of Ca^{2+} into neurons exposed to glutamate and the subsequent burst in ROS.[43] Patuletin, nepetin, and axillarin sustained GSH and boosted the enzymatic activities of GSH-px and glutathione disulfide reductase (GSSG-R).[43] The flavone calycopterin (CS 3.24) from *Dracocephalum kotschyi* Boiss. (family Lamiaceae Martinov) at a dose of $50\,\mu M$ protected PC12 cells against H_2O_2 with a decrease of ROS, pro-apoptotic Bax, caspase 3, calpain and caspase 12; and inactivation of MAPK p38, ERK1/2, and JNK.[3] The flavone jaceosidin (CS 3.25) from *Artemisia argyi* H. Lev & Vaniot (family Asteraceae Bercht. & J. Presl) inactivated ERK1/2 and MAPK p38, lowered iNOS, and consequently lowered the production of NO by BV-2 mouse microglial cells exposed to LPS with an IC$_{50}$ value equal to $27\,\mu M$.[44]

The flavanones (2S)-2′-methoxykurarinone (CS 3.26) and sophoraflavanone G (CS 3.27) from *Sophora flavescens* Aiton (family Fabaceae Lindl) at a dose of $40\,\mu M$ nullified the deleterious effects of glutamate against mouse hippocampal (HT-22) cells by commanding the production

■ **CS 3.21** Patuletin.

■ **CS 3.22** Nepetin.

■ **CS 3.23** Axillarin.

■ **CS 3.24** Calycopterin.

■ **CS 3.25** Jaceosidin.

■ **CS 3.26** (2S)-2′-methoxycurarinne.

■ **CS 3.27** Sophoraflavonone G.

heme oxidase (HO-1).[45] The flavan (+)-catechin (CS 3.28) shielded dopaminergic neurons against 6-hydroxydopamine (6-HODA) and rotenone at a dose of $40\,\mu M$.[5] Likewise, cianidanol at a dose of 500 nM protected hippocampal neurons against glutamate insults.[9] *Toxicodendron vernicifluum* (Stokes) F.A. Barkley (family Anacardiaceae R. Br.) produces the flavanones butin (CS 3.29) and eriodictyol (CS 3.30), which protected mouse hippocampal (HT-22) cells against glutamate by 59.6% and 22.3%, respectively, at a dose of $50\,\mu M$ with sustained enzymatic activity of SOD and GPx.[46] The same plant produces the chalcone butein (CS 3.31), which protected mouse hippocampal (HT-22) cells against glutamate by 57.1% at a dose of $50\,\mu M$ with sustained enzymatic activity of SOD and GPx, whereas isoliquiritigenin (CS 3.32) was inactive.[46] The chalcone verbenachalcone (CS 3.33) from a member of the genus *Verbena* L. (family Verbenaceae J.St.Hil) doubled the neuritogenic effects of 2 ng/mL of nerve growth factor (NGF) against PC12 cells.[47] From the same plant, littorachalcone (CS 3.34) doubled the neuritogenic effects of 2 ng/mL of NGF against PC12 cells.[48] The chalcone compound 2 (CS 3.35) at a dose of $6\,\mu M$ quadrupled the neurotrophic potencies of 2 ng/mL of NGF against PC12 cells.[48] The chalcone sappanchalcone (CS 3.36) from *Caesalpinia sappan* L. (family Fabaceae Lindl.) at a dose of $1\,\mu M$ protected 13.7% of cortical cells against glutamate insults.[47]

■ **CS 3.28** (+)-catechin.

■ **CS 3.29** Butin.

■ **CS 3.30** Eriodictyol.

■ **CS 3.31** Butein.

■ **CS 3.32** Isoliquiritigenin.

■ **CS 3.33** Verbenachalcone.

■ **CS 3.34** Littorachalcone.

■ **CS 3.35** Compound 2.

■ **CS 3.36** Sappanchalcone.

The isoflavonoids formononetin (CS 3.37) and calycosin (CS 3.38) isolated from *Astragalus mongholicus* Bunge (family Fabaceae Lindl.) sustained the viability of PC12 cells exposed to glutamate by 69.5% and 66.41% at a concentration of 0.05 μg/mL by maintaining the enzymatic activity of SOD and GPx.[2] Genistein (CS 3.39) from *Glycine max* (L.) Merr. (family Fabaceae Lindl.) protected dopaminergic neurons against MPP$^+$ at a dose of 40 μM. At the same dose, catechin shielded dopaminergic neurons against 6-hydroxydopamine (6-HODA) and rotenone.[5] Genistein at a dose of 50 nM protected motor neurons against microglial cytokine with a decrease in ROS, inactivation of caspase 8, reduction of pro-apoptotic Bax, protection of mitochondrial integrity, decrease of NF-κB, and consequent reduction of COX-2.[49] Along the same line, genistein at a dose of 16 mg/kg prolonged the survival of human SOD-1 (G93A) mice,[50] and at a dose of 10 mg/kg reduced the infarct volume of rodents subjected to experimental strokes by 12.1%, sustained the enzymatic activity of SOD and GPx, and blocked the phosphorylation of IκBα.[51] Daidzein (CS 3.40) and genistein from *Pueraria thomsonii* Benth (family Fabaceae Lindl.) protected PC12 cells against 6-hydroxydopamine (6-OHDA) with IC$_{50}$ values equal to 66.8 μM and 78.7 μM, respectively, with an inhibition of caspase 8.[52] The isoflavone glabridin (CS 3.41) from a dose of 25 mg/kg mitigated the locomotor disturbances incurred by experimental stroke in rodents with a reduction of infarct volume and an increase in cerebral SOD and GSH.[49] Furthermore, glabridin from *Glycyrrhiza glabra* L. (family Fabaceae Lindl.) sustained the viability of cortical neurons against an IC$_{50}$ equal to 4.1 μM by boosting the activity of the antioxidant enzyme SOD.[49]

■ **CS 3.37** Formononetin.

■ **CS 3.38** Calycosin.

■ **CS 3.39** Genistein.

■ **CS 3.40** Daidzein.

■ **CS 3.41** Glabridin.

3.1.2 *Citrus trifoliata* L.

History The plant was first described by Carl von Linnaeus in *Species Plantarum, Editio Secunda* published in 1763.

Synonyms *Aegle sepiaria* DC., *Citrus trifolia* Thunb., *Citrus trifoliata* subfo. *monstrosa* (T. Itô) Hiroë, *Citrus trifoliata* var. *monstrosa* T. Itô, *Citrus triptera* Desf., *Poncirus trifoliata* (L.) Raf., *Poncirus trifoliata* var. *monstrosa* (T. Itô) Swingle, *Pseudaegle sepiaria* (DC.) Miq.

Family Rutaceae Juss., 1789

Common Name Zhi (Chinese)

Habitat and Description It is a shrub which grows to 2 m high in China. The stems are terete and develop spines which are conspicuous,

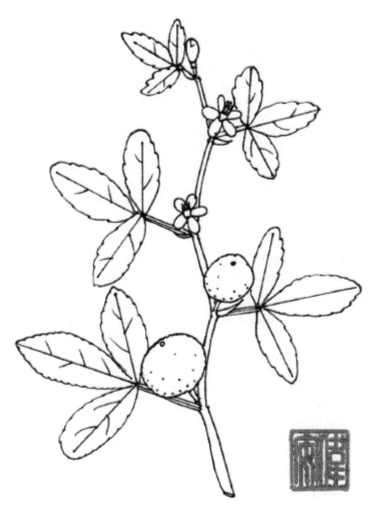

■ **FIGURE 3.2** *Citrus trifoliata* L.

stout, and grow to 5 cm long. The leaves are trifoliate. The rachis is flat and winged and bears 3 leaflets which are 2–5 cm × 1–3 cm, crenate, and present 3–5 pairs of secondary nerves and oil glands. The flowers are simple and axillary. The calyx is minute and 5-lobed. The corolla is pure white and develops 5 petals, which are 1.5–3 cm long, recurved, ephemeral, and spathulate. The androecium is conspicuous and includes 20 stamens. The ovary is ovoid and develop a sturdy style. The fruit is a pungent hesperidum which is globose and 5 cm across (Figure 3.2).

■ **CS 3.42** Isoimperatorin.

■ **CS 3.43** Esculetin.

■ **CS 3.44** Scopoletin.

■ **CS 3.45** Umbelliferone.

Medicinal Uses In China, the plant is used to treat allergies and spasms.

Phytopharmacology The plant contains the triterpenes 21α-methylmelianodiol, 21β-methylmelianodiol, hispidol A 25-Me ether, hispidol B 25-Me ether, 21α,25-dimethylmelianodiol, and 21β,25-dimethylmelianodiol; and coumarins such as isoschininallylol, auraptene, isoimperatorin (CS 3.42), bergapten, imperatorin, phellopterin, umbelliferone, isoschinilenol, poncimarin, and scopoletin[37,53]; anisocoumarin B, seselin, xanthyletin, isoangenomalin, 3-(1,1-dimethylallyl)-8-hydroxy-7-methoxycoumarin, poncitrin, 7,8-dihydrofurocoumarin-5,7-dimethoxycoumarin, 6,8-dimethoxycoumarin[54]; flavonoids including (2S)-poncirin, (2S)-naringin, (2S)-poncirenin,[55] atalantoflavone, citflavanone, 4,5,7-trihydroxyflavanone[54]; and an essential oil of α-pinene, myrcene, limonene, camphene, γ-terpinene, and 3-sulfanylhexan-1-ol.[56]

Proposed Research Pharmacological study of umbelliferone and derivatives for the treatment of neurodegenerative diseases.

Rationale 5-Lipoxygenase (5-LOX) in hippocampal and cerebellar neurons catalyzes the oxygenation of arachidonic acid at C5 to yield leukotriene A4 (LTA4), thus rending neurons vulnerable to oxidative insults.[57] Hence, inhibitors of 5-LOX may be useful for the treatment of neurodegenerative diseases, and such inhibitors are in fact coumarins. For instance, *Aesculus hippocastanum* L. (family Sapindaceae Juss.) produces esculetin (CS 3.43), which inhibited the enzymatic activity of 5-LOX and 12-lipoxygenase (12-LOX) with IC_{50} values equal 4×10^{-6} M and 2.5×10^{-2} M, respectively.[58] Likewise, scopoletin (CS 3.44) and umbelliferone (CS 3.45) from *Dystaenia takeshimana* inhibited the enzymatic activity of 5-LOX by 70% and 80.6% at a dose equal to 25 μg/mL.[59] The prenylated furanocoumarins imperatorin (CS 3.46), phellopterin (CS 3.47), and isoimperatorin from *Angelica apaensis* Shan & C.Q. Yuan (family Apiaceae Lindl.) inhibited the enzymatic activity of 5-LOX by 73.6%, 75.4%, and 56.6% at a dose of 25 μg/mL.[60] The prenylated coumarin psoralidin inhibited the enzymatic activities of 5-LOX with an IC_{50} value equal to 8.8 μM.[61] *Angelica pubescens* fo. *biserrata* Shan & C.Q. Yuan (family Apiaceae Lindl.) produces the prenylated coumarin osthenol (CS 3.48), which inhibited 5-LOX with an IC_{50} value equal to 43.1 μM.[62] Along the same line, osthole (CS 3.49) from *Angelica archangelica* L. (family Apiaceae Lindl.) inhibited the enzymatic activity of 5-LOX with an IC_{50} value equal to 36.2 μM.[63] The furanocoumarins psoralen (CS 3.50), xanthotoxin (CS 3.51), and (+)-marmesin (CS 3.52) from *Dystaenia takeshimana* (family Apiaceae Lindl.) inhibited the enzymatic activity of 5-lipoxygenase by 84.1%, 94.1%, and 56.7%, respectively, at a dose equal to $25\,\mu g/mL$.[59] In the same experiment, psoralen, xanthotoxin, scopoletin, umbelliferone, and (+)-marmesin inhibited the enzymatic activity of COX-2 by 88.2%, 60.9%, 62.1%, 61.5%, and 61.2%, respectively, at a dose equal to $25\,\mu g/mL$.[59]

COX-2 catalyzes the synthesis of prostaglandin G2 (PGG2) from arachidonic acid with concurrent generation of superoxide radicals (O_2^-), thus favoring oxidative insults.[64,65] Therefore, COX-2 inhibitors hinder neurodegeneration, and such inhibitors are coumarins like esculetin, herniarin, and scopoletin, which blunted the release of PGE2 from macrophages challenged with A23187 by 71.3%, 65.5%, and 77.3%, respectively, at a dose of $100\,\mu M$.[66,67] Along the same line, esculetin, esculetin 7-methyl ether (CS 3.53), and scopoletin from *Artemisia vulgaris* L. (family Asteraceae Bercht. & J. Presl) inhibited the enzymatic activity of monoamine oxidases (MAOs) with IC_{50} values equal to $30.1\,\mu mol$, $32.2\,\mu mol$, and $45\,\mu mol$, respectively.[68] Psoralidin (CS 3.54) inhibited the enzymatic activities of cyclo-oxygenase-1 (COX-1) with an IC_{50} value equal to $23\,\mu M$.[61] Other enzymes favoring neurodegeneration are monoamine oxidase A (MAO-A) and monoamine oxidase B (MAO-B), which catalyze the deamination of dopamine and generate H_2O_2.[69,70] The prenyloxylated coumarin geiparvarin (CS 3.55) from *Geijera parviflora* Lindl (family Rutaceae Juss.) inhibited the enzymatic activity of MAO-A and MAO-B with pIC_{50} values equal to 4.5 and 6.8, respectively.[71] Furthermore, the prenyloxylated coumarins 7-(6'R-hydroxy-3',7'-dimethyl-2'E,7'-octadienyloxy) coumarin and auraptene from *Dictamnus albus* L. (family Rutaceae Juss.) inhibited the enzymatic activity of MAO-B with IC_{50} values equal to $0.5\,\mu M$ and $0.6\,\mu M$, respectively.[72] 7-(6'R-hydroxy-3', 7'-dimethyl-2'E, 7'-octadienyloxy) coumarin (CS 3.56) and auraptene (CS 3.57) inhibited the enzymatic activity of MAO-A with IC_{50} values equal to $1.3\,\mu M$ and $34.6\,\mu M$, respectively.[72]

Other enzymes of neurological importance inhibited by coumarins are acetylcholinesterase (AChE) and β-secretase. In effect, the prenylated coumarin decursinol (CS 3.58) from *Angelica dahurica* (Fisch.) Benth. & Hook. f. (family Apiaceae Lindl.) inhibited the enzymatic activity of AChE with an IC_{50} value equal to $0.2\,\mu M$.[73] The same plant shelters isoimperatorin and imperatorin, and the prenyloxylated furanocoumarin oxypeucedanin (CS 3.59), which inhibited the enzymatic activity of AChE with IC_{50} values equal to $74.6\,\mu M$, $63.7\,\mu M$, and $89.1\,\mu M$, respectively.[73] Scopoletin inhibited the enzymatic activity of AChE with an IC_{50} value equal to $168\,\mu M$.[74] Of note, the prenyloxylated furanocoumarin imperatorin at a dose of $7\,mM$ inhibited the enzymatic activity of GABA-transaminase

■ **CS 3.46** Imperatorin.

■ **CS 3.47** Phellopterin.

■ **CS 3.48** Osthenol.

■ **CS 3.49** Osthole.

by 76%,[75] thus hindering glutamatergic excitotoxicity. The prenyloxy-lated coumarin umbelliprenin (CS 3.60) from *Ferulago campestris* Grec. (family Apiaceae Lindl.) inhibited the enzymatic activity of AChE with an IC_{50} equal to 380.1 μM.[76] Note that the prenyloxylated furanocou-marins 5-geranyloxy-8-methoxypsoralen (CS 3.61), bergamottin (CS 3.62), 8-geranyloxypsoralen (CS 3.63), 8-geranyloxy-5-methosypsoralen (CS 3.64), imperatorin, byakangelicol (CS 3.65), byakangelicine (CS 3.66), phellopterin, and kinidilin (CS 3.67) inhibited the enzymatic activity of β-secretase with IC_{50} values of 9.9 μM, 32.2 μM, 20.4 μM, 11.1 μM,

■ **CS 3.50** Psoralen.

■ **CS 3.51** Xanthotoxin.

■ **CS 3.52** (+)-marmesin.

■ **CS 3.53** Esculetin-7-methyl ether.

■ **CS 3.54** Psoralidin.

■ **CS 3.55** Geiparvarin.

■ **CS 3.56** 7-(6′R-hydroxy-3′,7′-dimethyl-2′E,7′-octadienyloxy) coumarin.

■ **CS 3.57** Auraptene.

■ **CS 3.58** Decursinol.

■ **CS 3.59** R-(+)-oxypeucedanin.

■ **CS 3.60** Umbelliprenin.

■ **CS 3.61** 5-geranyloxy-8-methoxypsoralen.

■ **CS 3.62** Bergamottin.

■ **CS 3.63** 8-geranyloxypsoralen.

■ **CS 3.64** 8-geranyloxy-5-methoxypsoralen.

91.8 μM, 104.9 μM, 219.7 μM, 143 μM, and 344 μM, respectively.[77] Given the ability of coumarins to inhibit the enzymatic activities of 5-LOX, MAOs, AChE and β-secretase, it is tempting to speculate that these benzopyrones may shelter polyenzymatic inhibitors for the treatment of Alzheimer's disease (AD) or Parkinson's disease (PD).

Furthermore, mounting lines of evidence have suggested that coumarins protect neurons against oxidative insults. Esculetin, for instance, protected hippocampal neurons against deprivation of oxygen and glucose,[78] and at a dose of 10 μM mitigated *N*-methyl-D-aspartate (NMDA), H_2O_2, and xanthine/xanthine oxidase insults against cortical cells by boosting the enzymatic activity of glutathione reductase and thus preserving GSH.[79] Furthermore, esculetin at a dose of 100 mg/kg reduced infarct size in rodents subjected to experimental stroke with concomitant increase of Bcl-2, decrease of pro-apoptotic Bax, and deactivation of caspase 3.[80] Scopoletin at a dose of 20 μg protected rodents against the dementia induced by scopolamine with probable enhancement of cholinergic neurotransmission.[74] Auraptene and 7-isopentenyloxycoumarin (CS 3.68) protected cortical cells against NMDA by 20% and 40%, respectively, at a dose equal to 100 μM.[81] *Eleutherococcus senticosus* (Rupr. ex Maxim.) Maxim. (family Araliaceae Juss.) produces isofraxidin (CS 3.69), which at a dose of 10 μM mitigated the deleterious effects of β-amyloid (Aβ_{25-35}) peptide against cortical neurons.[82] In addition, isofraxidin at a dose of 10 μg/mL negated ERK1/2 and MAPK p38 in macrophages exposed to LPS,[83] suggesting a blockade of apoptosis by inhibiting p53. Decursinol from *Angelica gigas* Nakai (family Apiaceae Juss.) at a dose of 0.004% of diet inhibited the dementia induced by Amyloid-β_{1-42} in rodents.[84] Along the same line, 4′-hydroxytigloyldecursinol (CS 3.70), 4′-hydroxydecursin (CS 3.71), (2′S,3′S)-epoxyangeloyldecursinol (CS 3.72), (2′R,3′R)-epoxyangeloyldecursinol, decursinol, and decursin (CS 3.73) from the same plant protected cortical cells against glutamate insults by 33.2%, 38.4%, 70%, 47.5%, 33.4%, and 39%, respectively, at a dose equal to 0.1 μM.[85] Furthermore, at a dose of 1 μM, decursinol and decursin protected 67.1% and 50% of cortical cells against glutamate by blocking the entry of extracellular Ca^{2+} into the cytoplasm.[86] In the same experiment, decursinol and decursin sustained the viability of cortical cells against NMDA and kainite by 40%, 25% and 60%, 30%, respectively.[86] Decursin and decursinol angelate (CS 3.74) at a dose of 10 μM protected PC12 cells against β-amyloid (Aβ_{25-35}) peptide with an increase in enzymatic activity of SOD, GPx, catalase (CAT), and GST, as a result of induction of Nrf2[87] and on probable account of ERK1/2 transient induction. Note that the incapacity of SOD to control ROS accounts for oxidative insults and apoptosis

■ CS 3.65 Byakangelicol.

■ CS 3.66 Byakangelicine.

■ **CS 3.67** Kinidiline.

in motor neurons during the course of amyotrophic lateral sclerosis (ALS), and agents able to boost the enzymatic activity of SOD are of therapeutic value. Auraptene at a dose of 25 mg/kg protected rodents against experimental stroke with inactivation of hippocampal glial cells and reduction of COX-2.[88] Likewise, osthole at a dose of 100 mg/kg attenuated the volume of cerebral infarct incurred in rodents with a reduction of MMP-9 in the hippocampus[89] by deactivating caspase 3 and JNK and activating ERK1/2.[90]

The stimulation by coumarins of ERK1/2 favors neuritogenesis.[91] For instance, scoparone (CS 3.75) at a dose of $200 \mu M$ evoked the growth of neurites from PC12 cells by increasing cytoplasmic Ca^{2+}, protein kinase A (PKA), protein kinase C (PKC), CaMKII, MEK1/2, and consequently ERK1/2.[92] Furthermore, scoparone at a dose of $200 \mu M$ increased cytoplasmic c-AMP and Ca^{2+}; activated PKA, Ca^{2+}/calmodulin kinase II (CaMK II), PKC, and synapsin I, hence stimulating tyrosine hydroxylase (TH),

■ **CS 3.68** 7-isopentenyloxycoumarin.

■ **CS 3.69** Isofraxidin.

■ **CS 3.70** 4′-hydroxytigloyldecursinol.

■ **CS 3.71** 4'-hydroxydecursin.

■ **CS 3.72** (2′,S)-(3′,S)-epoxyangeloxyl decursinol.

■ **CS 3.73** Decursin.

■ **CS 3.74** Decursinol angelate.

■ **CS 3.75** Scoparone.

■ **CS 3.76** Daphnetin.

aromatic L-amino acid decarboxylase (AADC), and CREB; and consequently enhanced the synthesis and secretion of dopamine from PC12 cells as well as neuritogenesis.[92–94] Auraptene at a dose of $30\,\mu$M induced MEK1/2, ERK1/2, and CREB and consequently the growth of neurites in PC12 cells.[95] Daphnetin (CS 3.76) from *Daphne tangutica* Maxim. (family Thymelaeaceae Juss.) inhibited the enzymatic activities of PKC and PKA by 99.1% and 98.6% at a dose equal to $200\,\mu$M,[96] but at a dose of $8\,\mu$mol/L evoked neuritogenesis of cortical neurons with an increase in microtubule-associated protein 2 (MAP2) and brain-derived neurotrophic factor (BDNF).[97]

Neuritogenesis is attenuated by the PKC inhibitor thrombin, which commands the aggregation of platelets,[98] and thrombolytic agents may be beneficial to prevent and delay Alzheimer's disease (AD). Of note, schinicoumarin (CS 3.77) and the prenyloxylated coumarins acetoxyauraptene (CS 3.78), schininallylol (CS 3.79), auraptene, collinin (CS 3.80), and (-)-acetoxycollinin (CS 3.81) from *Zanthoxylum schinifolium* Siebold & Zucc. (family Rutaceae Juss.) abrogated platelet aggregation induced by arachidonic acid at a dose of $100\,\mu$g/mL.[99]

■ **CS 3.77** Schinicoumarin.

■ **CS 3.78** Acetoxyauraptene.

■ **CS 3.79** Schininallylol.

■ **CS 3.80** Collinin.

■ **CS 3.81** (−)- acetoxycollinin.

3.1.3 *Polygala caudata* Rehder & E.H. Wilson

History The plant was first described by Alfred Rehder and Ernest Henry Wilson in *Plantae Wilsonianae* published in 1914.

Synonyms *Polygala comesperma* Chodat, *Polygala wattersii* Hance

Family Polygalaceae Hoffmanns. & Link, 1809

Common Name Wei ye yuan zhi (Chinese)

Habitat and Description It is a shrub which grows in China to a height 3 m tall. The stems are quadrangular and hairy. The leaves are simple, exstipulate, spiral, and crowded at apices of stems. The petiole is 1 cm long and channeled above. The blade is obovate 3–12 cm × 1–3 cm, thinly coriaceous, and presents 7 pairs of secondary nerves plus inter nerves, wavy at the margin, and caudate at apex. The inflorescence is an axillary or terminal raceme of tiny flowers which grows to 6 cm long. The calyx includes 5 sepals of irregular shape. The corolla is tubular, purplish, and 3-lobed. The androecium consists of 8 stamens attached to the corolla tube. The flower presents a tiny disc on which sits an ovoid ovary. The fruit is a 0.8 cm long capsule containing hairy seeds (Figure 3.3).

Medicinal Uses The plant is used to calm, and to treat hepatis and cough.

Phytopharmacology The plant engineers euxanthone, wubangziside A-C,3-dihydroxy-2-methoxyxanthone, lancerin, gentisein, mangiferin,[100–102] polycaudoside A, 2′-benzoylmangiferin, 1-methoxy-7-hydroxyxanthone, and 1,2,8-trihydroxyxanthone,[103] as well as the flavonoids dihydroquercetin and quercetin.[100]

Proposed Research Pharmacological study of gaudichaudione H and derivatives for the treatment of cancer and neurodegenerative diseases.

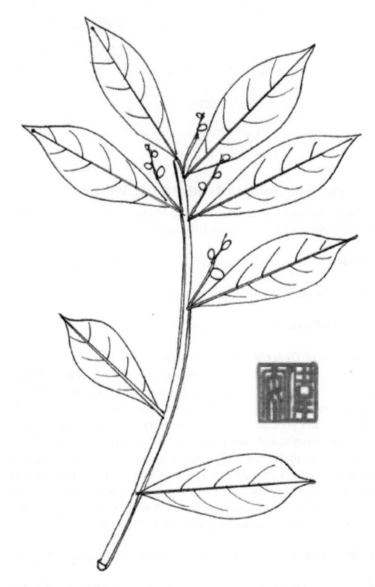

■ **FIGURE 3.3** *Polygala caudata* Rehder & E.H. Wilson.

Rationale In the presence of oxygen, MAO-A and MAO-B catalyze the replacement of the NH2 moiety of dopamine by a carbonyl with concomitant production of H_2O_2, thus favoring the accumulation of ROS and therefore apoptosis,[69,70] and inhibitors of such enzymes, like xanthones, favor neuroprotection. For instance, the synthetic 1,3-dihydroxy-2-methylxanthone

(CS 3.82) and 1,3-dihydroxy-4-methylxanthone (CS 3.83), both of which have a hydroxyl moiety in C3, inhibited the enzymatic activity of MAO-A with IC_{50} values equal to $3.7\,\mu M$ and $4.3\,\mu M$, respectively.[104] *Gentiana kochiana* Perr. et Song. (family Gentianaceae Juss.) elaborates gentiacaulein (CS 3.84) and gentiakochianin, which inhibited the enzymatic activity of MAO-A with IC_{50} values equal to $0.4\,\mu M$ and $164\,\mu M$, respectively, and MAO-B with IC_{50} values equal to $340\,\mu M$ and $63\,\mu M$, respectively,[105] suggesting that methoxylation in C3 does not nullify MAO-A activity and that ortho C7 and C8 hydroxyl moieties are not favorable. From *Gentianella amarella* var. *acuta* Herder (family Gentianaceae Juss.), bellidin (CS 3.85) at a dose of $10^{-5}\,M$ inhibited the enzymatic activities of AChE, MAO-A, and MAO-B by 17.5%, 90.5%, and 59%; whereas bellidifolin (CS 3.86) at a dose of $10^{-5}\,M$ inhibited the enzymatic activities of AChE, MAO-A, and MAO-B by 21.9%, 98.9%, and 65.2%, respectively,[106] substantiating further the contention that both C3 hydroxyl and methoxyl moieties are beneficial. Also, para C5 and C8 hydroxyl moieties *Cudrania tricuspidata* (Carriere) Bureaux

■ **CS 3.82** 1,3-dihydroxy-3-methylxanthone.

■ **CS 3.83** 1,3-dihydroxy-4-methylxanthone.

■ **CS 3.84** Gentiacaulein.

■ **CS 3.85** Bellidin.

■ **CS 3.86** Bellidifolin.

ex Lavalle (family Moraceae Gaudich.) elaborate the prenylated C3 hydroxy xanthone cudratricusxanthone A (CS 3.87), which inhibited the enzymatic activity of MAO with an IC_{50} value of $88.3\,\mu M$,[107] evidencing that prenylation does not favor MAO inhibition.

In the presence of oxygen, COX-1 and COX-2 catalyze the epoxidation of arachidonic acid (AA) into prostaglandin G2 (PGG_2) with simultaneous production of highly reactive superoxide radicals (O_2^-), which further load the cytoplasm of neurons and glial cells with ROS, thus favoring neuroapoptosis and neuroinflammation.[64,65]

Prenylated xanthones like γ-mangostin (CS 3.88) from *Garcinia mangostana* L. (family Clusiaceae Lindl.) blocked the production of PGE2 by C6 cells challenged by A23187 with an IC_{50} equal to $5\,\mu M$.[108,109] In addition, γ-mangostin lowered the enzymatic activity of COX-1 and COX-2 with IC_{50} values of $0.8\,\mu M$ and $2\,\mu M$, respectively,[110] and inhibited IKK.[111] Along the same line, γ-mangostin at a dose of $10\,\mu M$ reduced by 70% the expression of COX-2 in C6 cells challenged with LPS as a result of IKK inhibition, hence IκBα hyphosphorylation and NF-κB inhibition.[111] Inhibition of IKK may account for the synthesis inhibition of PGE2 by C6 cells challenging

■ **CS 3.87** Cudratricusxanthone A.

■ **CS 3.88** γ-mangostin.

A23187 by the prenylated xanthones garcinone B (CS 3.89) and patulone (CS 3.90) from *Hypericum patulum* Thunb. (family Hypericaceae Juss.) at a dose of $20\,\mu$M.[109] Indeed, gambogic acid (CS 3.91) from *Garcinia hanburyi* Hook. f. (family Clusiaceae Lindl.) abrogated the phosphorylation of IκBα by IKK in macrophages *(RAW 264.7)* at a dose of $1\,\mu$M, resulting in the blockade of NF-κB and therefore reduction of COX-2.[112] Prenyl moieties are not a prerequisite for activity as 1,3,6,7-tetrahydroxyxanthone (CS 3.92) from *Tripterospermum lanceolatum* (Hayata) H. Hara ex Satake (family Gentianaceae Juss.) inhibited the enzymatic activities of COX-2 and 12-LOX with IC$_{50}$ values equal to $19.6\,\mu$M and $1.2\,\mu$M, and inhibited the enzymatic activities of COX-1 and 5-LOX with IC$_{50}$ values equal to $16.2\,\mu$M

■ **CS 3.89** Garcinone B.

■ **CS 3.90** Patulone.

■ **CS 3.91** Gambogic acid.

■ **CS 3.92** 1,3,6,7-tetrahydroxyxanthone.

and 1.8 μM, respectively.[113] 2,6-Dihydroxy-1,7-dimethoxyxanthone (CS 3.93) and 3,4-dihydroxyxanthone (CS 3.94) from *Calophyllum membranaceum* Gardner & Champ. (family Calophyllaceae J. Agardh) inhibited the enzymatic activity of COX-2 with IC_{50} values equal to 2.9 μM and 1.8 μM, respectively.[114] 1,3,5-Trihydroxy-4-prenylxanthone from *Cudrania cochinchinensis* (Lour.) Kudô & Masam. (family Moraceae Gaudich) at a dose of 30 μM abated the production of NO by RAW 264.7 macrophages challenged with LPS as a consequence of a decrease of iNOS; hypophosphorylation of growth factor beta activated kinase-1 (TAK1), ERK1/2, JNK, IKK; and inhibition of NF-κB, Jun, and Fos.[115] The therapeutic significance of these results is that xanthones mitigate the inflammatory response of glial cells and may prevent fight, thus neurodegeneration.

The seco-xanthone garcinol (CS 3.95) from *Garcinia indica* Choisy (family Moraceae Gaudich.) protected DNA against oxidative insults with an IC_{50} value equal to 0.3 μM, inhibited xanthine oxidase activity with an IC_{50} value of 52 μM, and at a dose equal to 5 μM protected neurons against the cytokines secreted by astrocytes challenged with LPS by reducing iNOS and COX-2.[116] Of note, garcinol at a dose of 5 μM inhibited p300 and therefore PKC in dopaminergic neurons by 60%.[117] Because α-synuclein represses the expression of PKC, thus protecting dopaminergic N27 cells against MPP$^+$ insults via probable inhibition of p300 and NF-κB,[117] it is reasonable to infer that the aforementioned neuroprotective effects are owed to p300 blockade and PKC enhancement. α-Mangostin (CS 3.96) from *Garcinia mangostana* L. (family Clusiaceae Lindl.) at a dose of 50 nM protected cortical neurons against β-amyloid (Aβ_{1-40}) peptide and β-amyloid (Aβ_{1-42}) peptide by 90% and 85%, respectively, by docking into the peptides and thus hampering their aggregation.[118] α-Mangostin scavenged superoxide radical (O$_2{}^-$) and therefore at a dose of 12 μM protected cerebellar granule neurons against oxidative insults by 84.9%.[119]

Prenylated xanthones evoke the growth of neurites. In effect, the seco-xanthone garcinol at a dose of 5 μM enhanced the neuritogenic potencies of epidermal growth factor (EGF) against cortical via activation of ERK1/2[120] via probable activation of PKC. Euxanthone (CS 3.97) from *Polygala caudata* Rehder & E.H. Wilson. (family Polygalaceae Hoffmanns. & Link)

■ **CS 3.93** 2,6-dihydroxy-1,7-dimethoxyxanthone.

■ **CS 3.94** 3,4-dihydroxyxanthone.

■ **CS 3.95** Garcinol.

■ **CS 3.96** α-mangostin.

at a dose of $100\,\mu$M commanded the growth of neurites by 40% in neurons[121] via the stimulation of PKC[122] and induction of E2F-5.[123] The isoprenyl xanthones 1,4,5,6-tetrahydroxy-7,8-di(3-methylbut-2-enyl)xanthone (CS 3.98), 1,2,6-trihydroxy-5-methoxy-7-(3-methylbut-2-enyl)xanthone (CS 3.99), and 12b-hydroxy-des-D-garcigerrin A (CS 3.100) from *Garcinia xanthochymus* Hook. f. (family Clusiaceae Lindl.) at a dose equal to $10\,\mu$M augmented the neuritogenic activity of 2 ng/mL of NGF by 23.5%, 27.4%, and 28.8%, respectively,[124] suggesting the activation of ERK1/2. ERK1/2 activation might account for the neuritogenic activity of macluraxanthone (CS 3.101) from *Garcinia thorelii* Pierre (family Clusiaceae Lindl.) and garcinone A (CS 3.102) from *Cratoxylum cochinchinense* (Lour.) Blume (family Hypericaceae Juss.), which boosted the neuritogenic property of NGF against PC12 at doses equal to $3\,\mu$M and $10\,\mu$M, respectively.[125] Along the same line, *Garcinia xanthochymu*s engineers 1,3,5,6-tetrahydroxy-4,7,8-tri(3-methyl-2-butenyl)xanthone (CS 3.103),[126] and garciniaxanthone E (CS 3.104) at a dose of $3\,\mu$M and $10\,\mu$M enhanced the neuritogenic effect of 2 ng/mL of NGF against PC12 cells.[125]

■ **CS 3.97** Euxanthone.

■ **CS 3.98** 1,4,5,6-tetrahydroxy-7,8-di(3-methylbut-2-enyl)xanthone.

■ **CS 3.99** 1,2,6-trihydroxy-5-methoxy-7-(3-methylbut-2-enyl)xanthone.

■ **CS 3.100** 12b-hydroxy-des-D-garcigerrin A.

■ **CS 3.101** Macluroxanthone.

■ **CS 3.102** Garcinone A.

■ **CS 3.103** 1,3,5,6-tetrahydroxy-4,7,8-tri(3-methyl-2-butenyl)xanthone.

■ **CS 3.104** Garciniaxanthone E.

■ **CS 3.105** Gambogic amide.

Note that gambogic amide (CS 3.105) from *Garcinia hanburyi* Hook. f. (family Clusiaceae Lindl.) binds to and stimulates TrkA receptors with a K_d value equal to 75 nM, and activated phosphoinositide 3-kinase (PI3K) and Akt in hippocampal neurons at a dose of 50 nM and evoked the growth of neurites in PC12 cells.[126] In the same experiment, gambogic amide protected rodents against kainate at a dose of 2 mg/kg.[126] Cudratricusxanthone A induced JNK, evoked the expression of heme oxidase (HO-1) in mouse hippocampal (HT-22) cells, and therefore attenuated glutamate excitotoxicity at a dose of 10 μM,[110] suggesting the possibility of neuritogenesis.

REFERENCES

[1] Subarnas A, Oshima Y, Hikino H. Isoflavans and apterocarpan from *Astragalus mongholicus*. Phytochemistry 1991;30(8):2777–80.

[2] Yu D, Duan Y, Bao Y, Wei C, An L. Isoflavonoids from Astragalus mongholicus protect PC12 cells from toxicity induced by L-glutamate. J Ethnopharmacol 2005;98(1–2):89–94.

[3] Zhang LJ, Liu HK, Hsiao PC, Kuo LM, Lee IJ, Wu TS, et al. New isoflavonoid glycosides and related constituents from astragali radix (Astragalus membranaceus) and their inhibitory activity on nitric oxide production. J Agric Food Chem 2011;59(4):1131–7.

[4] He ZQ, Findlay JA. Constituents of *Astragalus membranaceus*. J Nat Prod 1991;54(3):810–15.

[5] Mercer LD, Kelly BL, Horne MK, Beart PM. Dietary polyphenols protect dopamine neurons from oxidative insults and apoptosis: investigations in primary rat mesencephalic cultures. Biochem Pharmacol 2005;69(2):339–45.

[6] Ha S, Lee K, Park P, Oh JA, Lee HR, Park SY, et al. Apigenin inhibits the production of NO and PGE_2 in microglia inhibits neuronal cell death in a middle cerebral artery occlusion-induced focal ischemia mice model. Neurochem Int 2008;52(4–5):878–86.

[7] Gresa-Arribas N, Serratosa J, Saura J, Solà C. Inhibition of CCAAT/enhancer binding protein δ expression by chrysin in microglial cells results in anti-inflammatory and neuroprotective effects. J Neurochem 2010;115(2):526–36.

[8] He XL, Wang YH, Bi MG, Du GH. Chrysin improves cognitive deficits and brain damage induced by chronic cerebral hypoperfusion in rats. Eur J Pharmacol 2012;680(1–3):41–8.

[9] Jang SW, Liu X, Yepes M, Shepherd KR, Miller GW, Liu Y, et al. A selective TrkB agonist with potent neurotrophic activities by 7,8-dihydroxyflavone. Proc Natl Acad Sci USA 2010;107(6):2687–92.

[10] Gao M, Zhang WC, Liu QS, Hu JJ, Liu GT, Du GH. Pinocembrin prevents glutamate-induced apoptosis in SH-SY5Y neuronal cells via decrease of bax/bcl-2 ratio. Eur J Pharmacol 2008;591(1–3):73–9.

[11] Liu R, Gao M, Yang ZH, Du GH. Pinocembrin protects rat brain against oxidation and apoptosis induced by ischemia-reperfusion both *in vivo* and *in vitro*. Brain Res 2008;1216:104–15.

[12] Cho J, Lee HK. Wogonin inhibits excitotoxic and oxidative neuronal damage in primary cultured rat cortical cells. Eur J Pharmacol 2004;485(1–3):105–10.

[13] Son D, Lee P, Lee J, Kim H, Kim SY. Neuroprotective effect of wogonin in hippocampal slice culture exposed to oxygen and glucose deprivation. Eur J Pharmacol 2004;493(1–3):99–102.

[14] Yeh CH, Yang ML, Lee CY, Yang CP, Li YC, Chen CJ, et al. Wogonin attenuates endotoxin-induced prostaglandin E2 and nitric oxide production via Src-ERK1/2-NFκB pathway in BV-2 microglial cells. Environ Toxicol 2013 in press.

[15] Moon HI, Cho SB, Lee JH, Lee YC, Lee JH, Lee CH, et al. Protective effects of methoxyflavone derivatives from black galingale against glutamate induced neurotoxicity in primary cultured rat cortical cells. Phytotherapy Res 2011;25(8):1215–17.

[16] Mu X, He G, Cheng Y, Li X, Xu B, Du G. Baicalein exerts neuroprotective effects in 6-hydroxydopamine-induced experimental parkinsonism *in vivo* and *in vitro*. Pharmacol Biochem Behav 2009;92(4):642–8.

[17] Chen YC, Chow JM, Lin CW, Wu CY, Shen SC. Baicalein inhibition of oxidative-stress-induced apoptosis via modulation of ERKs activation and induction of HO-1 gene expression in rat glioma cells C6. Toxicol Appl Pharmacol 2006;216(2):263–73.

[18] Liu C, Wu J, Gu J, Xiong Z, Wang F, Wang J, et al. Baicalein improves cognitive deficits induced by chronic cerebral hypoperfusion in rats. Pharmacol Biochem Behav 2007;86(3):423–30.

[19] Kang SS, Lee JY, Choi YK, Kim GS, Han BH. Neuroprotective effects of flavones on hydrogen peroxide-induced apoptosis in SH-SY5Y neuroblastoma cells. Bioorg Med Chem Lett 2004;14(9):2261–4.

[20] Heo HJ, Kim DO, Shin SC, Kim MJ, Kim BG, Shin DH. Effect of antioxidant flavanone, naringenin, from *Citrus junos* on neuroprotection. J Agric Food Chem 2004;52(6):1520–5.

[21] Kumar A, Prakash A, Dogra S. Naringin alleviates cognitive impairment, mitochondrial dysfunction and oxidative stress induced by D-galactose in mice. Food Chem Toxicol 2010;48(2):626–32.

[22] Kim HG, Ju MS, Ha SK, Lee H, Lee H, Kim SY, et al. Acacetin protects dopaminergic cells against 1-methyl-4-phenyl-1,2,3,6-tetrahydropyridine-induced neuroinflammation *in vitro* and *in vivo*. Biol Pharm Bull 2012;35(8):1287–94.

[23] Nishina A, Kimura H, Tsukagoshi H, Kozawa K, Koketsu M, Ninomiya M, et al. Neurite outgrowth in PC12 cells stimulated by components from Dendranthema × grandiflorum cv. "Mottenohoka" is enhanced by suppressing phosphorylation of p38MAPK. Evid Based Complement Alternat Med 2013:1–10.

[24] Zhang ZT, Cao XB, Xiong N, Wang HC, Huang JS, Sun SG, et al. Morin exerts neuroprotective actions in Parkinson disease models *in vitro* and *in vivo*. Acta Pharmacol Sin 2010;31(8):900–6.

[25] Lin CW, Wu MJ, Liu IY, Su JD, Yen JH. Neurotrophic and cytoprotective action of luteolin in PC12 cells through ERK-dependent induction of Nrf2-driven HO-1 expression. J Agric Food Chem 2010;58(7):4477–86.

[26] Jang S, Dilger RN, Johnson RW. Luteolin inhibits microglia and alters hippocampal-dependent spatial working memory in aged mice. J Nutr 2010;140(10):1892–8.

[27] Jang YJ, Kim J, Shim J, Kim J, Byun S, Oak MH, et al. Kaempferol attenuates 4-hydroxynonenal-induced apoptosis in PC12 cells by directly inhibiting NADPH oxidase. J Pharmacol Exp Ther 2011;337(3):747–54.

[28] Filomeni G, Graziani I, De Zio D, Dini L, Centonze D, Rotilio G, et al. Neuroprotection of kaempferol by autophagy in models of rotenone-mediated acute toxicity: possible implications for Parkinson's disease. Neurobiol Aging 2012;33(4):767–85.

[29] Park SE, Sapkota K, Kim S, Kim H, Kim SJ. Kaempferol acts through mitogen-activated protein kinases and protein kinase B/AKT to elicit protection in a model of neuroinflammation in BV-2 microglial cells. Br J Pharmacol 2011;164(3):1008–25.

[30] Campos-Esparza MR, Sánchez-Gómez MV, Matute C. Molecular mechanisms of neuroprotection by two natural antioxidant polyphenols. Cell Calcium 2009;45(4):358–68.

[31] Ansari MA, Abdul HM, Joshi G, Opii WO, Butterfield DA. Protective effect of quercetin in primary neurons against Abeta(1–42): relevance to Alzheimer's disease. J Nutr Biochem 2009;20(4):269–75.

[32] Suematsu N, Hosoda M, Fujimori K. Protective effects of quercetin against hydrogen peroxide-induced apoptosis in human neuronal SH-SY5Y cells. Neurosci Lett 2011;504(3):223–7.

[33]   Dok-Go H, Lee KH, Kim HJ, Lee EH, Lee J, Song YS, et al. Neuroprotective effects of antioxidative flavonoids, quercetin, (+)-dihydroquercetin and quercetin 3-methyl ether, isolated from *Opuntia ficus-indica* var. *saboten*. Brain Res 2003;965(1–2):130–6.

[34]   Ahmad A, Khan MM, Hoda MN, Raza SS, Khan MB, Javed H, et al. Quercetin protects against oxidative stress associated damages in a rat model of transient focal cerebral ischemia and reperfusion. Neurochem Res 2011;36(8):1360–71.

[35]   Zhang K, Ma Z, Wang J, Xie A, Xie J. Myricetin attenuated MPP^+-induced cytotoxicity by anti-oxidation and inhibition of MKK4 and JNK activation in MES23.5 cells. Neuropharmacology 2011;61(1–2):329–35.

[36]   Lee KW, Kang NJ, Rogozin EA, Kim HG, Cho YY, Bode AM, et al. Myricetin is a novel natural inhibitor of neoplastic cell transformation and MEK1. Carcinogenesis 2007;28:1918–27.

[37]   Xu GH, Kim JA, Kim SY, Ryu JC, Kim YS, Jung SH, et al. Terpenoids and coumarins isolated from the fruits of *Poncirus trifoliata*. Chem Pharm Bull (Tokyo) 2008;56(6):839–42.

[38]   Kumamoto T, Fujii M, Hou DX. Akt is a direct target for myricetin to inhibit cell transformation. Mol Cell Biochem 2009;332(1–2):33–41.

[39]   Maher P. The flavonoid fisetin promotes nerve cell survival from trophic factor withdrawal by enhancement of proteasome activity. Arch Biochem Biophys 2008;476(2):139–44.

[40]   Sagara Y, Vanhnasy J, Maher P. Induction of PC12 cell differentiation by flavonoids is dependent upon extracellular signal-regulated kinase activation. J Neurochem 2004;90(5):1144–55.

[41]   Zheng LT, Ock J, Kwon BM, Suk K. Suppressive effects of flavonoid fisetin on lipopolysaccharide-induced microglial activation and neurotoxicity. Int Immunopharmacol 2008;8(3):484–94.

[42]   Sung B, Pandey MK, Aggarwal BB. Fisetin, an inhibitor of cyclin-dependent kinase 6, down-regulates nuclear factor-kappaB-regulated cell proliferation, antiapoptotic and metastatic gene products through the suppression of TAK-1 and receptor-interacting protein-regulated IkappaBalpha kinase activation. Mol Pharmacol 2007;71(6):1703–14.

[43]   McDowell ML, Das A, Smith JA, Varma AK, Ray SK, Banik NL. Neuroprotective effects of genistein in VSC4.1 motoneurons exposed to activated microglial cytokines. Neurochem Int 2011;59(2):175–84.

[44]   Nam Y, Choi M, Hwang H, Lee MG, Kwon BM, Lee WH, et al. Natural flavone jaceosidin is a neuroinflammation inhibitor. Phytother Res 2013;27(3):404–11.

[45]   Jeong GS, Li B, Lee DS, Byun E, An RB, Pae HO, et al. Lavandulyl flavanones from *Sophora flavescens* protect mouse hippocampal cells against glutamate-induced neurotoxicity via the induction of heme oxygenase-1. Biol Pharm Bull 2008;31(10):1964–7.

[46]   Cho N, Choi JH, Yang H, Jeong EJ, Lee KY, Kim YC, et al. Neuroprotective and anti-inflammatory effects of flavonoids isolated from *Rhus verniciflua* in neuronal HT-22 and microglial BV2 cell lines. Food Chem Toxicol 2012;50(6):1940–5.

[47]   Moon HI, Chung IM, Seo SH, Kang EY. Protective effects of 3′-deoxy-4-O-methylepisappanol from *Caesalpinia sappan* against glutamate-induced neurotoxicity in primary cultured rat cortical cells. Phytother Res 2010;24(3):463–5.

[48] Li Y, Ishibashi M, Chen X, Ohizumi Y. Littorachalcone, a new enhancer of NGF-mediated neurite outgrowth, from *Verbena littoralis*. Chem Pharm Bull 2003;51(7):872–4.

[49] Yu XQ, Xue CC, Zhou ZW, Li CG, Du YM, Liang J, et al. *In vitro* and *in vivo* neuroprotective effect and mechanisms of glabridin, a major active isoflavan from *Glycyrrhiza glabra* (licorice). Life Sci 2008;82(1–2):68–78.

[50] Trieu VN, Uckun FM. Genistein is neuroprotective in murine models of familial amyotrophic lateral sclerosis and stroke. Biochem Biophys Res Commun 1999;258(3):685–8.

[51] Qian Y, Guan T, Huang M, Cao L, Li Y, Cheng H, et al. Neuroprotection by the soy isoflavone, genistein, via inhibition of mitochondria-dependent apoptosis pathways and reactive oxygen induced-NF-κB activation in a cerebral ischemia mouse model. Neurochem Int 2012;60(8):759–67.

[52] Lin CM, Lin RD, Chen ST, Lin YP, Chiu WT, Lin JW, et al. Neurocytoprotective effects of the bioactive constituents of *Pueraria thomsonii* in 6-hydroxydopamine (6-OHDA)-treated nerve growth factor (NGF)-differentiated PC12 cells. Phytochemistry 2010;71(17–18):2147–56.

[53] Guiotto A, Rodighiero P, Quintily U. Poncimarin, a new coumarin from *Poncirus trifoliata* L. Z Naturforsch C 1975;30(3):420–1.

[54] Feng T, Wang RR, Cai XH, Zheng YT, Luo XD. Anti-human immunodeficiency virus-1 constituents of the bark of *Poncirus trifoliata*. Chem Pharm Bull 2010;58(7):971–5.

[55] Han AR, Kim JB, Lee J, Nam JW, Lee IS, Shim CK, et al. A new flavanone glycoside from the dried immature fruits of *Poncirus trifoliata*. Chem Pharm Bull 2007;55(8):1270–3.

[56] Starkenmann C, Niclass Y, Escher S. Volatile organic sulfur-containing constituents in *Poncirus trifoliata* (L.) Raf. (Rutaceae). J Agric Food Chem 2007;55(11):4511–7.

[57] Manev H, Uz T, Sugaya K, Qu T. Putative role of neuronal 5-lipoxygenase in an aging brain. FASEB J 2000;14(10):1464–9.

[58] Neichi T, Koshihara Y, Murota S. Inhibitory effect of esculetin on 5-lipoxygenase and leukotriene biosynthesis. Biochim Biophys Acta 1983;753(1):130–2.

[59] Kim JS, Kim JC, Shim SH, Lee EJ, Jin W, Bae K, et al. Chemical constituents of the root of *Dystaenia takeshimana* and their anti-inflammatory activity. Arch Pharm Res 2006;29(8):617–23.

[60] Liu JH, Zschocke S, Reininger E, Bauer R. Comparison of *Radix Angelicae pubescentis* and substitutes—constituents and inhibitory effect on 5-lipoxygenase and cyclooxygenase. Pharm Biol 1998;36:207–16.

[61] Chi YS, Jong HG, Son KH, Chang HW, Kang SS, Kim HP. Effects of naturally occurring prenylated flavonoids on enzymes metabolizing arachidonic acid: cyclooxygenases and lipoxygenases. Biochem Pharmacol 2001;62(9):1185–91.

[62] Liu JH, Zschocke S, Reininger E, Bauer R. Inhibitory effects of *Angelica pubescens* f. *biserrata* on 5-lipoxygenase and cyclooxygenase. Planta Med 1998;64:525–9.

[63] Roos G, Waiblinger J, Zschocke S, et al. Isolation, identification and screening for COX-1- and 5-LO-inhibition of coumarins from *Angelica archangelica*. Pharm Pharmacol Lett 1997;4:157–60.

[64] Minghetti L. Cyclooxygenase-2 (COX-2) in inflammatory and degenerative brain diseases. J Neuropathol Exp Neurol 2004;63(9):901–10.

[65] Im JY, Kim D, Paik SG, Han PL. Cyclooxygenase-2-dependent neuronal death proceeds via superoxide anion generation. Free Radic Biol Med 2006;41(6):960–72.

[66] Silván AM, Abad MJ, Bermejo P, Sollhuber M, Villar A. Antiinflammatory activity of coumarins from *Santolina oblongifolia*. J Nat Prod 1996;59(12):1183–5.

[67] Kim HJ, Jang SI, Kim YJ, Chung HT, Yun YG, Kang TH, et al. Scopoletin suppresses pro-inflammatory cytokines and PGE2 from LPS-stimulated cell line, RAW 264.7 cells. Fitoterapia 2004;75(3–4):261–6.

[68] Lee1 SJ, Chung HY, Lee IK, Oh SU, You ID. Phenolics with inhibitory activity on mouse brain monoamine oxidase (MAO) from whole parts of *Artemisia vulgaris* L (Mugwort). Food Sci Biotechnol 2000;9(3):179–82.

[69] Youdim MB, Edmondson D, Tipton KF. The therapeutic potential of monoamine oxidase inhibitors. Nat Rev Neurosci 2006;7(4):295–309.

[70] Bortolato M, Chen K, Shih JC. Monoamine oxidase inactivation: from pathophysiology to therapeutics. Adv Drug Deliv Rev 2008;60(13–14):1527–33.

[71] Carotti A, Carrieri A, Chimichi S, Boccalini M, Cosimelli B, Gnerre C, et al. Natural and synthetic geiparvarins are strong and selective MAO-B inhibitors. Synthesis and SAR studies. Bioorg Med Chem Lett 2002;12(24):3551–5.

[72] Jeong SH, Han XH, Hong SS, Hwang JS, Hwang JH, Lee D, et al. Monoamine oxidase inhibitory coumarins from the aerial parts of *Dictamnus albus*. Arch Pharm Res 2006;29(12):1119–24.

[73] Kim DK, Lim JP, Yang JH, Eom DO, Eun JS, Leem KH. Acetylcholinesterase inhibitors from the roots of *Angelica dahurica*. Arch Pharm Res 2002;25(6):856–9.

[74] Hornick A, Lieb A, Vo NP, Rollinger J, Stuppner H, Prast H. Effects of the coumarin scopoletin on learning and memory, on release of acetylcholine from brain synaptosomes and on long-term potentiation in hippocampus. BMC Pharmacology 2008;8(Suppl. 1):A36.

[75] Choi SY, Ahn EM, Song MC, Kim DW, Kang JH, Kwon OS, et al. *In vitro* GABA-transaminase inhibitory compounds from the root of *Angelica dahurica*. Phytother Res 2005;19(10):839–45.

[76] Dall'Acqua S, Maggi F, Minesso P, Salvagno M, Papa F, Vittori S, et al. Identification of non-alkaloid acetylcholinesterase (AChE) inhibitors from *Ferulago campestris* (Besser) *Grecescu* (Apiaceae). Fitoterapia 2010;81(8):1208–12.

[77] Marumoto S, Miyazawa M. Structure-activity relationships for naturally occurring coumarins as *β*-secretase inhibitor. Bioorg Med Chem 2012;20(2):784–8.

[78] Arai K, Nishiyama N, Matsuki N, Ikegaya Y. Neuroprotective effects of lipoxygenase inhibitors against ischemic injury in rat hippocampal slice cultures. Brain Res 2001;904(1):167–72.

[79] Lee CR, Shin EJ, Kim HC, Choi YS, Shin T, Wie MB. Esculetin inhibits N-methyl-D-aspartate neurotoxicity via glutathione preservation in primary cortical cultures. Lab Anim Res 2011;27(3):259–63.

[80] Wang C, Pei A, Chen J, Yu H, Sun ML, Liu CF, et al. A natural coumarin derivative esculetin offers neuroprotection on cerebral ischemia/reperfusion injury in mice. J Neurochem 2012;121(6):1007–13.

[81] Epifano F, Molinaro G, Genovese S, Ngomba RT, Nicoletti F, Curini M. Neuroprotective effect of prenyloxycoumarins from edible vegetables. Neurosci Lett 2008;443(2):57–60.

[82] Tohda C, Ichimura M, Bai Y, Tanaka K, Zhu S, Komatsu K. Inhibitory effects of *Eleutherococcus senticosus* extracts on amyloid beta (25–35)-induced neuritic atrophy and synaptic loss. J Pharmacol Sci 2008;107(3):329–39.

[83] Niu X, Xing W, Li W, Fan T, Hu H, Li Y. Isofraxidin exhibited anti-inflammatory effects *in vivo* and inhibited TNF-α production in LPS-induced mouse peritoneal macrophages *in vitro* via the MAPK pathway. Int Immunopharmacol 2012;14(2):164–71.

[84] Yan JJ, Kim DH, Moon YS, Jung JS, Ahn EM, Baek NI, et al. Protection against beta-amyloid peptide-induced memory impairment with long-term administration of extract of *Angelica gigas* or decursinol in mice. Prog Neuropsychopharmacol Biol Psychiatry 2004;28(1):25–30.

[85] Kang SY, Lee KY, Sung SH, Kim YC. Four new neuroprotective dihydropyranocoumarins from *Angelica gigas*. J Nat Prod 2005;68(1):56–9.

[86] Kang SY, Kim YC. Decursinol and decursin protect primary cultured rat cortical cells from glutamate-induced neurotoxicity. J Pharm Pharmacol 2007;59(6):863–70.

[87] Li L, Li W, Jung SW, Lee YW, Kim YH. Protective effects of decursin and decursinol angelate against amyloid β-protein-induced oxidative stress in the PC12 cell line: the role of Nrf2 and antioxidant enzymes. Biosci Biotechnol Biochem 2011;75(3):434–42.

[88] Okuyama S, Minami S, Shimada N, Makihata N, Nakajima M, Furukawa Y. Anti-inflammatory and neuroprotective effects of auraptene, a citrus coumarin, following cerebral global ischemia in mice. Eur J Pharmacol 2013;699(1–3):118–23.

[89] Mao X, Yin W, Liu M, Ye M, Liu P, Liu J, et al. Osthole, a natural coumarin, improves neurobehavioral functions and reduces infarct volume and matrix metalloproteinase-9 activity after transient focal cerebral ischemia in rats. Brain Res 2011;1385:275–80.

[90] Chen T, Liu W, Chao X, Qu Y, Zhang L, Luo P, et al. Neuroprotective effect of osthole against oxygen and glucose deprivation in rat cortical neurons: involvement of mitogen-activated protein kinase pathway. Neuroscience 2011;183:203–11.

[91] Waetzig V, Herdegen T. The concerted signaling of ERK and JNKs is essential for PC12 cell neuritogenesis and converges at the level of target proteins. Mol Cell Neurosci 2003;24(1):238–49.

[92] Yang YJ, Lee HJ, Choi DH, Huang HS, Lim SC, Lee MK. Effect of scoparone on neurite outgrowth in PC12 cells. Neurosci Lett 2008;440(1):14–18.

[93] Yang YJ, Lee HJ, Huang HS, Lee BK, Choi HS, Lim SC, et al. Effects of scoparone on dopamine biosynthesis and L-DOPA-induced cytotoxicity in PC12 cells. J Neurosci Res 2009;87(8):1929–37.

[94] Yang YJ, Lee HJ, Lee BK, Lim SC, Lee CK, Lee MK. Effects of scoparone on dopamine release in PC12 cells. Fitoterapia 2010;81(6):497–502.

[95] Furukawa Y, Watanabe S, Okuyama S, Nakajima M. Neurotrophic effect of citrus auraptene: neuritogenic activity in PC12 cells. Int J Mol Sci 2012;13(5):5338–47.

[96] Yang EB, Zhao YN, Zhang K, Mack P. Daphnetin, one of coumarin derivatives, is a protein kinase inhibitor. Biochem Biophys Res Commun 1999;260(3):682–5.

[97]   Yan L, Zhou X, Zhou X, Zhang Z, Luo HM. Neurotrophic effects of 7,8-dihydroxycoumarin in primary cultured rat cortical neurons. Neurosci Bull 2012; 28(5):493–8.

[98]   Suidan HS, Stone SR, Hemmings BA, Monard D. Thrombin causes neurite retraction in neuronal cells through activation of cell surface receptors. Neuron 1992;8(2):363–75.

[99]   Chen IS, Lin YC, Tsai IL, Teng CM, Ko FN, Ishikawa T, et al. Coumarins and anti-platelet aggregation constituents from *Zanthoxylum schinifolium*. Phytochemistry 1995;39(5):1091–7.

[100]   Li W, Chan C, Leung H, Yeung H, Xiao P. Xanthones and flavonoids of *Polygala caudata*. Pharm Pharmacol Communications 1998;4(8):415–17.

[101]   Pan MD, Mao Q. Isolation and identification of wubangziside A and B from *Polygala caudata* Rehd et Wils. Acta Pharmaceutica Sinica 1984;19(12):899–903.

[102]   Pan MD, Mao Q. Isolation and identification of wubangziside C from *Polygala caudata* Rehd et Wils. Acta Pharmaceutica Sinica 1985;20(9):662–5.

[103]   Li W, Chan C-L, Leung H-W, Yeung HW, Xiao P. Xanthones from *Polygala caudata*. Phytochemistry 1999;51(7):953–8.

[104]   Chanmahasathien W, Li Y, Ishibashi M, Ruangrungsi N, Ohizumi Y. Xanthones with NGF-potentiating activity. Nat Med 2004;58(2):76–8.

[105]   Tomić M, Tovilović G, Butorović B, Krstić D, Janković T, Aljancić I, et al. Neuropharmacological evaluation of diethylether extract and xanthones of *Gentiana kochiana*. Pharmacol Biochem Behav 2005;81(3):535–42.

[106]   Urbain A, Marston A, Grilo LS, Bravo J, Purev O, Purevsuren B, et al. Xanthones from *Gentianella amarella* ssp. *acuta* with acetylcholinesterase and monoamine oxidase inhibitory activities. J Nat Prod 2008;71(5):895–7.

[107]   Hwang JH, Hong SS, Han XH, Hwang JS, Lee D, Lee H, et al. Prenylated xanthones from the root bark of *Cudrania tricuspidata*. J Nat Prod 2007;70(7):1207–9.

[108]   Jeong GS, An RB, Pae HO, Chung HT, Yoon KH, King DG, et al. Cudratricusxanthone a protects mouse hippocampal cells against glutamate-induced neurotoxicity via the induction of heme oxygenase-1. Planta Med 2008;74(11):1368–73.

[109]   Yamakuni T, Aoki K, Nakatani K, Kondo N, Oku H, Ishiguro K, et al. Garcinone B reduces prostaglandin E2 release and NF-kappaB-mediated transcription in C6 rat glioma cells. Neurosci Lett 2006;394(3):206–10.

[110]   Nakatani K, Nakahata N, Arakawa T, Yasuda H, Ohizumi Y. Inhibition of cyclooxygenase and prostaglandin E2 synthesis by gamma-mangostin, a xanthone derivative in mangosteen, in C6 rat glioma cells. Biochem Pharmacol 2002;63(1):73–9.

[111]   Nakatani K, Yamakuni T, Kondo N, Arakawa T, Oosawa K, Shimura S, et al. Gamma-Mangostin inhibits inhibitor-kappaB kinase activity and decreases lipopolysaccharide-induced cyclooxygenase-2 gene expression in C6 rat glioma cells. Mol Pharmacol 2004;66(3):667–74.

[112]   Palempalli UD, Gandhi U, Kalantari P, Vunta H, Arner RJ, Narayan V, et al. Gambogic acid covalently modifies IkappaB kinase-β subunit to mediate suppression of lipopolysaccharide-induced activation of NF-kappaB in macrophages. Biochem J 2009;419:401–9.

[113] Hsu M-F, Lin C-N, Lu M-C, Wang J-P. Inhibition of the arachidonic acid cascade by norathyriol via blockade of cyclooxygenase and lipoxygenase activity in neutrophils. Naunyn-Schmiedeberg's Arch Pharmacol 2004;369(5):507–15.

[114] Zou J, Jin D, Chen W, Wang J, Liu Q, Zhu X, et al. Selective cyclooxygenase-2 inhibitors from *Calophyllum membranaceum*. J Nat Prod 2005;68(10):1514–18.

[115] Chiou WF, Chen CC, Lin IH, Chiu JH, Chen YJ. 1,3,5-Trihydroxy-4-prenylxanthone represses lipopolysaccharide-induced iNOS expression via impeding posttranslational modification of IRAK-1. Biochem Pharmacol 2011;81(6):752–60.

[116] Liao CH, Ho CT, Lin JK. Effects of garcinol on free radical generation and NO production in embryonic rat cortical neurons and astrocytes. Biochem Biophys Res Commun 2005;329(4):1306–14.

[117] Jin H, Kanthasamy A, Ghosh A, Yang Y, Anantharam V, Kanthasamy AG. α-Synuclein negatively regulates protein kinase Cδ expression to suppress apoptosis in dopaminergic neurons by reducing p300 histone acetyltransferase activity. J Neurosci 2011;31(6):2035–51.

[118] Wang Y, Xia Z, Xu JR, Wang YX, Hou LN, Qiu Y, et al. A-mangostin, a polyphenolic xanthone derivative from mangosteen, attenuates β-amyloid oligomers-induced neurotoxicity by inhibiting amyloid aggregation. Neuropharmacology 2012;62(2):871–81.

[119] Pedraza-Chaverrí J, Reyes-Fermín LM, Nolasco-Amaya EG, Orozco-Ibarra M, Medina-Campos ON, González-Cuahutencos O, et al. ROS scavenging capacity and neuroprotective effect of alpha-mangostin against 3-nitropropionic acid in cerebellar granule neurons. Exp Toxicol Pathol 2009;61(5):491–501.

[120] Weng MS, Liao CH, Yu SY, Lin JK. Garcinol promotes neurogenesis in rat cortical progenitor cells through the duration of extracellular signal-regulated kinase signaling. J Agric Food Chem 2011;59(3):1031–40.

[121] Mak NK, Li WK, Zhang M, Wong RN, Tai LS, Yung KK, et al. Effects of euxanthone on neuronal differentiation. Life Sci 2000;66(4):347–54.

[122] Ha WY, Wu PK, Kok TW, Leung KW, Mak NK, Yue PY, et al. Involvement of protein kinase C and E2F-5 in euxanthone-induced neurite differentiation of neuroblastoma. Int J Biochem Cell Biol 2006;38(8):1393–401.

[123] Chanmahasathien W, Li Y, Satake M, Oshima Y, Ruangrungsi N, Ohizumi Y. Prenylated xanthones with NGF-potentiating activity from *Garcinia xanthochymus*. Phytochemistry 2003;64(5):981–6.

[124] Thull U, Kneubuhler S, Testa B, Borges MFM, Pinto MMM. Substituted xanthones as selective and reversible monoamine oxidase A (MAO-A) inhibitors. Pharm Res 1993;10(8):1187–90.

[125] Chanmahasathien W, Li Y, Satake M, Oshima Y, Ishibashi M, Ruangrungsi N, et al. Prenylated xanthones from *Garcinia xanthochymus*. Chem Pharm Bull (Tokyo) 2003;51(11):1332–4.

[126] Jang SW, Okada M, Sayeed I, Xiao G, Stein D, Jin P, et al. Gambogic amide, a selective agonist for TrkA receptor that possesses robust neurotrophic activity, prevents neuronal cell death. Proc Natl Acad Sci USA 2007;104(41):16329–34.

Topic **3.2**

Quinones

3.2.1 *Petroselinum crispum* (Mill.) Nyman ex A.W. Hill

History The plant was first described by Johann Mihály Fuss, in *Flora Transsilvaniae Excursoria* published in 1866.

Synonyms *Apium crispum* Mill., *Apium petroselinum* L., *Carum petroselinum* (L.) Benth. & Hook. f., *Petroselinum crispum* (Mill.) Mansf., *Petroselinum crispum* (Mill.) Nyman, *Petroselinum hortense* Hoffm., *Petroselinum hortense* var. crispum L.H. Bailey, *Petroselinum petroselinum* (L.) H. Karst., *Petroselinum sativum* Hoffm., *Petroselinum vulgare* Lag., *Selinum petroselinum* (L.) E.H.L. Krause

Family Apiaceae Lindl, 1836

Common Names Parsley, ou qin (Chinese)

Habitat and Description This common culinary herb grows to 1 m in Europe and China from a taproot. The leaves are spiral and exstipulate. The petiole is of variable length, to 10 cm long and sheathing at the base. The blade is dissected, 5–8 cm × 4–10 cm, serrate and aromatic. The inflorescence is an umbel of umbellules, which is 5 cm in diameter. The flowers are minute and white. The fruits are minute and dry achenes.

Medicinal Uses In China, the plant is diuretic and febrifuge.

Phytopharmacology The plant contains the coumarin oxypeucedadin[127]; the flavonoids apigenin, luteolin, and chrysoeriol; the flavonols kaempferol, quercetin, and isorhamnetin[128]; the polyacetylenes falcarinol, falcarindiol[129]— an essential oil sheltering notably 1,3,8-p-menthatriene, myristicin, apiol, myrcene, terpinolene, and 1-methyl-4-isopropenylbenzene[130,131]—a fixed oil containing petroselenic acid and phylloquinone.[132]

Proposed Research Pharmacological study of phylloquinone and derivatives for the treatment of amyotrophic lateral sclerosis (ALS) and spinal cord injuries.

Rationale An increasing amount of experimental evidence demonstrates that alkylated quinones sustain the viability of neurons and prompt neuritogenesis. For instance, the synthetic 2,4,4-trimethyl-3-(15-hydroxypentadecyl)-2-cyclohexen-1-one (CS 3.106) at a dose of 100 nM evoked the growth of neurites and neuron survival,[133] corrected nociception in rodents subjected to chronic nerve constriction injury,[134] and at 0.5 μM boosted cytoplasmic Ca^{2+} in synaptosomes,[135] implying a growth of neurites resulting from Ca^{2+} induced Ras, Raf, MEK1/2, and therefore extracellular signal-regulated kinase (ERK1/2). Along the same line, α-tocopherol quinone (CS 3.107) at a dose of 10 μM abated the production of nitric oxide (NO), tumor necrosis factor-α (TNF-α), and Interleukin-1β (IL-1β) by glial cells challenged with β-amyloid (Aβ_{1-42}) peptide,[136] and because this quinone is lacking in the cerebrospinal fluid of patients with ALS,[137] it is tempting to speculate that alkylated quinones may have value in treating amyotrophic lateral sclerosis, spinal cord injuries, and multiple sclerosis (MS).

Sargaquinoic acid (CS 3.108) from *Sargassum macrocarpum* C. Agardh (family Sargassaceae Kützing) at a dose of 6.2 μg/mL enhanced the neurotrophic effects of 10 ng m/L of nerve growth factor (NGF) toward rat pheochromocytoma (PC12) cells by 500%.[138] In serum-free medium, sargaquinoic acid at a dose equal to 1.5 μg/mL boosted the neuroprotective and neuritogenic effects of NGF at a dose of 0.08 ng/mL via the activation of phosphoinositide 3-kinase (PI3K),[139] Rac, and therefore JNK. Other alkylated quinones of neuropharmacological interest are vitamin K1 (CS 3.109) or phylloquinone, and vitamin K2 (CS 3.110) or menaquinone-4 at a dose of 100 μmol/L enhanced the neuritogenic effects of 10 ng/ mL of NGF against PC12 cells by activating protein kinase A (PKA).[140] Furthermore, phylloquinone and menaquinone-4 at a dose of 0.1 μM protected oligodendrocytes against glutamate insults by sustaining the levels of glutathione (GSH) and thus mitigating reactive oxygen species (ROS) insults.[141] Along the same line, pre-oligodendrocytes were protected against oxidative insults owed to arachidonic acid by phylloquinone and menaquinone-4 with IC_{50} values equal to 10 nM and 25 nM, respectively, by hindering 12-lipoxygenase (12-LOX),[142] which in fact has been shown as being of critical importance for neurodegeneration upon depletion of GSH and subsequent inhibition of glutathione peroxidase-4 (GPx-4).[143] Thus, several vitamin K derivatives were prepared, including Compound 1d (CS 3.111) and 2j (CS 3.112), which protected H22 neurons against glutamate with PC_{50} values equal to 61 nM and 88 nM, respectively.[144]

The aforementioned neurotrophic properties are in fact principally owed to the quinone or naphthoquinones moiety because 1,4-benzoquinone

■ **CS 3.106** 2,4,4-trimethyl-3-(15-hydroxypentadecyl)-2-cyclohexen-1-one.

■ **CS 3.107** α-tocopherol quinone.

■ **CS 3.108** Sargaquinoic acid.

■ **CS 3.109** Vitamin K1.

■ **CS 3.110** Vitamin K2.

■ **CS 3.111** Compound 1d.

■ **CS 3.112** Compound 2j.

■ **CS 3.113** 1,4-benzoquinone.

(CS 3.113) itself induced sprouting of neurites in PC12 cells exposed to NGF.[138] Likewise, lapachol (CS 3.114) and lawsone (CS 3.115) commanded the growth of neurites from PC12 cells by 265% and 329%, respectively.[138] Naphthazarin (CS 3.116) from members of the genus *Lomatia* R. Br. (family Proteaceae Juss.) at a dose of 1 mg/kg attenuated the locomotor impairments induced by MPTP in rodents and prevented nigro-striatal degeneration of dopaminergic neurons by preventing glial activation[145] via probable induction of heme oxidase (HO-1) and therefore attenuation of oxidative insults.[146] This assertion is supported by the demonstration that plumbagin (CS 3.117) from *Plumbago zeylanica* L. (family Plumbaginaceae Juss.) at a dose of $5\mu M$ induced ARE in SH-SY5Y cells, whereas menadione (CS 3.118) was inactive.[147] In the same experiment, *Plumbago zeylanica* L. (family Plumbaginaceae Juss.) evoked the

■ **CS 3.114** Lapachol.

transcription of HO-1 via ARE by Nrf2, without involving ROS and via activation of PI3K, protein kinase B (Akt), and ERK1/2[147], accounting probably for the fact that rodents subjected to experimental stroke were protected by this naphthoquinone at a dose of 3 mg/kg.[147] Plumbagin at a dose of 0.1 μM induced the conversion of embryonic neural progenitor cells into glia-restricted precursors via activation of STAT3.[148] In regard to STAT3, one should recall that interleukin-6 (IL-6) favors the survival of PC12 cells by activating JAK/STAT3, which translocates into the nucleus.[149] Another property of plumbagin is to generate ROS, thus inhibiting the phosphorylation of nuclear factor of kappa light polypeptide gene enhancer in B-cells inhibitor α (IκBα) by kappa light polypeptide gene enhancer in B-cells inhibitor (IκB) kinase (IKK) and therefore blocking nuclear factor kappa-light-chain-enhancer of activated B cells (NF-κB),[150] making naphthoquinones of interest to curb glial activation. In this light, β-lapachone (CS 3.119) from *Tabebuia avellanedae* Lorentz ex Griseb. (family Bignoniaceae Juss.) at a dose of 1 μM abated the production of NO and prostaglandin E2 (PGE2) by BV-2 cells challenged with lipopolysaccharide (LPS) as a result of inducible nitric oxide synthetase (iNOS), and cyclo-oxygenase-2 (COX-2) reduced expression.[151] In the same experiment, the production of β-lapachone induced the hypophosphorylation of IκBα and inactivation of mitogen-activated protein kinase (MAPK) p38, ERK1/2, and Akt.[151] Along the same line, isobutyrylshikonin (CS 3.120) and isovalerylshikonin (CS 3.121) at a dose of 4 μM abated the production of nitric oxide. IL-6, IL-1β, and TNF-α were hindered by BV-2 cells in the presence of β- (NO), TNF-α, and prostaglandin E2 (PGE2) by BV-2 cells challenged with LPS on account of IκBα hypophosphorylation, ERK1/2, and Akt inactivation.[151,152] The prenylated naphthoquinone shikonin (CS 3.122) (β-alkannin) from *Lithospermum erythrorhizon* Siebold & Zucc. (family Boraginaceae Juss.) at a dose of 50 mg/kg protected rodents against experimental stroke with an increase in the enzymatic activities of catalase (CAT), superoxide dismutase (SOD), and glutathione peroxidase (GPx) and a decrease in ROS.[153] In PC12 cells, the asterriquinone 1H5 (CS 3.123) at a dose of 100 μM activated TrkA, ERK1/2, and Akt, thus

■ **CS 3.115** Lawsone.

■ **CS 3.116** Naphthazarin.

■ **CS 3.117** Plumbagin.

■ **CS 3.118** Menadione.

■ **CS 3.119** β-lapachone.

■ **CS 3.120** Isobutyrylshikonin.

■ **CS 3.121** Isovalerylshikonin.

allowing survival in serum-free medium.[128] In the same experiment, 5E5 (CS 3.124) at a dose of $1 \mu M$ induced TrkA, Akt, ERK1/2, FRS2/SNT, and neurite outgrowth in PC12 cells exposed to NGF.[155]

Naphthoquinones not only are neurotrophic but also inhibit the enzymatic activity of enzymes involved in the pathophysiology of neurodegenerative diseases. In regard to Alzheimer's disease (AD), 2-hydroxy-1,4-naphthoquinone (CS 3.125), 5-hydroxy-1,4-naphthoquinone (CS 3.126), 5,8-dihydroxy-1,4-naphthoquinone (CS 3.127), and 2-methyl-5-hydroxy-1,4-naphthoquinone (CS 3.128) have the ability to inhibit the enzymatic activity of β-secretase with IC_{50} values equal to $5.9 \mu M$, $6.5 \mu M$, $9 \mu M$, and $12.9 \mu M$, respectively.[156] Along the same line, 1,4-naphthoquinone (CS 3.129), 5-hydroxy-1,4-naphthoquinone, and 6-hydroxy-1,4-naphthoquinone (CS 3.130) inhibited the formation of β-amyloid ($A\beta_{25-35}$) peptide aggregation with an IC_{50} value equal to $12.4 \mu M$, $11.1 \mu M$, and $2.7 \mu M$.[28] N-(3-chloro-1,4-dihydro-1,4-dioxo-2-naphthalenyl)-L-tryptophan (CS 3.131) prevented the aggregation of β amyloid polypeptides 1–40 ($A\beta_{1-40}$) and β amyloid polypeptides 1–42 ($A\beta_{1-42}$) by direct interaction, and at a dose of 50 mg/kg improved the cognition of rodents.[157] Impatienol (CS 3.132) and balsaminol (CS 3.133) from *Impatiens balsamina* L. (family Balsaminaceae A. Rich.) inhibited the enzymatic activity of COX-2 with IC_{50} values equal to 0.2 μM and 9.4 μM.[156] 2,3,6-Trimethyl-1,4-naphthoquinone (CS 3.134) from the aforementioned *Nicotiana tabacum* L. (family Solanaceae Juss.) inhibited the enzymatic activity of monoamine oxidase A (MAO-A) and monoamine oxidase B (MAO-B) with K_i values equal to 3 μM and 6 μM, respectively.[158] *In vivo*, 2,3,6-trimethyl-1,4-naphthoquinone at a dose of 400 mg/kg mitigated the locomotor disturbances incurred by MPTP.[159] 1,4-Naphthoquinone and menadione inhibited the enzymatic activity of MAO-B K_i values equal to 1.5 μM and 0.4 μM, respectively.[158] In the same experiment, 1,4-naphthoquinone and menadione inhibited the enzymatic activity of MAO-A with K_i values equal to 7.7 μM and 26.1 μM, respectively,[158] equal to 12.4 μM, 11.1 μM, and 2.7 μM.[155] N-(3-chloro-1,4-dihydro-1,4-dioxo-2-naphthalenyl)-L-tryptophan prevented the aggregation of β amyloid polypeptides 1–40 (Aβ_{1-40}) and β amyloid polypeptides 1–42 (Aβ_{1-42}) by direct interaction and at a dose of 50 mg/kg improved the cognition of rodents.[157]

■ **CS 3.122** Shikonin.

■ **CS 3.123** 1H5.

■ **CS 3.124** 5E5.

■ **CS 3.125** 2-hydroxy-1,4-naphthoquinone.

■ **CS 3.126** 5-hydroxy-1,4-naphthoquinone.

■ **CS 3.127** 5,8-dihydroxy-1,4-naphthoquinone.

■ **CS 3.128** 2-methyl-5-hydroxy-1,4-naphthoquinone.

■ **CS 3.129** 1,4-naphthoquinone.

■ **CS 3.130** 6-hydroxy-1,4-naphthoquinone.

■ **CS 3.131** N-(3-chloro-1,4-dihydro-1,4-dioxo-2-naphthalenyl)-L-tryptophan.

■ **CS 3.132** Impatienol.

■ **CS 3.133** Balsaminol.

■ **CS 3.134** 2,3,6-trimethyl-1,4-
naphthoquinone.

3.2.2 *Cassia javanica* L.

History The plant was first described by Carl von Linnaeus in *Systema Naturae* published in 1768.

Synonyms *Cassia bacillus* Gaertn., *Cassia megalantha* Decne., *Cassia nodosa* Buch.-Ham. ex Roxb., *Cathartocarpus javanicus* Pers.

Family Fabaceae Lindl., 1836

Common Name Qian cao (Chinese)

Habitat and Description It is a magnificent tree, which grows in China, Thailand, Vietnam, Laos, Malaysia, Burma, Indonesia, and India. The leaves are pinnate, spiral, and stipulate. The rachis is 15–40 cm long and bears 12 pairs of folioles, which are 2–8 cm × 1–4 cm and elliptic. The inflorescence is an axillary raceme. The calyx includes 5 sepals, which are ovate. The corolla consists of 5 petals, which are light pink, and 2.5–4.5 cm × 1–2 cm. The androecium presents 10 stamens. The fruit is an elongated and annelated pod which is 30–50 cm × 0.2–1.5 cm (Figure 3.4).

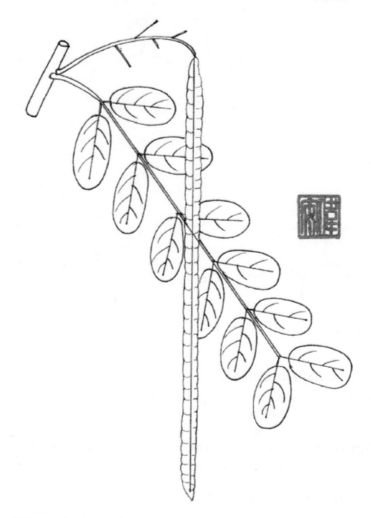

■ **FIGURE 3.4** *Cassia javanica* L.

Medicinal Uses In China, the plant is used as laxative.

Phytopharmacology The plant produces some flavonoids including leucocyanidin-4′-*O*-methyl ether-3-*O*-β-D-galactopyranoside, dihydrorhamnetin-3-O-β-D-glucopyranoside, quercetin-3′,4′,7-trimethyl ether-3-*O*-α-L-rhamnopyranoside, kaempferol-3-rhamnoglucoside, and quercetin[160]; kaempferol; kaempferol-3-methylether[161]; the anthraquinones emodin,[161] chrysophanol (CS 3.135), physcion (CS 3.136), 1,5-dihydroxy 4,7-dimethoxy 2-methyl anthraquinone 3-O-α-L(-) rhamnopyranoside, and

■ **CS 3.135** Chrysophanol.

■ **CS 3.136** Physcion.

1,3,6,7,8-pentahydroxy-4-methoxy 2-methyl anthraquinone[162]; and the triterpenoids α-amyrin and β-amyrin.[161]

Proposed Research Pharmacological study of emodin and derivatives for the treatment of neurodegenerative diseases.

Rationale There is evidence indicating that anthraquinones have the ability to inhibit the activity of enzymes involved in the pathophysiology of AD, Parkinson's disease (PD), and ALS. In effect, some mutagenic and, therefore, carcinogenic mycotoxins including norsolorinic acid (CS 3.137) and averufin (CS 3.138) isolated from the mold *Emericella navahoensis* M. Chr. & States (family Trichocomaceae E. Fisch) inhibited the enzymatic activity of monoamine oxidase (MOA) with IC_{50} values equal to $0.3\,\mu M$ and $92.4\,\mu M$, respectively.[163] Along the same line, averythrin (CS 3.139), averufin dimethylether (CS 3.140), emodin (CS 3.141), islandicin (CS 3.142), and versiconol (CS 3.143) inhibited the enzymatic activity of MOA with IC_{50} values equal to $3.7\,\mu M$, $303\,\mu M$, $50\,\mu M$, $148.2\,\mu M$, and $52.8\,\mu M$, respectively.[163] Solorinic acid (CS 3.144) isolated from the lichen *Solorina crocea* (L.) Ach. (family Peltigeraceae Dumort.) inhibited the enzymatic activity of MAO with an IC_{50} value equal to $14.3\,\mu M$.[164] This finding supports the notion that hydroxyl moieties in C1, C3, C6, and C8 as well as an alkyl in C2 favor activity and anthraquinones against MOA, whereas methoxylations or alkylations of C1, C3, C6, and C8 abate the activity.

Emodin from *Rheum tanguticum* Maxim. ex Balf. (family Polygonaceae Juss.) inhibited the enzymatic activity of 5-lipoxygenase (5-LOX) with an IC_{50} value equal to $4.3\,\mu M$, whereby chrysophanol and physcion were inactive.[162] The synthetic anthracenone compounds 4f, 4g, and 4h (CS 3.145–3.147) inhibited the enzymatic activity of 5-LOX with IC_{50} values equal to $2\,\mu M$, $3\,\mu M$, and $2\,\mu M$, respectively.[166] *Kniphofia foliosa* Hochst. (family Xanthorrhoeaceae Dumort.) shelters knipholone (CS 3.148), which inhibited the production of leukotriene by human neutrophile granulocytes with an IC_{50} value equal to $4.2\,\mu M$,[167] reinforcing further the contention that a benzyl moiety at the vicinity of C4 favors the activity. In addition,

■ **CS 3.137** Norsolorinic acid.

■ **CS 3.138** Averfulin.

■ **CS 3.139** Averythrin.

■ **CS 3.140** Averufin dimethyl ether.

■ **CS 3.141** Emodin.

knipholone inhibited the synthesis of 12-hydroxyeicosatetraenoic acid (12-HETE) by platelets by 28.6% at a dose of $10\,\mu g/mL$,[167] and up to $50\,\mu g/mL$ did not interfere with the enzymatic activity of cyclo-oxygenase-1 (COX-1) and cyclo-oxygenase-2 (COX-2).[167] In regard to 12-LOX, the anthracenones anthranilin and danthron inhibited the synthesis of 12-HETE by 36% and 44%, respectively, in epidermal cells exposed to arachidonic acid at a dose of $100\,\mu M$.[168] Along the same line, the synthetic anthracenones 10a, 10f, 10g, 10i, 12e, 14c (CS 3.149–3.154) and anthralin (CS 3.155) inhibited the production of 12-HETE by 12-LOX in mouse epidermal

■ **CS 3.142** Islandicin.

■ **CS 3.143** Versiconol.

■ **CS 3.144** Solorinic acid.

■ **CS 3.145** Compound 4f.

■ **CS 3.146** Compound 4g.

■ **CS 3.147** Compound 4h.

homogenates with IC_{50} values equal to $7\,\mu M$, $6\,\mu M$, $6\,\mu M$, $4\,\mu M$, $5\,\mu M$, $11\,\mu M$, and $9\,\mu M$, respectively,[169] implying that a benzyl moiety near C3 and C3 favor the activity. Rhein from *Rheum officinale* Baill. (family Polygonaceae Juss.) and aloe-emodin from *Aloe vera* (L.) Burm. f. (family Xanthorrhoeaceae Dumort.) inhibited the peroxidation of linoleic acid by 15-lipoxygenase (15-LOX) with IC_{50} values equal to $64\,\mu mol/L$ and $65\,\mu mol/L$, respectively.[170] It therefore seems plausible that anthraquinones may circumvent neuroinflammation and neuroapoptosis as the production of 12-hydroxyeicosatetraenoic acid (12-HETE) by 12-LOX from arachidonic acid (AA) evokes mitochondrial insults.

Emodin and rhein inhibited the production of NO by RAW 264.7 challenged with LPS with IC_{50} values equal to $60.7\,\mu M$ and $67.3\,\mu M$.[171] In the same experiment, emodin hindered the expression of COX-2 and iNOS, whereas rhein solely abrogated iNOS expression.[171] Furthermore, aloe-emodin, emodin, and rhein nullified the production of NO by macrophages (RAW 264.7) challenged with LPS at a dose of $30\,\mu M$ via inhibition of iNOS.[172] In the same experiment, aloe-emodin, emodin, and chrysophanol abated the production of PGE2 by RAW 264.7 macrophages challenged with LPS on account of COX-2 inhibition at a dose of $30\,\mu M$ (12/42). In CaCo cells, $100\,\mu M$ of emodin inactivated ERK1/2, JNK, and MAPK p38.[172]

Emodin at a dose of $30\,\mu M$ reduced the excitatory postsynaptic potential (EPSP) of CA1 pyramidal neurons in rat hippocampus by blocking the secretion of glutamate from Schaffer collateral/commissural terminals via the activation of adenosine A1 receptors.[173] One must recall here that Schaffer collaterals are involved in the apoptosis of CA1 pyramidal

■ **CS 3.148** Knipholone.

■ **CS 3.149** Compound 10a.

■ **CS 3.150** Compound 10f.

■ **CS 3.151** Compound 10g.

■ **CS 3.152** Compound 10i.

■ **CS 3.153** Compound 12e.

■ **CS 3.154** Compound 14e.

■ **CS 3.155** Anthralin.

■ **CS 3.156** Danthron.

neurons during stroke, and, therefore, emodin and congeners may be of value in this instance.[174] Danthron (CS 3.156) at a dose of $30 \mu M$ protected cortical neurons against β-amyloid ($A\beta_{25-35}$) peptide, Fe^{3+}, buthionine sulfoximine (BSO), sodium nitroprussite, and H_2O_2 but was inactive against N-methyl-D-aspartate (NMDA) and kainate.[175] Furthermore, emodin protected cortical neurons against amyloid β_{25-35} peptides at a dose equal to $20 \mu M$ via activation of Akt, ERK1/2, inactivation of JNK1/2, and induction of Bcl-2 at a dose of $20 \mu M$.[176] The activation of Akt may account for the fact that aloe-emodine at a dose of $15 \mu M$ mitigated apoptosis of retina ganglion cells exposed to N-methyl-D-aspartate (NMDA) to 50% with expression of SOD.[177] Furthermore, anthraquinone-2-sulfonic acid (CS 3.157) protected cortical neurons against H_2O_2 at a dose of $50 \mu M$ by activating Akt and inhibiting caspase-3.[178] The synthetic anthraquinone derivatives AQ22, AQ23, AQ24, and AQ32 (CS 3.158–3.161) antagonized NMDA receptors expressed by oocytes with IC_{50} values equal to $22 \mu M$, $6.5 \mu M$, $4 \mu M$, and $8.5 \mu M$, respectively.[179] Along the same line, AQ33, AQ34, AQ33b, AQ343, and AQ444 (CS 3.162–3.166) blocked N-methyl-D-aspartate (NMDA) receptors expressed by oocytes with IC_{50} values equal to $5.6 \mu M$, $7.1 \mu M$, $5.1 \mu M$, $0.3 \mu M$, and $0.5 \mu M$, respectively,[180] suggesting that a polyamine chain in C2 with C5 fragments favors activity.

By virtue of their planaricity and therefore intercalating properties, anthraquinones bind to amyloid peptides and inhibit their

■ **CS 3.157** Anthraquinone-2-sulfonic acid.

■ **CS 3.158** AQ22.

■ **CS 3.159** AQ23.

■ **CS 3.160** AQ24.

■ **CS 3.161** AQ32.

■ **CS 3.162** AQ33.

■ **CS 3.163** AQ34.

■ **CS 3.164** AQ33b.

■ **CS 3.165** AQ343.

■ **CS 3.166** AQ444.

■ **CS 3.167** Mitoxantrone.

aggregations. For instance, emodin inhibited the polymerization of the tau constructs with an IC_{50} value equal to $1.9\,\mu M$ in N2a cells, and at a dose of $15\,\mu M$ abated the formation of neurofibrillary tangles.[181] Another example is the synthetic mitoxantrone (CS 3.167), which intercalated in and stabilized the tau splicing regulatory element (SRE) RNA structure with an IC_{50} value equal to 700 nM, thus mitigating the inclusion of exon 10 into the mature tau protein and therefore neurodegeneration in AD.[182] Of further interest against Alzheiner's disease (AD), rhein (CS 3.168) and aloe-emodin (CS 3.169) inhibited the enzymatic activity of acetylcholinesterase (AChE) by 18.1% and 57.2% at a dose of 1 mg/mL.[183] Obtusifolin (CS 3.170) from *Cassia obtusifolia* L. (family Fabaceae Lindl.) inhibited the enzymatic activity of AChE with an IC_{50} value equal to $18.5\,\mu M$ and mitigated the dementia incurred by scopolamine at a dose of 0.5 mg/kg via inhibition of AChE,[184] suggesting that a methoxyl moiety in C1 with ortho hydroxy is beneficial.

■ **CS 3.168** Rhein.

■ **CS 3.169** Aloe-emodin.

■ **CS 3.170** Obtusifolin.

REFERENCES

[127] Chaudhary SK, Ceska O, Tetu C. Oxypeucedanin, a major furocoumarin in parsley, *Petroselinum crispum*. Planta Med 1986;6:462–4.

[128] Hempel J, Pforte H, Raab B, Engst W, Böhm H, Jacobasch G. Flavonols and flavones of parsley cell suspension culture change the antioxidative capacity of plasma in rats. Die Nahrung 1999;43(3):201–4.

[129] Nitz S, Spraul MH, Drawert F. C17 polyacetylenic alcohols as the major constituents in roots of *Petroselinum crispum* mill. ssp. tuberosum. J Ag Food Chem 1990;38(7):1445–7.

[130] Simon JE, Quinn J. Characterization of essential oil of parsley. J Ag Food Chem 1988;36(3):467–72.

[131] Ellenbracht F, Barz W, Mangold HK. Unusual fatty acids in the lipids from organs and cell cultures of *Petroselinum crispum*. Planta 1980;150(2):114–19.

[132] Bolton-Smith C, Price RJG, Fenton ST, Harrington DJ, Shearer MJ. Compilation of a provisional UK database for the phylloquinone (vitamin K1) content of foods. Br J Nutr 2000;83(4):389–99.

[133] Gonzalez de Aguilar JL, Girlanda-Junges C, Coowar D, Duportail G, Loeffler JP, Luu B. Neurotrophic activity of 2,4,4-trimethyl-3-(15-hydroxypentadecyl)-2-cyclohexen-1-one in cultured central nervous system neurons. Brain Res 2001;920(1–2):65–73.

[134] Tamura Y, Monden M, Suzuki H, Yamada M, Koyama K, Shiomi H. Beneficial action of 2,4,4-trimethyl-3-(15-hydroxypentadecyl)-2-cyclohexen-1-one, a novel

long-chain fatty alcohol, on diabetic hypoalgesia and neuropathic hyperalgesia. J Pharmacol Sci 2006;102(2):248–52.

[135] Jover E, Gonzalez de Aguilar JL, Luu B, Lutz-Bucher B. Effect of a cyclo-hexenonic long-chain fatty alcohol on calcium mobilization. Eur J Pharmacol 2005;516(3):197–203.

[136] Yang SG, Wang WY, Ling TJ, Feng Y, Du XT, Zhang X, et al. α-Tocopherol qui-none inhibits β-amyloid aggregation and cytotoxicity, disaggregates preformed fibrils and decreases the production of reactive oxygen species, NO and inflam-matory cytokines. Neurochem Int 2010;57(8):914–22.

[137] Tohgi H, Abe T, Saheki M, Yamazaki K, Takahashi S. Alpha-Tocopherol quinone level is remarkably low in the cerebrospinal fluid of patients with sporadic amyo-trophic lateral sclerosis. Neurosci Lett 1996;207(1):5–8.

[138] Tsang CK, Sagara A, Kamei Y. Structure-activity relationship of a neurite out-growth-promoting substance purified from the brown alga, *Sargassum macrocar-pum*, and its analogues on PC12D cells. J Appl Phycol 2001;13(4):349–57.

[139] Tsang CK, Kamei Y. Sargaquinoic acid supports the survival of neuronal PC12D cells in a nerve growth factor-independent manner. Eur J Pharmacol 2004;488(1–3):11–18.

[140] Tsang CK, Kamei Y. Novel effect of vitamin K (1) (phylloquinone) and vitamin K (2) (menaquinone) on promoting nerve growth factor-mediated neurite out-growth from PC12D cells. Neurosci Lett 2002;323(1):9–12.

[141] Li J, Lin JC, Wang H, Peterson JW, Furie BC, Furie B, et al. Novel role of vita-min K in preventing oxidative injury to developing oligodendrocytes and neu-rons. J Neurosci. 2003;23(13):5816–26.

[142] Li J, Wang H, Rosenberg PA. Vitamin K prevents oxidative cell death by inhib-iting activation of 12-lipoxygenase in developing oligodendrocytes. J Neurosci Res 2009;87(9):1997–2005.

[143] Seiler A, Schneider M, Förster H, Roth S, Wirth EK, Culmsee C, et al. Glutathione peroxidase 4 senses and translates oxidative stress into 12/15-lipox-ygenase dependent- and AIF-mediated cell death. Cell Metab 2008;8(3):237–48.

[144] Josey BJ, Inks ES, Wen X, Chou CJ. Structure-activity relationship study of vitamin K derivatives yields highly potent neuroprotective agents. J Med Chem 2013;56(3):1007–22.

[145] Choi SY, Son TG, Park HR, Jang YJ, Oh SB, Jin B, et al. Naphthazarin has a protective effect on the 1-methyl-4-phenyl-1,2,3,4-tetrahydropyridine-induced Parkinson's disease model. J Neurosci Re 2012;90(9):1842–9.

[146] Son TG, Kawamoto EM, Yu QS, Greig NH, Mattson MP, Camandola S. Naphthazarin protects against glutamate-induced neuronal death via activation of the Nrf2/ARE pathway. Biochem Biophys Res Commun 2013;433(4):602–6.

[147] Son TG, Camandola S, Arumugam TV, Cutler RG, Telljohann RS, Mughal MR, et al. Plumbagin, a novel Nrf2/ARE activator, protects against cerebral ischemia. J Neurochem 2010;112(5):1316–26.

[148] Luo Y, Mughal MR, Ouyang TG, Jiang H, Luo W, Yu QS, et al. Plumbagin pro-motes the generation of astrocytes from rat spinal cord neural progenitors via activation of the transcription factor Stat3. J Neurochem 2010;115(6):1337–49.

[149] Wu YY, Bradshaw RA. Induction of neurite outgrowth by Interleukin-6 is accompanied by activation of Stat3 signaling pathway in a variant PC12 cell (E2) line. J Biol Chem 1996;271(22):13023–32.

[150] Sandur SK, Ichikawa H, Sethi G, Kwang SA, Aggarwal BB. Plumbagin (5-hydroxy-2-methyl-1,4-naphthoquinone) suppresses NF-kappaB activation and NF-kappaB-regulated gene products through modulation of p65 and IkappaBalpha kinase activation, leading to potentiation of apoptosis induced by cytokine and chemotherapeutic agents. J Biol Chem 2006;281(25):17023–33.

[151] Moon DO, Choi YH, Kim ND, Park YM, Kim GY. Anti-inflammatory effects of beta-lapachone in lipopolysaccharide-stimulated BV2 microglia. Int Immunopharmacol. 2007;7(4):506–14.

[152] Nam KN, Son MS, Park JH, Lee EH. Shikonins attenuate microglial inflammatory responses by inhibition of ERK, Akt, and NF-kappaB: neuroprotective implications. Neuropharmacology 2008;55(5):819–25.

[153] Wang Z, Liu T, Gan L, Wang T, Yuan X, Zhang B, et al. Shikonin protects mouse brain against cerebral ischemia/reperfusion injury through its antioxidant activity. Eur J Pharmacol 2010;643(2–3):211–17.

[154] Lin B, Pirrung MC, Deng L, Li Z, Liu Y, Webster NJ. Neuroprotection by small molecule activators of the nerve growth factor receptor. J Pharmacol Exp Ther 2007;322(1):59–56.

[155] Bermejo-Bescós P, Martín-Aragón S, Jiménez-Aliaga KL, Ortega A, Molina MT, Buxaderas E, et al. In vitro antiamyloidogenic properties of 1,4-naphthoquinones. Biochem Biophys Res Commun 2010;400(1):169–74.

[156] Scherzer-Attali R, Farfara D, Cooper I, Levin A, Ben-Romano T, Trudler D, et al. Naphthoquinone-tryptophan reduces neurotoxic Aβ*56 levels and improves cognition in Alzheimer's disease animal model. Neurobiol Dis 2012;46(3):663–72.

[157] Oku H, Ishiguro K. Cyclooxygenase-2 inhibitory 1,4-naphthoquinones from *Impatiens balsamina* L. Biol Pharm Bull 2002;25(5):658–60.

[158] Coelho Cerqueira E, Netz PA, Diniz C, Petry do Canto V, Follmer C. Molecular insights into human monoamine oxidase (MAO) inhibition by 1,4-naphthoquinone: evidences for menadione (vitamin K3) acting as a competitive and reversible inhibitor of MAO. Bioorg Med Chem 2011;19(24):7416–24.

[159] Castagnoli KP, Steyn SJ, Petzer JP, Van der Schyf CJ, Castagnoli Jr. N. Neuroprotection in the MPTP Parkinsonian C57BL/6 mouse model by a compound isolated from tobacco. Chem Res Toxicol 2001;14(5):523–7.

[160] Tiwari RD, Singh J. The flavonoids of *Cassia javanica* flowers. Phytochemistry 1979;18(5):906.

[161] El-Sayyad SM, Ross SA. A phytochemical study of some *Cassia* species cultivated in Egypt. J Nat Prod 1983;46(3):431–2.

[162] Tiwari RD, Sharma MN. Anthraquinone pigments from *Cassia javanica* seeds 1981;43(12):381–3.

[163] Yamazaki M, Satoh Y, Maebayashi Y, Horie Y. Monoamine oxidase inhibitors from a fungus, *Emericella navahoensis*. Chem Pharm Bull 1988;36(2):670–5.

[164] Okuyama E, Hossain CF, Yamazaki M. Monoamine oxidase inhibitors from a lichen, *Solorina rocea* (L.) Ach. Jpn J Pharmacol 1991;4(2):159–62.

[165] Jin JH, Ngoc TM, Bae K, Kim YS, Kim HP. Inhibition of experimental atopic dermatitis by rhubarb (rhizomes of *Rheum tanguticum*) and 5-lipoxygenase inhibition of its major constituent, emodin. Phytother Res 2011;25(5):755–9.

[166] Müller K, Altmann R, Prinz H. 10-Benzoyl-1, 8-dihydroxy-9 (10 H)-anthracenones: Synthesis and biological properties. Eur J Med Chem 1998; 33(3):209–14.

[167] Wube AA, Bucar F, Asres K, Gibbons S, Adams M, Streit B, et al. Knipholone, a selective inhibitor of leukotriene metabolism. Phytomedicine 2006;13(6):452–6.

[168] Bedford CJ, Young JM, Wagner BM. Anthralin inhibition of mouse epidermal arachidonic acid lipoxygenase *in vitro*. J Invest Dermatol 1983;81(6):566–71.

[169] Müller K, Altmann R, Prinz H. 2-Arylalkyl-substituted anthracenones as inhibitors of 12-lipoxygenase enzymes. 2. Structure-activity relationships of the linker chain. Eur J Med Chem 2002;37(1):83–9.

[170] Malterud KE, Farbrot TL, Huse AE, Sund RB. Antioxidant and radical scavenging effects of anthraquinones and anthrones. Pharmacology 1993;47(Suppl. 1):77–85.

[171] Wang CC, Huang YJ, Chen LG, Lee LT, Yang LL. Inducible nitric oxide synthase inhibitors of Chinese herbs III. *Rheum palmatum*. Planta Med. 2002;68(10):869–74.

[172] Choi RJ, Ngoc TM, Bae K, Cho HJ, Kim DD, Chun J, et al. Anti-inflammatory properties of anthraquinones and their relationship with the regulation of P-glycoprotein function and expression. Eur J Pharm Sci 2013;48(1–2):272–81.

[173] Gu JW, Hasuo H, Takeya M, Akasu T. Effects of emodin on synaptic transmission in rat hippocampal CA1 pyramidal neurons *in vitro*. Neuropharmacology 2005;49(1):103–11.

[174] Onodera H, Sato G, Kogure K. Lesions to Schaffer collaterals prevent ischemic death of CA1 pyramidal cells. Neurosci Lett 1986;68(2):169–74.

[175] Kwon YS, Koh JY, Song DK, Kim HC, Kwon MS, Choi YS, et al. Danthron inhibits the neurotoxicity induced by various compounds causing oxidative damages including beta-amyloid (25–35) in primary cortical cultures. Biol Pharm Bull 2004;27(5):723–6.

[176] Liu T, Jin H, Sun QR, Xu JH, Hu HT. Neuroprotective effects of emodin in rat cortical neurons against beta-amyloid-induced neurotoxicity. Brain Res 2010;1347:149–60.

[177] Lin HJ, Lai CC, Lee CPD, Fan SS, Tsai Y, Huang SY, et al. Aloe-emodin metabolites protected N-methyl-d-aspartate-treated retinal ganglion cells by Cu-Zn superoxide dismutase. J Ocul Pharmacol Ther 2007;23(2):152–71.

[178] Jackson TC, Verrier JD, Kochanek PM. Anthraquinone-2-sulfonic acid (AQ2S) is a novel neurotherapeutic agent. Cell Death Dis 2013;4:e451.

[179] Kashiwagi K, Tanaka I, Tamura M, Sugiyama H, Okawara T, Otsuka M, et al. Anthraquinone polyamines: novel channel blockers to study N-methyl-D-aspartate receptors. J Pharmacol Exp Ther 2004;309(3):884–93.

[180] Kashiwagi K, Williams K, Igarashi K. Anthraquinone polyamines: novel channel blockers of N-methyl-D-aspartate receptors. Amino Acids 2007;33(2):299–304.

[181] Pickhardt M, Gazova Z, von Bergen M, Khlistunova I, Wang Y, Hascher A, et al. Anthraquinones inhibit tau aggregation and dissolve Alzheimer's paired helical filaments *in vitro* and in cells. J Biol Chem 2005;280(5):3628–35.

[182] Zheng S, Chen Y, Donahue CP, Wolfe MS, Varani G. Structural basis for stabilization of the tau pre-mRNA splicing regulatory element by novantrone (mitoxantrone). Chem Biol 2009;16(5):557–66.

[183] Orhan I, Tosun F, Şener B. Coumarin, anthraquinone and stilbene derivatives with anticholinesterase activity. J Biosci 2008;63(5–6):366–70.

[184] Kim DH, Hyun SK, Yoon BH, Seo JH, Lee KT, Cheong JH, et al. Gluco-obtusifolin and its aglycon, obtusifolin, attenuate scopolamine-induced memory impairment. J Pharmacol Sci 2009;111(2):110–16.

Topic **3.3**

Lignans

3.3.1 *Machilus thunbergii* Sieb. et Zucc

History The plant was first described by Philipp Franz Balthasar von Siebold and Joseph Gerhard Zuccarini in *Abhandlungen der Mathematisch-Physikalischen Classe der Königlich Bayerischen Akademie der Wissenschaften* published in 1846.

Synonyms *Litsea coreana* H. Lév., *Machilus arisanensis* (Hayata) Hayata, *Machilus kwashotensis* Hayata, *Machilus macrophylla* var. arisanensis Hayata, *Machilus nanshoensis* Kaneh., *Machilus taiwanensis* Kamik,. *Machilus thunbergii* var. kwashotensis (Hayata) Yamam., *Machilus thunbergii* var. trochodendroides Masam., *Persea arisanensis* (Hayata) Kosterm., *Persea thunbergii* (Siebold & Zucc.) Kosterm.

Family Lauraceae Juss., 1789

Common Name Hong nan (Chinese)

Habitat and Description It is an ornamental tree which grows in China, Korea, and Japan. The leaves are simple, exstipulate, spiral, and gathered at apices of stems. The petiole is 1–3 cm and reddish at first. The leaf blade is obovata, dull green and glossy above, coriaceous, 4–10 cm × 3 cm × 6 cm, and presenting 11 pairs of secondary nerves. The inflorescence is an axillary raceme which is 10 cm long, red, and fleshy. The perianth includes 6 tepals, which are 0.5 cm long. The androecium includes 9 stamens. The fruit is a purplish berry, which is 1 cm in diameter seated on a persistent perianth (Figure 3.5)

Medicinal Uses In Japan, the plant is used to heal stomach ulcers and to treat allergies.

Phytopharmacology The plant accumulates series of lignans which include machilin A, B, and E[185]; machilin F-I; nectandrin A; nectandrin B[186]; meso-dihydroguaiaretic acid; nordihydroguaiaretic acid; (+)-guaiacin; (−)-isoguaiacin; isoguaiacin dimethylether; licarin A[187];

■ **FIGURE 3.5** *Machilus thunbergii* Sieb. et Zucc.

(−)-acuminatin; (−)-sesamin; (+)-galbelgin[188]; and it contains flavonoids including kaempferol and quercetin.[189]

Proposed Research Pharmacological study meso-dihydroguaiaretic acid and derivatives for the treatment of neurodegenerative diseases.

Rationale A substantial amount of data has been accumulated on the beneficial role of lignans against microglial activation by virtue of their ability to abrogate the transcription of cytokines and chemokines. For instance, piperkadsin C (CS 3.171) and futoquinol (CS 3.172) from the aforementioned *Piper kadsura* (Choisy) Ohwi (family Piperaceae Giseke) reduced the production of nitric oxide (NO) by microglial BV-2 cells challenged with lipopolysaccharides (LPS) with IC_{50} values equal to 14.6 μM and 16.8 μM, respectively, by inhibiting the enzymatic activity of inducible nitric oxide synthetase (iNOS).[190] Likewise, obovatol (CS 3.173) from *Magnolia officinalis* Rehder & E.H. Wilson (family Magnoliaceae Juss.) inhibited the production of NO by macrophage (RAW 264.7)

■ **CS 3.171** Piperkadsin C.

■ **CS 3.172** Futoquinol.

■ **CS 3.173** Obovatol.

cells at a dose of $5\,\mu$M by reducing the expression of iNOS and attenuating the expression of iNOS and cyclo-oxygenase-2 (COX-2) by hindering the phosphorylation of nuclear factor of kappa light polypeptide gene enhancer in B-cells inhibitor α (IκBα) and thus nuclear factor kappa-light-chain-enhancer of activated B cells (NF-κB), inactivating extracellular signal-regulated kinase (ERK1/2), JNK,[191] and the enzymatic activity of peroxidase 2(Prx2).[192] Macelignan (CS 3.174) from *Myristica fragrans* Houtt. (family Myristicaceae R. Br.) at a dose of $10\,\mu$M hindered the production of tumor necrosis factor-α (TNF-α), interleukin-6 (IL-6), and NO by microglial cells with a decrease in iNOS and COX-2,[193] suggesting the inhibition of AP-1 and NF-κB. Along the same lines, *Magnolia fargesii* (Finet & Gagnep.) W.C. Cheng (family Magnoliaceae Juss.) engineers (+)-eudesmin (CS 3.175), (+)-magnolin (CS 3.176), and (+)-yangambin (CS 3.177); and epimagnolin B (CS 3.178) inhibited the production of NO by BV-2 glial cells challenged with LPS with IC_{50} values equal to $30\,\mu$M, $20.5\,\mu$M, $28.6\,\mu$M, and $10.9\,\mu$M, respectively.[194] In microglial BV-2 cells exposed to LPS, epimagnolin B blocked the degradation of IκBα and therefore reduced the expression of iNOS—hence a reduction of NO—and attenuated the expression of COX-2—hence a reduction in prostaglandin E2 (PGE2) at a dose of $50\,\mu$M.[194] Arctigenin (CS 3.179) from *Arctium lappa* L. (family Asteraceae Bercht. & J. Presl) at a dose of $1\,\mu$M deactivated ERK1/2, mitogen-activated protein kinase (MAPK) p38, and JNK in RAW 264.7 cells challenged with LPS via inhibition of mitogen-activated protein kinase (MAPK) kinase 1/2 (MKK1/2), mitogen-activated protein

■ **CS 3.174** Macelignan.

■ **CS 3.175** (+)-eudesmin.

■ **CS 3.176** (+)-magnolin.

■ **CS 3.177** (+)-yangambine.

■ **CS 3.178** Epimagnolin.

■ **CS 3.179** Arctigenin.

■ **CS 3.180** Sesamin.

kinase (MAPK) kinase 3/6 (MKK3/6), and mitogen-activated protein kinase (MAPK) kinase 4/7 (MKK4/7). In the same experiment, arctigenin at a dose of $1\,\mu$M inhibited the transcriptional activity of activator protein-1 (AP-1) and thus the production of TNF-α.[195] Sesamin (CS 3.180) at a dose of $100\,\mu$M inhibited the production of NO by microglia challenged with thrombin by inactivating MAPKs p44/42.[196]

Attenuating the activation of glial cells abates neuroinflammation, thus protecting neurons as exemplified with 4-*O*-methylhonokiol (CS 3.181), which at a dose of $1\,$mg/kg alleviated the dementia induced by LPS poisoning in rodents by repressing the expression of iNOS and COX-2 in the cortex and hippocampus.[197] Along the same lines, 4-*O*-methylhonokiol

■ **CS 3.181** 4-O-methylhonokiol.

■ **CS 3.182** Honokiol.

■ **CS 3.183** Magnolol.

■ **CS 3.184** Deoxyschizandrin.

(CS 3.182) at the same dose improved the cognition of rodents poisoned with amyloid β_{1-42} peptide by preventing neuroapoptosis in the *hippocampus* as a result of increased Bcl-2 expression, reduced pro-apoptotic Bcl-2-associated X protein (Bax), deactivation of MAPK p38, and glial cells inhibition.[198] In addition, 4-*O*-methylhonokiol inhibited the enzymatic activity of β-secretase with an IC_{50} value equal to 10.3 μM and, in rodents poisoned with amyloid β_{1-42} peptide, reduced the expression of β-secretase in the cortex and hippocampus.[199] Furthermore, 4-*O*-methylhonokiol at a dose of 1 mg/kg ameliorated the cognition of PS2 mutant rodents, reduced $A\beta_{1-42}$ in the cerebral cortex and hippocampus, deactivated ERK1/2, and assuaged glial cells.[200] The clinical significance of these data are quite clear in that they support a role for 4-*O*-methylhonokiol as a neuroprotective lead for Alzheimer's disease (AD). Honokiol delayed the onset of seizures in rodents poisoned with NMDA and reduced lethality by maintaining the cerebral levels of glutathione (GSH) and the enzymatic activity of glutathione peroxidase (GPx) at a dose of 3 mg/kg[201]; and magnolol (CS 3.183) at a dose of 80 mg/kg delayed the onset of seizures and decreased the mortality of rodents poisoned with pentylenetetrazol (PTZ).[202] The aforementioned elevation of GSH may very well result from a decrease in reactive oxygen species (ROS) because this lignan weakened PSD95-nNOS interaction, thus decreasing the enzymatic activity of neuronal nitric oxide synthetase (nNOS) and the production of NO at a dose of 10 μM in cortical neurons deprived of both oxygen and glucose by.[203] Note that 10 mg/kg of magnolol and 1 mg/kg of honokiol improved the cognition of SAMP8 mice with elevation of cholinergic neurons in the forebrain and activation of protein kinase B (Akt)[204] as a probable consequence of ROS reduction. Deoxyschizandrin (CS 3.184), gomisin N (CS 3.185), and wuweizisu C (CS 3.186) from *Schisandra chinensis* (Turcz.) Baill. (family Schisandraceae Blume) protected cortical cells against glutamate insults by 60%, 70%, and 55%, respectively, at a dose equal to 5 μM by sustaining GSH and the enzymatic activity of GPx.[205] In the same experiment, wuweizisu C and gomisin N elicited specific protection against kainate with a blockade of Ca^{2+} influx.[205] In addition, deoxyschizandrin, gomisin N, and wuweizisu C did not prevent Ca^{2+} influx evoked by NMDA but reduced the production of NO.[205] From the same plant, schizandrin B and schizandrin C (CS 3.187 and 3.188) protected rat pheochromocytoma (PC12) cells against amyloid β_{25-35} peptide at a dose of 25 μM and 100 μM, respectively, by abating ROS levels, decreasing pro-apoptotic Bax, and inactivating caspase-3[206] via the probable activation of Akt.

Kadsura polysperma Y.C. Yang (family Schisandraceae Blume) shelters polysperlignans A and D (CS 3.189 and 3.190) protected PC12 cells

■ **CS 3.185** Gomisin N.

■ **CS 3.186** Wuweizisu C.

■ **CS 3.187** Schizandrin B.

■ **CS 3.188** Schizandrin C.

against H_2O_2 by 46.4% and 50%, respectively, at a dose of $10\,\mu$M, whereas tiegusanin I (CS 3.191) protected PC12 cells against amyloid β_{25-35} peptide by 60.7% at a dose of $10\,\mu$M.[207] *Kadsura ananosma* Kerr (family Schisandraceae Blume) produces ananolignan F and ananolignan L (CS 3.192 and 3.193), which protected SH-SY5Y cells against H_2O_2 by 90% and 79.2%, respectively.[208] L-3-n-Butylphthalide (CS 3.194) from *Apium graveolens* L. (family Apiaceae Lindl.) protected SH-SY5Y cells and hippocampal neurons against amyloid β_{25-35} peptide[209] and H_2O_2 at a dose of $10\,\mu$M with an increase of Bcl-2 by inducing protein kinase C (PKC) at a dose of $10\,\mu$M.[210] Of further interest against AD, l-3-n-butylphthalide inhibited the phosphorylation of tau protein induced by amyloid β_{25-35}

■ **CS 3.189** Polysperlignan A.

■ **CS 3.190** Polysperlignan D.

■ **CS 3.191** Tiegusanin I.

■ **CS 3.192** Ananolignan F.

■ **CS 3.193** Ananolignan L.

peptide at a dose of $10\,\mu M$.[209] The aforementioned *Saururus chinensis* (Lour.) Baill. (family *Saururaceae Rich.* ex T. Lestib.) produces sauchinone (CS 3.195), which abated apoptosis of C6 cells induced by staurosporine (ST) at a dose of $10\,\mu M$ via the activation of Bcl-2 and therefore inactivation of caspase 3.[211] In fact, sauchinone at a dose of $10\,\mu M$ protected cortical neurons against oxygen glucose deprivation by increasing the enzymatic activity of catalase (CAT).[212]

Machilus thunbergii Sieb. et Zucc. (family Lauraceae Juss.) engineers (−)-isoguaiacin, meso-dihydroguaiaretic acid (CS 3.196), licarin A, and (+)-guaiacin, which protected cortical neurons against glutamate insults by 35%, 58.5%, 54.2%, and 36.1%, respectively, at a dose equal to $1\,\mu M$.[213]

■ **CS 3.194** L-3-n-butylphthalide.

■ **CS 3.195** Sauchinone.

■ **CS 3.196** Meso-dihydroguaiaretic acid.

■ **CS 3.197** Nordihydroguaiaretic acid.

Along the same lines, meso-dihydroguaiaretic acid, nordihydroguaiaretic acid (CS 3.197), (+)-guaiacin (CS 3.198), (−)-isoguaiacin (CS 3.199), and licarin A (CS 3.200) protected cortical neurons against glutamate with IC_{50} values equal to 0.2×10^{-6}M, 2.1×10^{-6}M, 38.4×10^{-6}M, 56.2×10^{-6}M, and 0.3×10^{-6}M, respectively; whereas machilin A (CS 3.201) and isoguaiacin dimethyl ether (CS 3.202) were inactive.[214] In the same experiment, meso-dihydroguaiaretic acid and licarin A maintained GSH and sustained the enzymatic activities of superoxide dismutase (SOD), GSH peroxidase (GSH-px), and GSH reductase (GSH-R),[214] suggesting a reduction in ROS. This contention is supported by the fact that meso-dihydroguaiaretic acid protected cortical neurons against staurosporine at a dose of $1\,\mu$M by blocking Ca^{2+} influx and subsequent ROS generation and sustaining the enzymatic activity of SOD and thus preserving mitochondrial integrity.[215] Arctigenin, from *Torreya nucifera* (L.) Siebold & Zucc. (family Taxaceae Gray), at a dose of $1\,\mu$M protected cortical neurons against glutamate by blocking the kainate receptor with an IC_{50} value equal to 83 nM.[216] Sesamin at a dose of $20\,\mu$M protected PC12 cells against L-DOPA–induced apoptosis with inhibition of ROS; activation of ERK1/2, Bad, and Bcl-2; and elevation of SOD activity.[217] Macelignan protected mouse hippocampal (HT-22) cells against glutamate insults at a dose of $1\,\mu$M via decrease in ROS.[193] 9,9′-*O*-di-(E)-Feruloyl-meso-5,5′-dimethoxysecoisolariciresinol (CS 3.203), 9,9′-*O*-di-(E)-sinapinoyl-meso-5,5′-dimethoxysecoisolariciresinol (CS 3.204), and 9,9′-*O*-di-(E)-feruloyl-meso-secoisolariciresinol (CS 3.205) isolated from *Lindera obtusiloba* Blume (family Lauraceaea Juss.) protected HT-22 cells against glutamate by 46%, 45.5%, and 70%, respectively, at a dose of $1\,\mu$M.[218] Of note, nordihydroguaiaretic acid from *Larrea tridentata* (Sessé & Moc. ex DC.) Coville (family Zygophyllaceae R. Br.) mitigated cortical neuron apoptosis incurred by oxygen and glucose deprivation at a dose equal to $20\,\mu$M via deactivation of JNK as a result of 12-lipoxygenase (12-LOX) and 15-lipoxygenase (15-LOX) inhibition.[219] There is increasing evidence to support the notion that lignans induce neuritogenesis. In this light,

■ **CS 3.198** (+)-guaiacin.

■ **CS 3.199** (−)-isoguaiacin.

■ **CS 3.200** Licarin A.

■ **CS 3.201** Machilin.

■ **CS 3.202** (−)-isoguaiacin dimethyl ether.

■ **CS 3.203** 9,9′-*O*-di-(E)-feruloyl-meso-5,5′-dimethoxysecoisolariciresinol.

■ **CS 3.204** 9,9′-O-di-(E)-sinapinoyl-meso-5,5′-dimethoxysecoisolariciresinol.

■ **CS 3.205** 9,9′-O-di-(E)-feruloylmeso-secoisolariciresinol.

■ **CS 3.206** Americanoic acid A methyl ester.

■ **CS 3.207** (−)-talaumidin.

■ **CS 3.208** Isodunnianol.

Phytolacca americana L. (family Phytolaccaceae R. Br.) produces americanoic acid A methyl ester (CS 3.206), which induced neurite outgrowth in cortical neurons at a dose equal to $0.1\,\mu$M.[220] (−)-Talaumidin (CS 3.207) from *Aristolochia arcuata* Mast. (family Aristolochiaceae Juss.) induced the growth of neurites by hippocampal neurons at a dose of $10\,\mu$M.[221] Likewise isodunnianol (CS 3.208) from *Illicium fargesii* Finet & Gagnep. (family Schisandraceae Blume) induced the growth of neurites from cortical neurons at a dose of $10\,\mu$M,[222] and magnolol and honokiol at a dose of $1\,\mu$M induced the growth of neurites in cortical neurons.[223] Clovanemagnolol (CS 3.209) from *Magnolia obovata* Thunb. (family Magnioliaceae Juss.) at a dose of $0.01\,\mu$M commanded the growth of neurites from hippocampal neurons.[224] Honokiol induced the neurite outgrowth and protection of cortical neurons at a dose of $10\,\mu$M.[223]

These ideas raise interesting questions regarding how neuritogenesis is induced by lignans. Obovatol induced the growth of neurites from embryonic neural cells at a dose equal to $10\,\mu$M by enhancing the production of nerve growth factor (NGF) and BDNF as a result of ERK1/2 induction solely.[225] Along the same lines, 4-*O*-methylhonokiol at a dose of $10\,\mu$M boosted neuritogenesis in embryonic neuronal cells via the activation of ERK1/2, and imposed the secretion of NGF and brain-derived neurotrophic factors (BDNF) as a result of activation of ERK1/2.[226] Schizandrin (CS 3.210) at a dose of $3\,\mu$mol/mL induced the growth of neurites from hippocampal neurons by increasing cytoplasmic

■ **CS 3.209** Clovanemagnolol.

■ **CS 3.210** Schizandrin.

Ca^{2+}, activating CaMIIK and therefore CREB, and activating PKC and MEK1/2 and therefore ERK1/2.[227] (+)- and (−)-Syringaresinol (CS 3.211 and 3.212) from *Magnolia officinalis* Rehder & E.H. Wilson (family Magnoliaceae Juss.) commanded neurite outgrowth in PC12 cells and Neuro2a cells at a dose of $24\,\mu$M via the probable inhibition of cyclic AMP phosphodiesterase (PDE), hence an increase in cAMP and activation of protein kinase A (PKA)[228] and induction of CREB and ERK1/2. SC-1 (CS 3.213) evoked the sprouting of neurites from PC12 cells at a dose equal to $10\,\mu$M by transient activation of MEK1/2 and thus ERK1/2, whereas sesamin was inactive at the same dose.[229] However, sesamin

■ **CS 3.211** (−)-Syringaresinol.

■ **CS 3.212** (+)-Syringaresinol.

■ **CS 3.213** SC-1.

from *Sesamum indicum* L. (family Pedaliaceae R. Br.) at a dose of 50 μM evoked an increase of dopamine in PC12 cells with an increase in tyrosine hydroxylase (TH) as a result of increased cAMP, activation of PKA, and subsequent induction of CREB.[217]

REFERENCES

[185] Shimomura H, Sashida Y, Oohara M. Lignans from *Machilus thunbergii*. Phytochemistry 1987;26(5):1513–15.

[186] Shimomura H, Sashida Y, Oohara M. Lignans from *Machilus thunbergii*. Phytochemistry 1988;27(2):634–6.

[187] Choong JM, So RK, Kim J, Young CK. Meso-dihydroguaiaretic acid and licarin A of *Machilus thunbergii* protect against glutamate-induced toxicity in primary cultures of a rat cortical cells. Br J Pharmacol 2005;146(5):752–9.

[188] Yu YU, Kang SY, Park HY, Sung SH, Lee EJ, Kim SY, et al. Antioxidant lignans from *Machilus thunbergii* protect CCl4-injured primary cultures of rat hepatocytes. J Pharm Pharmacol 2000;52(9):1163–9.

[189] Karikome H, Mimaki Y, Sashida Y. A butanolide and phenolics from *Machilus thunbergii*. Phytochemistry 1991;30(1):315–19.

[190] Kim KH, Choi JW, Ha SK, Kim SY, Lee KR. Neolignans from *Piper kadsura* and their anti-neuroinflammatory activity. Bioorg Med Chem Lett 2010;20(1):409–12.

[191] Choi MS, Lee SH, Cho HS, Kim Y, Yun YP, Jung HY, et al. Inhibitory effect of obovatol on nitric oxide production and activation of NF-kappaB/MAP kinases in lipopolysaccharide-treated RAW 264.7 cells. Eur J Pharmacol 2007;556(1–3):181–9.

[192] Ock J, Han HS, Hong SH, Lee SY, Han Y-M, Kwon B-M, et al. Obovatol attenuates microglia-mediated neuroinflammation by modulating redox regulation. Br J Pharmacol 2010;159(8):1646–62.

[193] Jin DQ, Lim CS, Hwang JK, Ha I, Han JS. Anti-oxidant and anti-inflammatory activities of macelignan in murine hippocampal cell line and primary culture of rat microglial cells. Biochem Biophys Res Commun 2005;331(4):1264–9.

[194] Kim JY, Lim HJ, Lee DY, Kim JS, Kim DH, Lee HJ, et al. *In vitro* anti-inflammatory activity of lignans isolated from *Magnolia fargesii*. Bioorg Med Chem Lett 2009;19(3):937–40.

[195] Cho MK, Jang YP, Kim YC, Kim SG. Arctigenin, a phenylpropanoid dibenzylbutyrolactone lignan, inhibits MAP kinases and AP-1 activation via potent MKK inhibition: the role in TNF-alpha inhibition. Int Immunopharmacol 2004;4(10–11):1419–29.

[196] Ohnishi M, Monda A, Takemoto R, Matsuoka Y, Kitamura C, Ohashi K, et al. Sesamin suppresses activation of microglia and p44/42 MAPK pathway, which confers neuroprotection in rat intracerebral hemorrhage. Neuroscience 2013;232:45–52.

[197] Lee YJ, Choi DY, Choi IS, Kim KH, Kim YH, Kim HM, et al. Inhibitory effect of 4-*O*-methylhonokiol on lipopolysaccharide-induced neuroinflammation, amyloidogenesis and memory impairment via inhibition of nuclear factor-kappaB *in vitro* and *in vivo* models. J Neuroinflamm 2012;9:35.

[198] Lee YK, Choi IS, Ban JO, Lee HJ, Lee US, Han SB, et al. 4-*O*-methylhonokiol attenuated β-amyloid-induced memory impairment through reduction of oxidative damages via inactivation of mitogen-activated protein kinase (MAPK) p38 MAP kinase. J Nutr Biochem 2011;22(5):476–86.

[199] Lee JW, Lee YK, Lee BJ, Nam SY, Lee SI, Kim YH, et al. Inhibitory effect of ethanol extract of *Magnolia officinalis* and 4-*O*-methylhonokiol on memory impairment and neuronal toxicity induced by beta-amyloid. Pharmacol Biochem Behavior 2010;95(1):31–40.

[200] Lee YJ, Choi IS, Park MH, Lee YM, Song JK, Kim YH, et al. 4-*O*-Methylhonokiol attenuates memory impairment in presenilin 2 mutant mice through reduction of oxidative damage and inactivation of astrocytes and the ERK pathway. Free Radic Biol Med 2011;50(1):66–77.

[201] Cui HS, Huang LS, Sok DE, Shin J, Kwon BM, Youn UJ, et al. Protective action of honokiol, administered orally, against oxidative stress in brain of mice challenged with NMDA. Phytomedicine 2007;14:696–700.

[202] Chen CR, Zhou XZ, Luo YJ, Huang ZL, Urade Y, Qu WM. Magnolol, a major bioactive constituent of the bark of *Magnolia officinalis*, induces sleep via the benzodiazepine site of GABA(A) receptor in mice. Neuropharmacology 2012;63(6):1191–9.

[203] Hu Z, Bian X, Liu X, Zhu Y, Zhang X, Chen S, et al. Honokiol protects brain against ischemia-reperfusion injury in rats through disrupting PSD95-nNOS interaction. Brain Res 2013;1491:204–12.

[204] Matsui N, Takahashi K, Takeichi M, Kuroshita T, Noguchi K, Yamazaki K, et al. Magnolol and honokiol prevent learning and memory impairment and cholinergic deficit in SAMP8 mice. Brain Res 2009;1305:108–17.

[205] Kim SR, Lee MK, Koo KA, Kim SH, Sung SH, Lee NG, et al. Dibenzocyclooctadiene lignans from *Schisandra chinensis* protect primary cultures of rat cortical cells from glutamate-induced toxicity. J Neurosci Res 2004;76(3):397–405.

[206] Song JX, Lin X, Wong RN, Sze SC, Tong Y, Shaw PC, et al. Protective effects of dibenzocyclooctadiene lignans from *Schisandra chinensis* against beta-amyloid and homocysteine neurotoxicity in PC12 cells. Phytother Res 2011;25(3):435–43.

[207] Dong K, Pu JX, Zhang HY, Du X, Li XN, Zou J, et al. Dibenzocyclooctadiene lignans from *Kadsura polysperma* and their antineurodegenerative activities. J Nat Prod 2012;75:249–56.

[208] Yang JH, Zhang HY, Wen J, Du X, Chen JH, Zhang HB, et al. Dibenzocyclooctadiene lignans with antineurodegenerative potential from *Kadsura ananosma*. J Nat Prod 2011;74:1028–35.

[209] Peng Y, Xing C, Lemere CA, Chen G, Wang L, Feng Y, et al. l-3-n-Butylphthalide ameliorates beta-amyloid-induced neuronal toxicity in cultured neuronal cells. Neurosci Lett 2008;434(2):224–9.

[210] Peng Y, Hu Y, Feng N, Wang L, Wang X. L-3-n-butyl-phthalide alleviates hydrogen peroxide-induced apoptosis by PKC pathway in human neuroblastoma SK-N-SH cells. Naunyn Schmiedeberg's Arch Pharmacol 2011;383(1):91–9.

[211] Song H, Choong Y, Moon A. Sauchinone, a lignan from *Saururus chinensis*, inhibits staurosporine-induced apoptosis in C6 rat glioma cells. Biol Pharm Bull 2003;26(10):1428–30.

[212] Choi IY, Yan H, Park Y-K, Kim W-K. Sauchinone reduces oxygen-glucose deprivation-evoked neuronal cell death via suppression of intracellular radical production. Arch Pharm Res 2009;32(11):1599–606.

[213] Ma CJ, Sung SH, Kim YC. Neuroprotective lignans from the bark of *Machilus thunbergii*. Planta Med 2004;70(1):79–80.

[214] Ma CJ, Kim SR, Kim J, Kim YC. Meso-dihydroguaiaretic acid and licarin A of *Machilus thunbergii* protect against glutamate-induced toxicity in primary cultures of a rat cortical cells. Br J Pharmacol 2005;146(5):752–9.

[215] Ma CJ, Lee MK, Kim YC. Meso-dihydroguaiaretic acid attenuates the neurotoxic effect of staurosporine in primary rat cortical cultures. Neuropharmacology 2006;50(6):733–40.

[216] Jang YP, Kim SR, Choi YH, Kim J, Kim SG, Markelonis GJ, et al. Arctigenin protects cultured cortical neurons from glutamate-induced neurodegeneration by binding to kainate receptor. J Neurosci Res 2002;68(2):233–40.

[217] Zhang M, Lee HJ, Park KH, Park HJ, Choi HS, Lim SC, et al. Modulatory effects of sesamin on dopamine biosynthesis and L-DOPA-induced cytotoxicity in PC12 cells. Neuropharmacology 2012;62(7):2219–26.

[218] Lee KY, Kim SH, Jeong EJ, Park JH, Kim SH, Kim YC, et al. New secoisolariciresinol derivatives from lindera obtusiloba stems and their neuroprotective activities. Planta Med 2010;76(3):294–7.

[219] Liu Y, Wang H, Zhu Y, Chen L, Qu Y, Zhu Y. The protective effect of nordihydroguaiaretic acid on cerebral ischemia/reperfusion injury is mediated by the JNK pathway. Brain Res 2012;1445:73–81.

[220] Takahasi H, Yanagi K, Ueda M, Nakade K, Fukuyama Y. Structures of 1,4-benzodioxane derivatives from the seeds of *Phytolacca americana* and their neuritogenic activity in primary cultured rat cortical neurons. Chem Pharm Bull 2003;51(12):1377–81.

[221] Esumi T, Hojyo D, Zhai H, Fukuyama Y. First enantioselective synthesis of (−)-talaumidin, a neurotrophic diaryltetrahydrofuran-type lignan. Tetrahedron Lett 2006;47(24):3979–83.

[222] Moriyama M, Huang J-M, Yang C-S, Hioki H, Kubo M, Harada K, et al. Structure and neurotrophic activity of novel sesqui-neolignans from the pericarps of *Illicium fargesii*. Tetrahedron 2007;63(20):4243–9.

[223] Fukuyama Y, Nakade K, Minoshima Y, Yokoyama R, Zhai H, Mitsumoto Y. Neurotrophic activity of honokiol on the cultures of fetal rat cortical neurons. Bioorg Med Chem Lett 2002;12:1163–6.

[224] Khaing Z, Kang D, Camelio AM, Schmidt CE, Siegel D. Hippocampal and cortical neuronal growth mediated by the small molecule natural product clovanemagnolol. Bioorg Med Chem Lett 2011;21(16):4808–12.

[225] Lee YK, Song JK, Choi IS, Jeong JH, Moon DC, Yun YP, et al. Neurotrophic activity of obovatol on the cultured embryonic rat neuronal cells by increase of neurotrophin release through activation of ERK pathway. Eur J Pharmacol 2010;649(1–3):168–76.

[226] Lee YK, Choi IS, Kim YH, Kim KH, Nam SY, Yun YW, et al. Neurite outgrowth effect of 4-O-methylhonokiol by induction of neurotrophic factors through ERK activation. Neurochem Res 2009;34(12):2251–60.

[227] Yang S-H, Jeng C-J, Chen C-H, Chen Y, Chen Y-C, Wang S-M. Schisandrin enhances dendrite outgrowth and synaptogenesis in primary cultured hippocampal neurons. J Sci Food Agric 2011;91(4):694–702.

[228] Chiba K, Yamazaki M, Umegaki E, Li MR, Xu ZW, Terada S, et al. Neuritogenesis of herbal (+)- and (−)-syringaresinol separated by chiral HPLC in PC12h and Neuro2a cells. Biol Pharm Bull 2002;25(6):791–3.

[229] Hamada N, Fujita Y, Tanaka A, Naoi M, Nozawa Y, Ono Y, et al. Metabolites of sesamin, a major lignan in sesame seeds, induce neuronal differentiation in PC12 through activation of ERK1/2 pathway. J Neural Transm 2009;116(7):841–52.

Index of Natural Products

Note: Page numbers followed by "*f*" refers to figures.

Index of Pharmacological Terms

Index of Plants

Note: Page numbers followed by "*f*" refers to figures.

Subject Index

Printed and bound by CPI Group (UK) Ltd, Croydon, CR0 4YY

08/05/2025

01864979-0004